Fuzzy Multiple Objective Decision Making

Fuzzy Multiple Objective Decision Making

Gwo-Hshiung Tzeng
Jih-Jeng Huang

CRC Press
Taylor & Francis Group
Boca Raton London New York

CRC Press is an imprint of the
Taylor & Francis Group, an **informa** business
A CHAPMAN & HALL BOOK

CRC Press
Taylor & Francis Group
6000 Broken Sound Parkway NW, Suite 300
Boca Raton, FL 33487-2742

First issued in paperback 2019

ISBN-13: 978-1-4665-5461-0 (hbk)
ISBN-13: 978-0-367-37964-3 (pbk)

Library of Congress Cataloging-in-Publication Data

Huang, Jih-Jeng, author.
Fuzzy multiple objective decision making / Jih-Jeng Huang, Gwo-Hshiung Tzeng.
pages cm
Includes bibliographical references and index.
ISBN 978-1-4665-5461-0 (hardback)
1. Multiple criteria decision making. 2. Fuzzy sets. 3. Management science. I. Tzeng, Gwo-Hshiung, author. II. Title.

T57.95.H83 2013
658.4'033--dc23 2013014250

Visit the Taylor & Francis Web site at
http://www.taylorandfrancis.com

and the CRC Press Web site at
http://www.crcpress.com

Contents

SECTION II Applications of Multi-Objective Decision Making

Preface

Operations research has been adapted by management science scholars to manage realistic problems for a long time. Among these methods, mathematical programming models play a key role in optimizing systems. However, traditional mathematical programming focuses on single objective optimization rather than multi-objective optimization as we encounter real situations. Hence, the concept of multi-objective programming was proposed by Kuhn, Tucker, and Koopmans in 1951 and since then has served as the mainstream technique of mathematical programming.

Multi-objective programming (MOP) can be considered a natural extension of single objective programming by simultaneously optimizing multi-objectives in mathematical programming models. However, the optimization of multi-objectives triggers the issue of Pareto solutions and complicates the derived answers. In addition, more scholars incorporate the concepts of fuzzy sets and evolutionary algorithms to multi-objective programming models and enrich the field of multi-objective decision making (MODM).

The contents of this book are divided into two parts methodologies and applications. In the first part, we introduce the most popular methods that are used to calculate the solution of MOP in the field of MODM. Furthermore, we included three new topics of MODM: multi-objective evolutionary algorithms (MOEAs), expanding De Novo programming to changeable spaces including decision spaces and objective spaces, and network data envelopment analysis (NDEA). In Part II covering applications, we propose different kinds of practical applications of MODM. These applications can provide readers the insights for better understanding of MODM in depth. This book may be useful for the following groups based on their specific objectives:

- Undergraduate and graduate students who wish to extend their knowledge of the methods of MODM or publish papers in journals of operations research and management science
- Practitioners who seek to make effective decisions by using MODM methods

Finally, we hope all our readers will be satisfied with this book and reap great rewards from it. Suggestions and corrections are very welcome and appreciated.

Gwo-Hshiung Tzeng and Jih-Jeng Huang

Biographical Note

Gwo-Hshiung Tzeng was born in 1943 in Taiwan. In 1967, he received a bachelor's degree in business management from the Tatung Institute of Technology (now Tatung University), Taiwan. In 1971, he received a master's degree in urban planning from Chung Hsing University (now Taipei University), Taiwan. In 1977, he received a Ph.D. course in management science from Osaka University, Osaka, Japan.

Gwo-Hshiung Tzeng was an associate professor at Chiao Tung University, Taiwan, from 1977 to 1981, a research associate at Argonne National Laboratory from July 1981 to January 1982, a visiting professor in the Department of Civil Engineering at the University of Maryland, College Park, MD, from August 1989 to August 1990, a visiting professor in the Department of Engineering and Economic System, Energy Modeling Forum at Stanford University, from August 1997 to August 1998, a professor at Chaio Tung University from 1981 to 2003, and a chair professor at Chiao Tung University. He was named a National Distinguished Chair Professor (highest honor offered by the Ministry of Education Affairs, Taiwan) and Distinguished Research Fellow (highest honor Offered by NSC, Taiwan) in 2000. His current research interests include statistics, multivariate analysis, networks, routing and scheduling, multiple criteria decision making, fuzzy theory, application of hierarchical structure analysis to technology management, energy, the environment, transportation systems, transportation investment, logistics, locations, urban planning, tourism, technology management, electronic commerce, global supply chain, etc. He was awarded a Highly Cited Paper (March 13, 2009) ESI "Compromise solution by MCDM methods: A comparative analysis of VIKOR and TOPSIS" as published in the "European Journal of Operational Research" on July 16th, 156(2), 445–455, 2004, which recently has been identified by Thomson Reuters' Essential Science Indicators SM as one of the most cited papers in the field of Economics and Business.

He received the MCDM Edgeworth-Pareto Award from the International Society on Multiple Criteria Decision Making (June 2009), the world Pinnacle of Achievement Award in 2005, and the National Distinguished Chair Professor Award (highest honor offered) of the Ministry of Education Affairs of Taiwan; additionally, he is a three time recipient of a distinguished research award and was twice named a distinguished research fellow (highest honor offered) of the National Science Council of Taiwan. He organized a Taiwan affiliate chapter of the International Association of Energy Economics in 1984 and he was the Chairman of the Tenth

International Conference on Multiple Criteria Decision Making, July 19–24, 1992, in Taipei, the Co-Chairman of the 36th International Conference on Computers and Industrial Engineering, June 20–23, 2006, Taipei, Taiwan, and the Chairman of the International Summer School on Multiple Criteria Decision Making 2006, July 2–14, Kainan University, Taiwan. He is a member of IEEE, IAEE, ISMCDM, World Transport, the Operations Research Society of Japan, the Society of Instrument and Control Engineers Society of Japan, the City Planning Institute of Japan, the Behavior Metric Society of Japan, and the Japan Society for Fuzzy Theory and Systems and participates in many societies of Taiwan. He is editor-in-chief of the *International Journal of Information Systems for Logistics and Management.*

1 Introduction

1.1 PROFILE OF MULTIPLE CRITERION DECISION MAKING

The decision-making process involves a series of steps: identifying the problem, constructing the preferences, evaluating the alternatives, and determining the best alternative (Simon, 1977; Keeney and Raiffa, 1993; Kleindorfer et al., 1993). Generally speaking, three kinds of formal analysis can be employed to solve decision-making problems (Bell et al., 1988; Kleindorfer et al., 1993): (1) descriptive analysis focuses on the problems that decision makers actually face; (2) prescriptive analysis considers the methods used by decision makers to improve their decisions; and (3) normative analysis is concerned with the ideal resolution of problems.

In this book, we limit our topics to normative analysis and prescriptive analysis since descriptive analysis (or behavior decision research) falls within the fields of psychology, marketing, and consumer research (Kahneman and Tversky, 2000). On the other hand, normative analysis and prescriptive analysis are concentration areas in the decision science, economics, and operations research (OR) fields.

Decision making is extremely intuitive for solving a single criterion problem because we need to choose only the alternative with the highest preference rating. However, when decision makers evaluate the alternatives based on multiple criteria, many problems such as weights of criteria, preference dependence, and conflicts among criteria seem to complicate the decision problems and should be overcome by more sophisticated methods.

In order to deal with multiple criteria decision making (MCDM) problems, the first steps are identifying the problems and determining how many attributes or criteria exist in the problems. Next, we must collect the appropriate data or information in which the preferences of decision makers can be correctly reflected and considered. This step is known as constructing the preferences. The next step is building a set of possible alternatives or strategies to guarantee that the goal will be reached—evaluate the alternatives. After that, we can select an appropriate method that helps us evaluate, rank, and improve the possible alternatives or strategies. This step involves finding and determining the best alternative.

To facilitate systematic research in the field of MCDM, Hwang and Yoon (1981) suggested that the MCDM problems can be classified into two main categories: multiple attribute decision making (MADM) and multiple objective decision making (MODM) based on different purposes and data types. MADM is applied in the evaluation phase, which is usually associated with a limited number of predetermined alternatives and discrete preference ratings. MODM is especially suitable for the design and planning steps and allows a user to achieve the optimal or aspired goals by considering the various interactions of the given constraints.

However, conventional MCDM considers only discrete decision problems and lacks a general paradigm for specific real-world problems such as group decisions and uncertain preferences.

Most of the MCDM problems in the real world, therefore, should be regarded naturally as fuzzy problems (Zadeh, 1965; Bellman and Zadeh, 1970) consisting of goals, aspects (or dimensions), attributes (or criteria), and possible alternatives (or strategies). More precisely, we can classify MCDM problems within the fuzzy environment into two categories: fuzzy multiple attribute decision making (FMADM) and fuzzy multiple objective decision making (FMODM) problems based on the concepts of MADM and MODM. Figures 1.1 and 1.2 illustrate the profiles of MCDM.

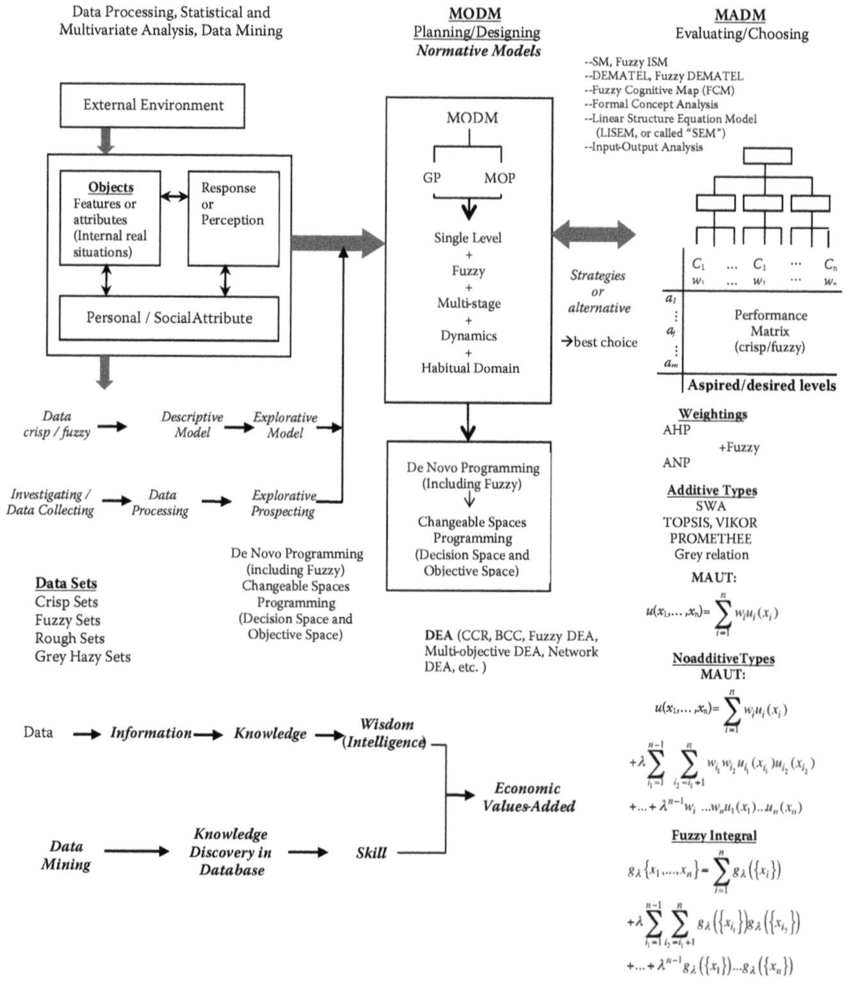

FIGURE 1.1 Profile of multiple criteria decision making.

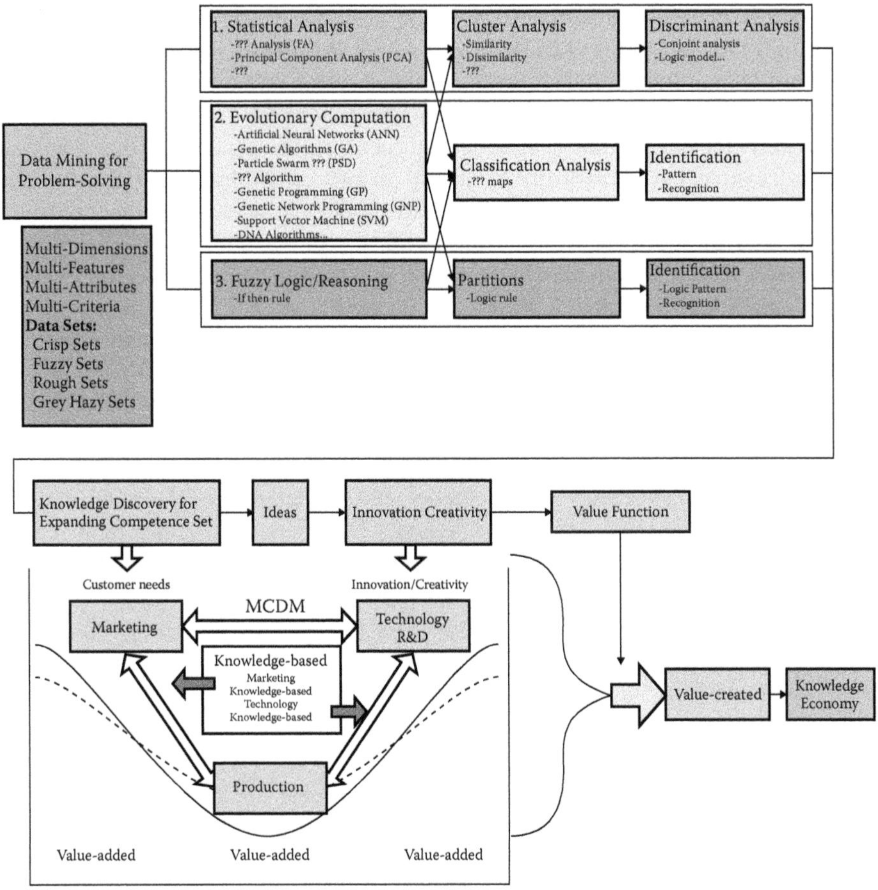

FIGURE 1.2 Data mining concepts of intelligent computation in knowledge economy.

1.2 HISTORICAL DEVELOPMENT OF MULTIPLE ATTRIBUTE DECISION MAKING

The historical origination of MADM can be traced back to a correspondence between Nicolas Bernoulli (1687–1759) and Pierre Rémond de Montmort (1678–1719) discussing the St. Petersburg paradox:

> A game is played by flipping a fair coin until it comes up tails, and the total number of flips, n, determines the prize, which equals \$2 × n. If the coin comes up heads in the first time, it is flipped again, and so on. The problem arises that how much are you willing to pay for this game?

According to the expected value theory, the result can be calculated that $EV = \Sigma_{n=1}^{\infty}(1/2)^n \times 2^n$ and shows that the expected value will go to infinity. However, the result is obviously against our intuition since no one is willing to pay more than $1,000 for this game. The answer of the St. Petersburg paradox was unavailable until Daniel Bernoulli (1700–1782) published his influential research on utility theory in 1738. We ignore the concrete discussions for describing the solution of the St. Petersburg paradox in detail but focus on the conclusion that a human makes decisions based on utility value, not on expected value. The implication is that a human chooses the alternative with the highest utility value while confronting the MADM problems.

In 1947, von Neumann and Morgenstern published a famous book on the theory of games and economic behavior and devised a detailed mathematical theory of economic and social organization based on the game theory. It is no doubt that their great work opened the door to MADM. In brief, the methods for dealing with MADM problems can be divided into multiple attribute utility theory (MAUT) and outranking methods, particularly the ELECTRE (Benayoun et al., 1966; Roy, 1968) and PROMETHEE (Brans et al., 1984) methods.

On the basis of Bernoulli's utility theory, MAUT determines the decision maker's preferences that may be represented as a hierarchical structure by using an *appropriate* utility function. By evaluating the utility function, a decision maker (DM) can easily determine the best alternative with the highest utility value. Although many papers have been proposed for determining the appropriate utility function of MAUT (Fishburn, 1970), the main criticism of MAUT concentrates on the unrealistic assumption-preferential independence (Grabisch, 1995; Hillier, 2000).

Preferential independence means that the preference outcome of one criterion over another criterion is not influenced by the remaining criteria. However, it should be highlighted that the criteria are usually interactive in practical MCDM problems. In order to overcome the non-additive issue, the Choquet integral was proposed (Choquet, 1953; Sugeno, 1974). The Choquet integral can represent a certain kind of interaction among criteria using the concept of redundancy and support/synergy. However, another critical problem of the Choquet integral concerns correct determination of fuzzy measures.

Conversely, instead of building complex utility functions, outranking methods compare the preference relations among alternatives to acquire the information for choosing the best alternative. Although outranking methods were proposed to overcome the empirical difficulties experienced with the utility function in handling practical problems, the main criticisms of outranking methods noted the absence of axiomatic foundations such as the classical aggregate problems, the structural problems, and the non-compensatory problems (Bouyssou and Vansnick, 1986).

In 1965, fuzzy sets (Zadeh, 1965; Bellman and Zadeh, 1970) were proposed to confront the problems of linguistic or uncertain information and act as a generalization of conventional set theory. After successful applications in the field of automatic control, fuzzy sets were incorporated into MADM recently for dealing with the MADM problems involving subjective uncertainty. Figure 1.3 illustrates the holistic development of MADM.

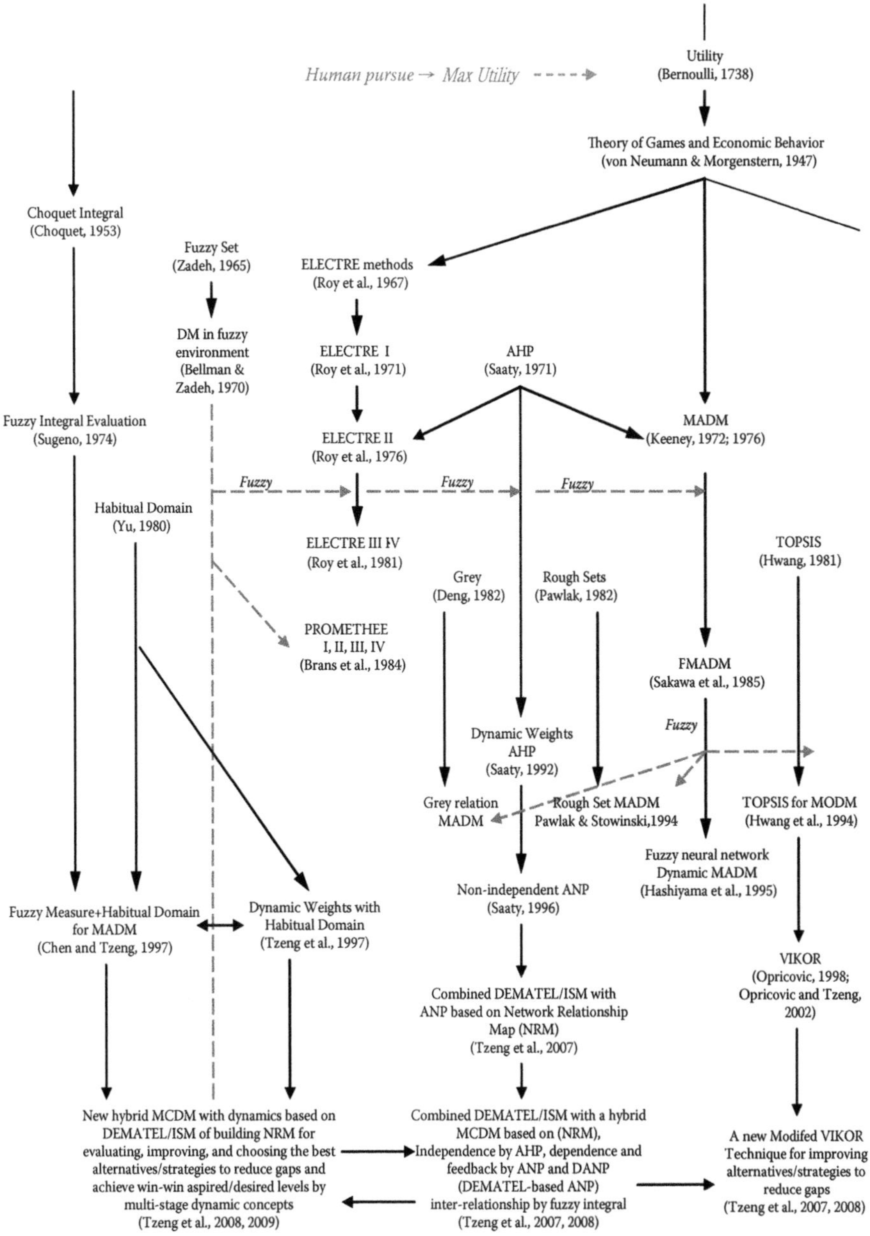

FIGURE 1.3 Development of multiple attribute decision making.

1.3 HISTORICAL DEVELOPMENT OF MULTIPLE OBJECTIVE DECISION MAKING

Multiple objective decision making (MODM) aims to resolve optimal design problems in which several (conflicting) objectives must be achieved simultaneously. MODM is characterized by a set of conflicting objectives and a set of well-defined constraints. Therefore, it is naturally associated with mathematical programming methods for dealing with optimization problems. However, two main difficulties involving the trade-off and scale issues complicate MODM problems through the mathematical programming model.

The trade-off problem arises because a final optimal solution is reached usually through mathematical programming and multiple objectives must transform into a weighted single objective. Therefore, a process of obtaining the trade-off information between the considered objectives should be identified first. Note that if trade-off information is unavailable, Pareto solutions should be derived. The scaling problem, on the other hand, arises when the number of dimensions increases beyond capacity and summons the "curse of dimensionality," i.e., the computational cost increases tremendously. Recently, many evolution algorithms, such as genetic algorithms (Holland, 1975), genetic programming (Koza, 1992), and evolution strategy (Rechenberg, 1973), have been suggested to handle this problem.

Since Kuhn and Tucker (1951) published multiple objectives using the vector optimization concept, and Yu (1973) proposed the compromise solution method to cope with MODM problems, considerable work has been done on various applications such as transportation investment and planning, econometric and development planning, financial planning, business management, investment portfolio selection, land use planning, water resource management, public policy, and environmental issues. The theoretical work is extended from simple multiple objective programming to multi-level, multi-objective programming and multi-stage multi-objective programming to confront very complicated real-world problems.

On the other hand, conventional MODM seems to ignore the problem of subjective uncertainty. Since the objectives and constraints may involve linguistic and fuzzy variables, the fuzzy numbers should be incorporated into MODM for dealing with more extensive problems. After Bellman and Zadeh (1970) proposed the concept of decision making under fuzzy environments, many distinguished works such as Hwang and Yoon (1981), Zimmermann (1978), Sakawa (1983, 1984a, and 1984b), and Lee and Li (1993) led to studies of fuzzy multiple objective linear programming (FMOLP).

FMOLP formulates the objectives and the constraints as fuzzy sets based on their individual linear membership functions. The decision set is defined by the intersection of all fuzzy sets and the relevant hard constraints. A crisp solution is generated by selecting the optimal solution, such that it has the highest degree of membership in the decision set. For further discussions, readers can refer to Zimmermann (1978), Werners (1987), and Martinson (1993). Figure 1.4 depicts the holistic development of MADM.

In addition, the trend of MODM has changed gradually from the win–lose strategy to the win–win strategy. Under the win–lose strategy, a firm can optimize its system only via its given resources and restricted capabilities. That is, the firm faces the traditional optimal problems. However, more firms are interested in creating

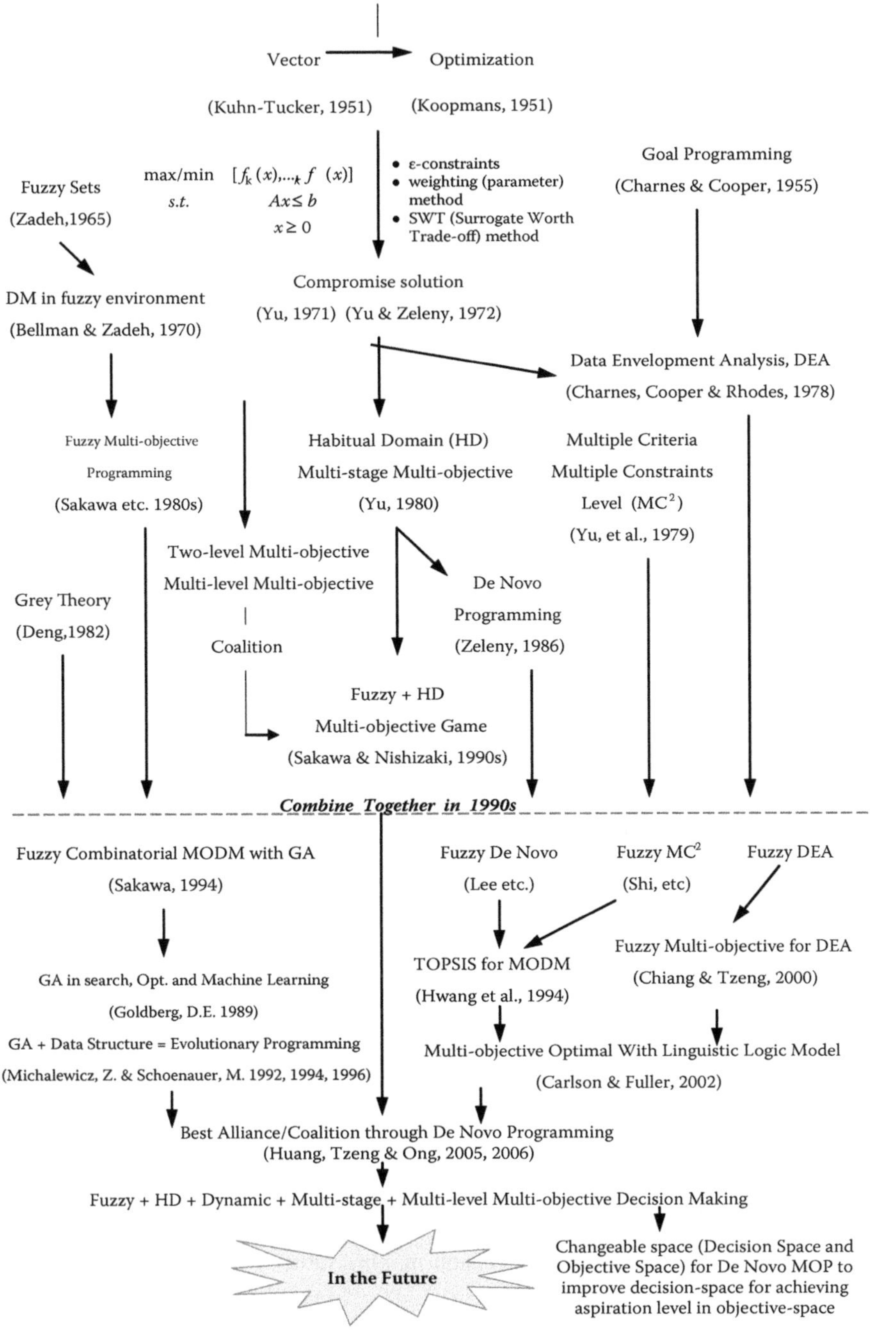

FIGURE 1.4 Development of multiple objective decision making.

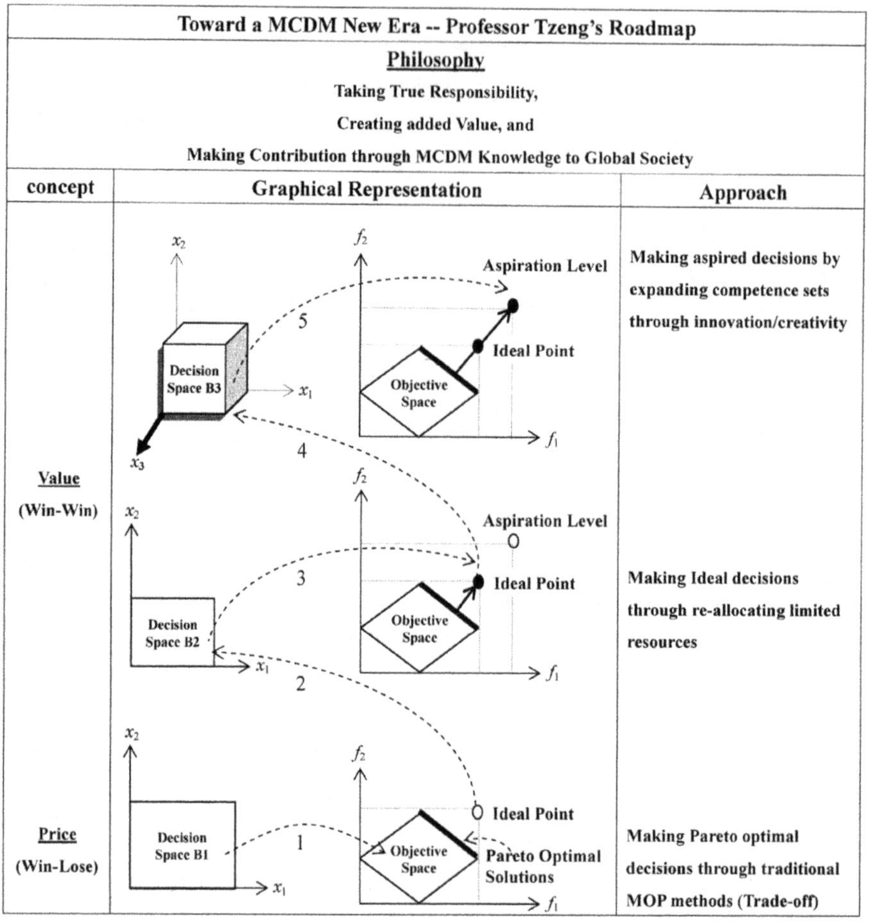

FIGURE 1.5 Concept of win–win strategy.

added value and achieving better parameters through flexible resources and expanding competence sets. Hence, the win–win strategy is another area of focus for this chapter. The win–win strategy is shown in Figure 1.5.

1.4 INTRODUCTION TO FUZZY SETS

In this section, we concentrate on the basic concepts of fuzzy sets and the arithmetic operations of fuzzy numbers in decision making rather than introduce all related topics in detail.

1.4.1 Basic Concepts

In contrast to classical set theory for coping with Boolean logic problems, fuzzy sets were proposed to represent the influences of the elements belonging to specific sets.

Instead of using the characteristic function for mapping purposes, a fuzzy subset $\tilde{A}$ of a universal set X can be defined by its membership function $\mu_{\tilde{A}}(x)$ as

$$\tilde{A} = \{(x, \mu_{\tilde{A}}(x)) \mid x \in X\}, \tag{1.1}$$

where $x \in X$ denotes the elements belonging to the universal set, and

$$\mu_{\tilde{A}}(x): X \to [0,1]. \tag{1.2}$$

Given a discrete finite set $X = \{x_1, x_2, \ldots, x_n\}$, a fuzzy subset $\tilde{A}$ of X can also be represented as

$$\tilde{A} = \sum_{i=1}^{n} \mu_{\tilde{A}}(x_i) / x_i. \tag{1.3}$$

For a continuous case, a fuzzy set $\tilde{A}$ of X can be represented as

$$\tilde{A} = \int_X \mu_{\tilde{A}}(x) / x. \tag{1.4}$$

Next, we present some definitions that will be used in the FMADM models as follows.

Definition 1.1. Consider a fuzzy subset $\tilde{A}$ of a set X; the support of $\tilde{A}$ is a crisp set of X defined by

$$\text{supp}(\tilde{A}) = \{x \in X \mid \mu_{\tilde{A}}(x) > 0\}. \tag{1.5}$$

Definition 1.2. The α-cut of a fuzzy subset $\tilde{A}$ of X can be defined by

$$\tilde{A}(\alpha) = \{x \in X \mid \mu_{\tilde{A}}(x) \geq \alpha\}, \qquad \forall \alpha \in [0,1]. \tag{1.6}$$

Definition 1.3. Let $\tilde{A}$ represent a fuzzy subset of a set X; the height of $\tilde{A}$ is the least upper bound (*sup*) of $\mu_{\tilde{A}}(x)$ and is defined by

$$h(A) = \sup_{x \in X} \mu_{\tilde{A}}(x). \tag{1.7}$$

Definition 1.4. A fuzzy subset $\tilde{A}$ of a set X is said to be normal if and only if its height is unity and called subnormal if and only if its height is not unity.

Fuzzy sets were originally proposed to deal with the problems of subjective uncertainty that arise from using linguistic variables to represent a problem or the event. Note that a linguistic variable is a variable that is expressed by words or sentences in a natural or artificial language. For example, linguistic variables with triangular fuzzy numbers may take on effect values such as very high (very good), high (good), fair, low (poor), and very low (very poor) as shown in Figure 1.6 to indicate the membership functions of the expression values [$\mu_{\tilde{A}}(x)$].

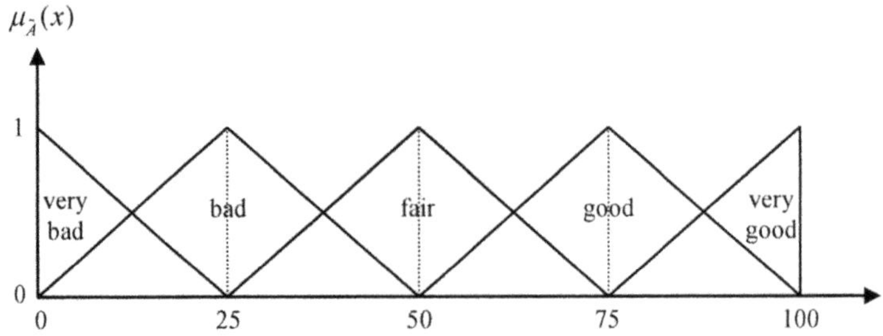

FIGURE 1.6 Membership function of five levels of linguistic variables.

The adoption of linguistic variables has become more widespread recently. These variables are used to assess the linguistic ratings given by the evaluators. Furthermore, linguistic variables are also employed to measure the achievement of the performance for each criterion. Since the linguistic variables can be defined by the corresponding membership function and fuzzy interval, we can naturally manipulate the fuzzy numbers to deal with the FMADM problems.

1.4.2 Fuzzy Arithmetic Operations

Fuzzy arithmetic operations involve adding, subtracting, multiplying, and dividing fuzzy numbers. Generally, these fuzzy arithmetic operations are based on the extension principle and α-*cut* arithmetic. For more detailed discussions of fuzzy arithmetic operations, readers can refer to Dubois et al. (2000); Dubois and Prade (1987); Dubois et al. (1993); Dubois and Prade (1988); Kaufmann and Gupta (1985, 1988); and Mares (1994). In this section, we briefly introduce the fuzzy arithmetic operations according to the extension principle and α-*cut* arithmetic, respectively.

1.4.2.1 Extension Principle

Let $\tilde{m}$ and $\tilde{n}$ be two fuzzy numbers and z denote a specific event. The membership functions of the four basic arithmetic operations for $\tilde{m}$ and $\tilde{n}$ can be defined by

$$\mu_{\tilde{m}+\tilde{n}}(z) = \sup_{x,y}\{\min(\tilde{m}(x),\tilde{n}(y)) \mid x+y=z\}; \quad (1.8)$$

$$\mu_{\tilde{m}-\tilde{n}}(z) = \sup_{x,y}\{\min(\tilde{m}(x),\tilde{n}(y)) \mid x-y=z\}; \quad (1.9)$$

$$\mu_{\tilde{m}\times\tilde{n}}(z) = \sup_{x,y}\{\min(\tilde{m}(x),\tilde{n}(y)) \mid x\times y=z\}; \quad (1.10)$$

$$\mu_{\tilde{m}\div\tilde{n}}(z) = \sup_{x,y}\{\min(\tilde{m}(x),\tilde{n}(y)) \mid x\div y=z\}. \quad (1.11)$$

The procedures of calculating two fuzzy numbers, $\tilde{m}$ and $\tilde{n}$, based on the extension principle can be illustrated by the following example:

X	1	3	5	7	9
Y	1	3	5	7	9
$\mu_{\tilde{m}}(x)$	0.2	0.4	0.6	1.0	0.8
$\mu_{\tilde{n}}(y)$	0.1	0.3	1.0	0.7	0.5

$\mu_{\tilde{m}+\tilde{n}}(10) = \sup\{0.2,0.4,0.6,0.3,0.1\} = 0.6$
$\mu_{\tilde{m}-\tilde{n}}(-2) = \sup\{0.2,0.4,0.6,0.5\} = 0.6$
$\mu_{\tilde{m}\times\tilde{n}}(9) = \sup\ \{0.2,0.3,0.1\} = 0.3$
$\mu_{\tilde{m}\div\tilde{n}}(3) = \sup\{0.1,0.3\} = 0.3$

Next, we provide another method to derive the fuzzy arithmetic operations based on the concept of α-*cut.*

1.4.2.2 α-Cut Arithmetic

Let $\tilde{m} = [m^l, m^m, m^u]$ and $\tilde{n} = [n^l, n^m, n^u]$ be two fuzzy numbers in which the superscripts l, m, and u denote the infimum, the mode, and the supremum, respectively. The standard fuzzy arithmetic operations can be defined using the concepts of α-*cut* as follows:

$$\tilde{m}(\alpha) + \tilde{n}(\alpha) = [m^l(\alpha) + n^l(\alpha), m^u(\alpha) + n^u(\alpha)]; \tag{1.12}$$

$$\tilde{m}(\alpha) - \tilde{n}(\alpha) = [m^l(\alpha) - n^u(\alpha), m^u(\alpha) - n^l(\alpha)]; \tag{1.13}$$

$$\tilde{m}(\alpha) \div \tilde{n}(\alpha) \approx [m^l(\alpha), m^u(\alpha)] \times [1/n^u(\alpha), 1/n^l(\alpha)]; \tag{1.14}$$

$$\tilde{m}(\alpha) \times \tilde{n}(\alpha) \approx [M, N] \tag{1.15}$$

where (α) denotes the α-*cut* operation, ≈ is the approximation operation, and

$$M = \min\{m^l(\alpha)n^l(\alpha), m^l(\alpha)n^u(\alpha), m^u(\alpha)n^l(\alpha), m^u(\alpha)n^u(\alpha)\};$$

$$N = \max\{m^l(\alpha)n^l(\alpha), m^l(\alpha)n^u(\alpha), m^u(\alpha)n^l(\alpha), m^u(\alpha)n^u(\alpha)\}.$$

An example is also given to illustrate the computation of fuzzy numbers. Let two fuzzy numbers $\tilde{m} = [3,5,8]$ and $\tilde{n} = [2,4,6]$. Then

$$\tilde{m}(\alpha) + \tilde{n}(\alpha) = [(3+2\alpha) + (2+2\alpha), (8-3\alpha) + (6-2\alpha)];$$

$$\tilde{m}(\alpha) - \tilde{n}(\alpha) = [(3+2\alpha) - (2+2\alpha), (8-3\alpha) - (6-2\alpha)];$$

$$\tilde{m}(\alpha) \div \tilde{n}(\alpha) \approx \left[\frac{(3+2\alpha)}{(6-2\alpha)}, \frac{(8-3\alpha)}{(2+2\alpha)}\right];$$

$$\tilde{m}(\alpha) \times \tilde{n}(\alpha) \approx [(3+2\alpha)(2+2\alpha), (8-3\alpha)(6-2\alpha)].$$

1.4.3 Ranking Fuzzy Numbers

Since the fuzzy arithmetic operations based on the α-*cut* arithmetic result in a fuzzy interval, determining the optimal alternative is not always obvious and involves the problem of ranking fuzzy numbers or *defuzzification*. In previous works, the procedure of defuzzification has been proposed to locate the best non-fuzzy performance (BNP) value. Defuzzified fuzzy ranking methods generally can be divided into four categories: (1) preference relation, (2) fuzzy mean and spread, (3) fuzzy scoring, and (4) linguistic methods (Chen and Hwang, 1992).

Although more than 30 defuzzified methods have been proposed over the past 20 years, only the center of area (CoA) is described in this book because of its simplicity and usefulness. Consider the preference ratings of an alternative with n attributes represented using fuzzy numbers. The BNP values of the alternative using CoA can be formulated as:

$$S = \frac{\sum_{i=1}^{n} x_i \mu(x_i)}{\sum_{i=1}^{n} \mu(x_i)} \tag{1.16}$$

where x_i denotes the preference ratings of the ith attribute and $\mu(x_i)$ is the corresponding membership function.

1.5 OUTLINE OF THE BOOK

This book is divided into two parts. Part I discusses concepts and theories of multiple objective decision making. In Chapter 2, we introduce multi-objective evolutionary algorithms (MOEAs) that are used widely for solving all kinds of MODM problems. Next, we

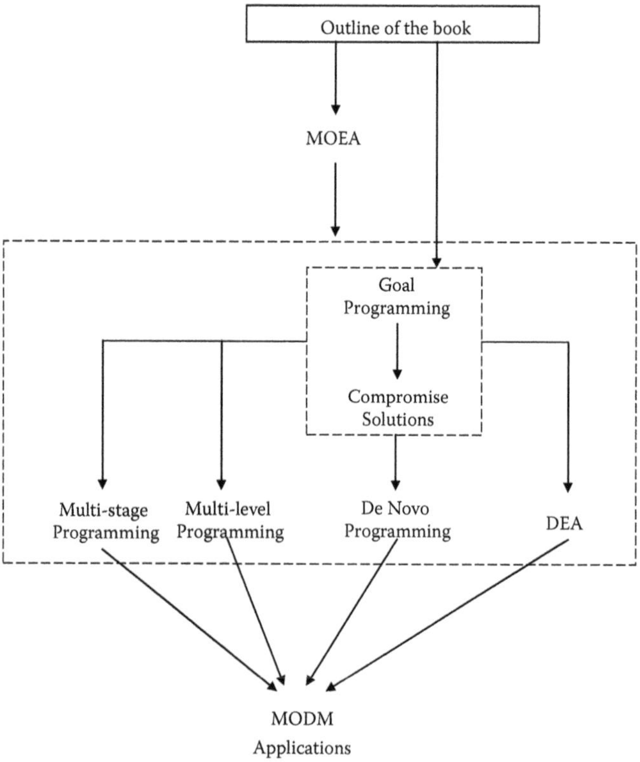

FIGURE 1.7 Book outline.

propose goal programming and compromise solutions in Chapter 3 and Chapter 4, respectively; these are the most popular methods used in MODM. In Chapter 5, we introduce the concept of de novo programming used to reallocate or reshape available resources for eliminating trade-offs between objectives. In Chapters 6 and 7, we introduce multi-stage and multi-level programming for solving network and hierarchical MODM problems, respectively. In Chapter 8, we propose data envelopment analysis (DEA) for evaluating the efficiency of decision-making units (DMUs) based on input and output data.

Part II (Chapters 9 through 16) covers applications of MODM such as strategic alliances, supply chain management, portfolio selection, optimal planning and design of systems, and production efficiency. The outline is depicted in Figure 1.7.

Section I

Concepts and Theory of Multi-Objective Decision Making

2 Multi-Objective Evolutionary Algorithms

In this chapter, we will introduce the use of genetic algorithms for solving multi-objective programming (MOP) problems and the concrete contents of genetic algorithms (GAs).

2.1 CONCEPTS OF GENETIC ALGORITHMS

In traditional optimization methods, gradients and derivatives are usually used to guide the search for an optimal solution. However, when the objective function is not differentiable or the dimensionality of the search space is quite large, these techniques usually perform poorly. GAs are now considered alternative methods to solve such optimization problems. GAs were pioneered by Holland (1975) and the concept is to mimic the natural evolution of a population by allowing solutions to reproduce, creating new solutions that then compete for survival in the next iteration. The fitness improves over generations and the best solution is finally achieved.

Holland (1975) tested the ability of GAs for dealing with different kinds of objective functions. These experimental results indicate that GAs are very robust and perform better than traditional optimization methods. In addition, Bosworth et al. (1972) performed one of the earlier studies of GAs to conclude that GAs are not sensitive to increasing dimensionality and noise. These results were consistent with the experimental results of De Jong's study (1975). GAs were widely used later for various optimization problems, and many revised GAs have been proposed to improve their capabilities, such as GENOCOP (Michalewicz, 1992) or increase their applications, such as genetic programming (Koza, 1992).

Let us consider the following notations for describing the procedures of GAs. The initial population $P(0)$ is encoded randomly by strings. In each generation t, the more fit elements are selected for the mating pool and then processed by three basic genetic operators (reproduction, crossover, and mutation) to generate new offspring. On the basis of the principle of survival of the fittest, the best chromosome of a candidate solution is obtained. The pseudo code of GAs illustrates the computation procedures as shown in Figure 2.1.

The power of the GA lies in its ability of simultaneous searching a population of points in parallel instead of a single point. Therefore, a GA can find the approximate optimum quickly without falling into a local optimum. In addition GAs do not have the limitation of differentiability, as do other optimization techniques.

```
procedure
GA
begin
          t – o
          initialize  P(o)
          evaluate    P(t)
          while not satisfy stopping rule do
          begin
                    t– t + 1
                    select      P(t)  from  P(t-1)
                    alter       P(t)
                    evaluate    P(t)
end
          end
```

FIGURE 2.1 Genetic algorithm procedures.

2.2 GA PROCEDURES

In order to illustrate how GAs can be used to find optimal decision variables, we first consider the following multi-objective programming (MOP) problem.

2.2.1 String Representation

To represent the solutions of a MOP problem in GAs, we should encode these solutions as chromosomes. Usually, each element of a chromosome consists of a binary string (bit) or real-number string. For binary encoding, the precision of a solution depends on the number of bits used. However, the binary encoding for a function optimization problem may suffer critical drawbacks because of the existence of Hamming cliffs (Ludvig et al., 1997). For example, the 1000000 and 0111111 pair belong to neighboring points in phenotype space but have maximum Hamming distance in genotype space. Hence, the binary strings do not preserve the locality of points in the phenotype space.

In contrast, real-number strings code possible solutions as a vector of real numbers of the same length as the solution vector, and are more suitable for dealing with function optimization problems (Eshelman and Schaffer, 1993; Walters and Smith, 1995; Michalewicz, 1996). This is because the topological structure of the genotype space for real-number encoding is identical to that of the phenotype space. Hence, it is easy to form effective genetic operators by borrowing useful techniques from conventional methods (Gen and Cheng, 2000). In addition, compared with binary encoding, real-number encoding is capable of representing very large domains or unknown domains (Michalewicz, 1996).

2.2.2 Population Initialization

Usually, the initial population $P(0)$ is selected at random. Each genotype in the population can be initialized to present the degree of variance from the uniform distribution. Note that there is no standard to determine the size $P(0)$ of the initial population. Bhandari et al. (1996) showed that as the number of iterations extends to infinity, the elitist model of GAs will provide the optimal string for any population size.

2.2.3 Fitness Computation

The fitness function can be considered as the link between the GAs and the problem to be solved. For MOP problems, the fitness function of GAs is clearly identical to the objective functions of a model. Therefore, we can easily use the values of chromosomes to calculate the fitness of each offspring. However, in MOP problems, the ideal solution (best value) of each objective is usually unavailable due to the trade-offs among objectives. Therefore, the best offspring of a MOP problem is a Pareto set rather than a unique solution.

2.2.4 Genetic Operators

In most kinds of different GAs, three common genetic operators (selection, crossover, and mutation) are used to generate offspring, i.e., solutions. Next, we will discuss the content of each genetic operator.

2.2.4.1 Selection

The selection operator chooses chromosomes from the mating pool using the concept of "survival of the fittest" applicable to natural genetic systems. Thus, the best chromosomes receive more copies and the worst die off. The first method of the selection operator is called wheel selection. The probability of variable selection is proportional to its fitness value in the population, according to the formula given by

$$P(x_i) = \frac{f(x_i)}{\sum_{j=1}^{N} f(x_j)} \tag{2.1}$$

where $f(x_i)$ represents the fitness value of the ith chromosome and N is the population size. Hence, each offspring's chance of being selected is directly proportional to its fitness. However, this method cannot ensure that the worst offspring will not be selected.

In contrast, we can use tournament selection to retain better solutions and cast off worse solutions in a population. In the tournament selection, chromosomes are selected to compete according to their fitness values and the better ones are chosen and placed in the mating pool. Tournament selection has been shown to provide better convergence and computational time complexity properties when compared to other selection operators (Goldberg and Deb, 1991).

In sum, the purpose of selection is to replace the worst chromosomes with the selected chromosomes. Therefore, we can guarantee survival of the best solution in each generation via this elitist strategy procedure.

2.2.4.2 Crossover

The goal of crossover, also called recombination, is to exchange information between two parent chromosomes to produce two new offspring for the next population. The common forms of crossover are one-point, two-point, *n*-point, and uniform (Syswerda, 1989). One-point and two-point crossovers are the specific forms of *n*-point crossovers when crossover points are equal to one and two, respectively.

We can use the example of two-point crossover, probably the most popular of different crossover operators, to demonstrate how the technique performs. Let a crossover probability be P_c. The proceeding in two-point crossover occurs when two parent chromosomes are swapped after two randomly selected points between [1, *N*–1], creating two children. If the parent chromosomes are selected by

$$\begin{array}{l} \text{parent1} = 1 \quad 0 \quad 0 \;\big|\; 1 \quad 1 \quad 0 \quad 1 \;\big|\; 1 \quad 0 \quad 0 \\ \text{parent2} = 0 \quad 1 \quad 1 \;\big|\; 0 \quad 0 \quad 0 \quad 1 \;\big|\; 1 \quad 1 \quad 0 \end{array}$$

two offspring will be produced as

$$\begin{array}{l} \text{offspring1} = 1 \quad 0 \quad 0 \quad 0 \quad 1 \quad 0 \quad 1 \quad 1 \quad 0 \quad 0 \\ \text{offspring2} = 0 \quad 1 \quad 1 \quad 1 \quad 0 \quad 0 \quad 1 \quad 1 \quad 1 \quad 0 \end{array}$$

On the other hand, uniform crossover randomly selects *n* number of points, which is less than the chromosome length and then each selected point is swapped over. The greatest advantage of uniform crossover is the capability to combine any schemata that do not disagree at any single position, i.e., uniform crossover allows any pattern to be swapped. Note that a schema is a template that identifies a subset of chromosomes with similarities at certain genes. This characteristic can solve the problems of *n*-point crossovers, including one-point or two-point crossovers that cannot generate specific schemata.

2.2.4.3 Mutation

Mutation is a random process by which one genotype is replaced by another to generate a new chromosome. Each genotype has the probability of changing mutation P_m from 0 to 1 or vice versa in a binary string. If real-number strings are considered, the selected valued can be replaced by a random value between 0 and 9. These mutations can be considered the errors of duplicating DNA. Sometimes these errors can result in good features and better offspring.

In conclusion, GAs differ from traditional optimal techniques in several ways (Buckles and Petry, 1992). First, GAs optimize the trade-off between exploring new points in the search space and exploiting the information discovered thus far. GAs also have the property of implicit parallelism in that their effect is equivalent to an extensive search of hyperplanes of the given space without directly searching all hyperplane values. Furthermore, GAs operate on several solutions simultaneously, gathering information from current search points to direct subsequent searches and try to avoid the problems of local optimization.

2.3 MULTI-OBJECTIVE EVOLUTIONARY ALGORITHMS (MOEAs)

Several disadvantages of classical MOP algorithms have been proposed (Deb, 2001; Coello et al., 2002) to support the requirements of genetic algorithms:

1. The convergence to a classical optimal solution depends on the chosen initial solution.
2. Most classical algorithms tend to get stuck to suboptimal solutions.
3. An algorithm efficient in solving one optimization problem may not be efficient in solving a different optimization problem.
4. Classical algorithms usually can deal only with MOP problems with convex sets.
5. Classical algorithms are not efficient in handling problems involving discrete search spaces.

The concept of genetic algorithms was first used by Schaffer (1984, 1985) and Schaffer and Grefenstette (1985) to solve MOP problems in the mid-1980s. They developed the vector evaluation genetic algorithm (VEGA) to solve two-objective optimization problems of machine learning. Later, more and more GA-based models were proposed, for example, the vector-optimized evolution strategy (VOES; Kursawe, 1990), weight-based genetic algorithms (WBGAs; Hajela and Lin, 1993), random weighted genetic algorithms (RWGAs; Murata and Ishibuchi, 1995), multi-objective genetic algorithms (MOGAs; Fonseca and Fleming, 1993), non-dominated sorting genetic algorithms (NSGAs; Deb, 1994), and niched Pareto genetic algorithms (NPGAs; Horn et al., 1994).

Here we introduce the concept of elitist non-dominated sorting genetic algorithms (NSGA-II; Deb et al., 2000a and b) for MOP used. The NSGA-II uses an explicit diversity-preserving mechanism to derive the Pareto solutions of MOP problems. The offspring and parent populations are combined to sort a set of non-dominated items instead of using only offspring populations. Although the procedure may require more time to deal with the problem, it allows a global non-domination check among the offspring and parent solutions.

The NSGA-II utilizes crowded tournament selection, recombination, and mutation operators to create an offspring population, just like other genetic-based algorithms. The crowded tournament selection operator assumes every solution i has two attributes, i.e., a non-domination rank (r_i) and local crowding distance (d_i), in the population. Next, we may say that solution i wins a tournament with another solution j if $r_i < r_j$ or $r_i = r_j$ and $d_i > d_j$. This procedure ensures that the chosen solution comes from a selection of better non-dominated solutions. In addition, the calculation of the crowding distance (d_i) can be derived from the estimate of the perimeter of the cuboid formed by using the nearest neighbors as the vertices.

The NSGA-II has a number of pros and cons. First, compared with some genetic-based algorithms such as MOGA, NSGA, and NPGA, no extra niching parameter is needed. Second, the NSGA-II exhibits a convergence proof to the Pareto optimal solutions. On the other hand, the procedure of the non-dominated sorting requires more time than other methods. From the viewpoint of the computational complexity, the non-dominated sorting of a population requires at most $O(MN^2)$, if M denotes the number of objectives and N is the size of the population.

2.3.1 Numerical Example

In this section, we demonstrate how MOEA can be used for dealing with MOP problems. Let us consider the following two-objective programming problem:

$$\begin{aligned} \max \quad & 3x_1^2 - 5x_2 \\ \max \quad & 2x_1 + 3x_2^2 \\ s.t. \quad & x_1 + x_2 \leq 50, \\ & x_2, x_2 \geq 0. \end{aligned}$$

We can depict the decision space (left side) and outcome space (right side) of the above problem as shown in Figure 2.2. The outcome space indicates the trade-offs between the objectives and forms the Pareto solutions. Therefore, the optimal solutions of the problem lie on the specific point of the frontier line, according to the preference of the decision maker.

If the goal of the decision maker is to obtain the Pareto solutions of a problem, MOEA is a good choice. In this section, we employ the NSGA-II to derive the frontier set of the problem as follows.

Let the variable coding be real-number strings; population size equals 200, number of generation is 2000, crossover rate is 0.8, and mutation rate is 0.1. We can then

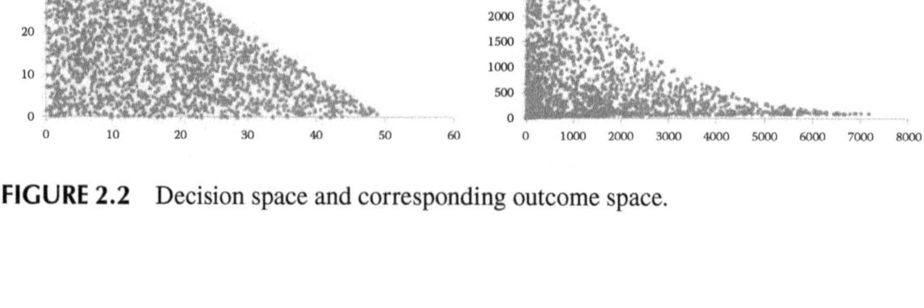

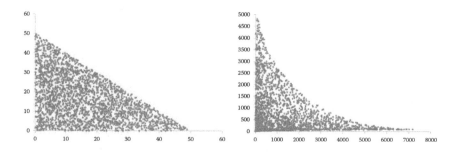

FIGURE 2.2 Decision space and corresponding outcome space.

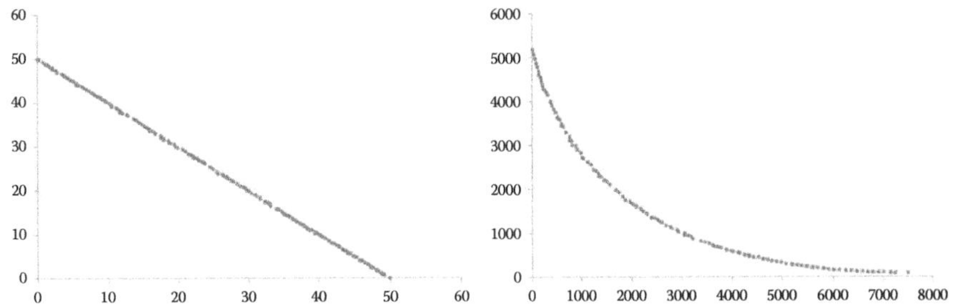

FIGURE 2.3 Optimal decision variable solutions and corresponding objective values by NSGA-II.

run the NSGA-II to derive the optimal solutions of decision variables and the corresponding values of the objective functions, as shown in Figure 2.3.

Comparing Figures 2.1 and 2.2, we can see that the NSGA-II simply derives the Pareto solutions of the objective fucntions and shows the usefulness of MOEA for dealing with MOP problems, especially large-scale MOP problems.

3 Goal Programming

The purpose of multiple objective decision making (MODM) is to achieve the efficient frontier of multiple objective programming (MOP). Traditionally, the weighting method and ε-constraint method have been used widely. The weighting method transforms a set of objectives into a single objective by multiplying each objective with a user-supplied weight. On the other hand, the ε-constraint method keeps only one of the objectives and restricts the rest of the objectives within some user-specified values to derive the efficient frontier. Cohon (1978) developed an algorithm to generate the efficient set systematically. However, goal programming (GP) is the best known method for dealing with MODM problems.

GP is an analytical approach devised to address decision making problems in which targets have been assigned to all the attributes and the decision maker is interested in minimizing the non-achievements of the corresponding goals (Romero, 2001). Generally, goal programming deals with the following MOP problems:

$$\begin{aligned} \max \quad & f_i(x), \quad i = 1,\ldots,n \\ s.t. \quad & g(x) \leq 0, \\ & x \geq 0, \end{aligned}$$

where $f_i(\boldsymbol{x})$ denotes the ith objective function.

3.1 GOAL SETTING

Three kinds of goal settings are utilized in multi-objective optimization problems: (1) minimize all the objective functions; (2) maximize all the objective functions; and (3) minimize some and maximize others. However, we can use duality principle (Reklaitis et al., 1983; Rao, 1984) to convert a minimization problem into a maximization problem by multiplying the objective function by −1. Hence, we can simplify mixed types of objectives and include all types of objectives into maximization problems.

Unlike single-objective problems, multi-objective programming problems do not usually allow all objectives to be optimized due to trade-offs or conflicts among objectives. Hence, there exists a set of non-dominated solutions (the efficient set) such that all points belonging to non-dominated solutions are regarded as indifferent. If we want to determine an optimal solution from the set of non-dominated solutions, we must measure which point is nearest to the ideal. Assume a two-objective programming maximization problem:

$$\begin{aligned} \max \quad & f_1(\boldsymbol{x}), f_2(\boldsymbol{x}) \\ s.t. \quad & g(\boldsymbol{x}) \leq 0 \ \text{(or write } \boldsymbol{Ax} \leq \boldsymbol{b}\text{)}, \\ & x \geq 0. \end{aligned} \tag{3.1}$$

Figure 3.1 illustrates these concepts in detail.

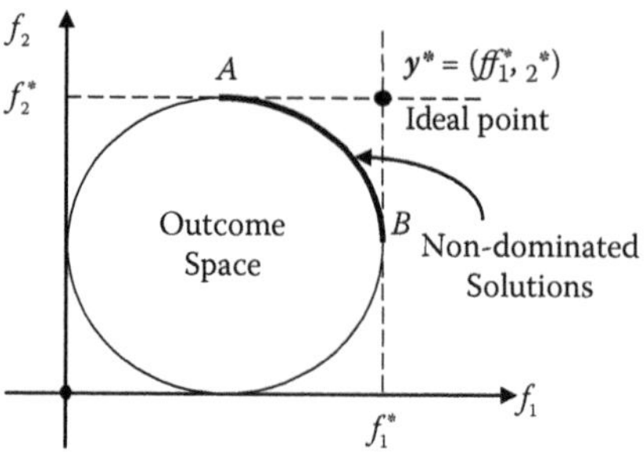

FIGURE 3.1 Concepts of multi-objective programming.

The ideal point is composed of individual optimal objective values. In this example, $y^* = (f_1^*, f_2^*)$. However, the ideal point vector usually corresponds to a non-existent solution because of the trade-off between objectives. Therefore, the problem of optimization in multi-objective programming is transformed into finding a feasible solution (location on the non-dominated set) that is nearest to the ideal points.

Three kinds of targets or goal settings (the one-lower upper goal, one-sided upper goal, and two-sided goal) can be consider in goal programming. A one-sided lower goal sets a lower limit that decision makers do not want to fall under, that is, $f_i(\boldsymbol{x}) + d_i^- = T_i$, where d_i^- is the underachievement derivational variable of the ith objective and T_i denotes the target of the ith objective. A one-sided upper goal sets an upper limit that decision makers do not want to exceed, that is, $f_i(\boldsymbol{x}) - d_i^+ = T_i$, where, where d_i^+ is the overachievement derivational variable of the ith objective. A two-sided goal sets an exact target that decision makers do not want to miss on either side, that is, $f_i(\boldsymbol{x}) + d_i^- - d_i^+ = T_i$.

Initially conceived as an application of single objective linear programming by Charnes and Cooper (1955, 1961), goal programming gained popularity in the 1960s and 1970s based on the works of Ijiri (1965), Lee (1972), and Ignizio (1976). A key element of a GP model is the achievement function that represents a mathematical expression of the unwanted deviation variables. Each type of achievement function leads to a different GP variant.

Tamiz and others (1995) show that around 65% of GP applications reported used lexicographic achievement functions, 21% utilized weighted achievement functions, and the remaining applications involved other types of achievement functions such as a min–max structure in which maximum deviation is minimized.

The next section introduces three kinds of goal programming: weighted goal programming, lexicographic goal programming, and min–max goal programming, to deal with multi-objective programming problems.

3.2 WEIGHTED GOAL PROGRAMMING

Goal programming was proposed by Charnes and Cooper (1961) to deal with linear multi-objective programming problems. The idea of goal programming is to seek a solution that is nearest to the ideal point by considering the relative weights of importance of objectives. That is why this method is called weighted goal programming. Thus, a decision maker should first assign the targets or goals to each objective and then minimize the "distance" from the targets to the objectives. Usually we can use L_p-norm to define the distance between the targets and the objectives to find the solution.

To achieve various kinds of goal programming, the normalization issue should be considered first. It is obvious that different goals may have different scales or measures and the differences will cause the problem of incommensurability. Therefore, these measures should be normalized as a common unit so that deviational variables can be summed up directly. Several normalization methods can be used (Tamiz et al., 1998):

Percentage normalization — Each target value is divided by 100 to ensure that all deviations are measured on a percentage scale. Clearly, this method is not suitable if the target value is equal to zero.

Euclidean normalization — The normalization constant uses the Euclidean norm i.e., $L_{p=2}$, as the denominator of objectives.

Summation normalization — This method uses the absolute value of the objective coefficients, i.e., $L_{p=1}$, to normalize objectives.

Zero–one normalization — The normalization constant in this method is equal to the distance of the target value minus the worst value of an objective.

Another issue we should consider in goal programming is the selection of preferential weights to express the decision maker's preferences with respect to each goal. First, we can directly integrate the analytic hierarchy and network processes (AHP and ANP) into goal programming to obtain the preferential weights of the objectives (Gass, 1986 and 1987). Second, we can use the interactive MCDM method to elicit the weight data for the objectives (Lara and Romero, 1992). In this chapter, we focus on the issue of goal programming. Readers interested in AHP and ANP may refer to the authors' 2011 book titled *Multiple Attribute Decision Making: Methods and Applications.*

A generalization of weighted goal programming can be described as:

$$
\begin{aligned}
\mathit{min} \quad & \left[\sum_{i=1}^{n}\left(\frac{w_i^- d_i^- + w_i^+ d_i^+}{k_i}\right)^p\right]^{1/p}, \quad p \geq 1 \\
\mathit{s.t.} \quad & g(\boldsymbol{x}) \leq 0 \ \ (\text{or write } \boldsymbol{Ax} \leq \boldsymbol{b}) \\
& f_i(\boldsymbol{x}) + d_i^- - d_i^+ = t_i, \qquad i = 1,\ldots,n, \\
& d_i^- \cdot d_i^+ = 0, \\
& d_i^-, \ \ d_i^+ \geq 0, \\
& x \geq 0,
\end{aligned}
\tag{3.2}
$$

where w_i^- and w_i^+ denote the weight factors for positive and negative deviations of the ith objective, respectively, d_i^- and d_i^+ denote the positive deviation representing overachievement of the ith goal and negative deviation representing underachievement of the ith goal, respectively, $k_i = b_i/100$ (or $k_i = x_i^* - x_i^-$, x_i^* is positive ideal point or set aspiration level, x_i^- is negative ideal point or set the worst value) denotes the normalization constant, and t_i denotes the target value of the ith goal set by the decision maker.

Example 3.1 Consider the following two-objective programming problem:

$$\begin{aligned} \max \quad & f_1(\boldsymbol{x}) = 3x_1 + 5x_2 \\ \max \quad & f_2(\boldsymbol{x}) = 6x_1 + 4x_2 \\ s.t. \quad & x_1 + x_2 \le 30, \\ & 2x_1 + 3x_2 \le 60 \\ & x_1, x_2 \ge 0. \end{aligned}$$

If we set $p = 2$ and $f_1(\boldsymbol{x})$ to be two times more important than $f_2(\boldsymbol{x})$, we can formulate the following weighted goal programming:

$$\begin{aligned} \min \quad & \left[\left(\frac{2d_1^-}{15} \right)^2 + \left(\frac{d_2^-}{18} \right)^2 \right]^{1/2} \\ s.t. \quad & 3x_1 + 5x_2 + d_1^- - d_1^+ = 150, \\ & 6x_1 + 4x_2 + d_2^- - d_2^+ = 180, \\ & x_1 + x_2 \le 30, \\ & 2x_1 + 3x_2 \le 60 \\ & d_1^- \cdot d_1^+ = 0, d_2^- \cdot d_2^+ = 0, \\ & d_1^-, d_2^-, x_1, x_2 \ge 0. \end{aligned}$$

Solving the above problem, we can obtain $d_1^- = 58.38$, $d_2^- = 16.22$, $x_1 = 25.14$ and $x_2 = 3.24$. Then, we can calculate $f_1(\boldsymbol{x}) = 91.62$ and $f_2(\boldsymbol{x}) = 163.78$, respectively.

3.3 LEXICOGRAPHY GOAL PROGRAMMING

The second model, known as the lexicography achievement model, consists of an ordered vector whose dimension coincides with the Q number of priority levels established in the model. Each component in this vector represents the unwanted deviation variables of the goals placed in the corresponding priority level (Ignizio, 1976).

$$\begin{aligned} lexmin \quad & \sum_{j=1}^{l} p_j \left[\sum_{i=1}^{n} \left(\frac{w_i^- d_i^- + w_i^+ d_i^+}{k_i} \right)^p \right]^{1/p}, \quad p \ge 1 \\ s.t. \quad & g(\boldsymbol{x}) \le 0 \ \text{(or writing } \boldsymbol{Ax} \le \boldsymbol{b}) \\ & f_i(\boldsymbol{x}) + d_i^- - d_i^+ = t_i, \quad i = 1, \ldots, n, \\ & d_i^- \cdot d_i^+ = 0, \\ & d_i^-, d_i^+ \ge 0, \\ & \boldsymbol{x} \ge 0, \\ & p_1 \succ p_2 \succ \ldots \succ p_l, \quad l \le n, \end{aligned} \tag{3.3}$$

where $\succ$ denotes the priority operator. If $p_i \succ p_j$, the ith objective should be achieved before the jth objective is considered. Lexicographic achievement functions imply a non-compensatory structure of preferences. In other words, there are no finite trade-offs among goals placed in different priority levels (Romero, 1991).

Example 3.2 Consider the following two-objective programming problem:

$$\begin{aligned} \text{First priority:} \quad & \max \quad f_1(\boldsymbol{x}) = 3x_1 + 5x_2 \\ & \qquad\quad f_2(\boldsymbol{x}) = 7x_1 + 3x_2 \\ \text{Second priority:} \quad & \max \quad f_3(\boldsymbol{x}) = 6x_1 + 4x_2 \\ & \qquad\quad f_4(\boldsymbol{x}) = -3x_1 + 5x_2 \\ & \text{s.t.} \quad x_1 + x_2 \le 30, \\ & \qquad\quad 2x_1 + 3x_2 \le 60, \\ & \qquad\quad x_1, x_2 \ge 0 \end{aligned}$$

where all objectives have equal importance and $f_1(\boldsymbol{x})$ and $f_2(\boldsymbol{x})$ are prior to $f_3(\boldsymbol{x})$ and $f_4(\boldsymbol{x})$. We can then formulate the first age of lexicography goal programming:

$$\begin{aligned} \min \quad & \frac{d_1^-}{15} + \frac{d_2^-}{13.5} \\ \text{s.t.} \quad & 3x_1 + 5x_2 + d_1^- - d_1^+ = 150, \\ & 7x_1 + 3x_2 + d_2^- - d_2^+ = 135, \\ & x_1 + x_2 \le 30, \\ & 2x_1 + 3x_2 \le 60, \\ & d_i^-, d_i^+, x_1, x_2 \ge 0. \end{aligned}$$

We can solve the above problem to obtain the solution as $d_1^- = 22.5$ and $d_2^- = 0$. The explanation is that $f_1(\boldsymbol{x})$ is short by 22.5 to achieve its goal and $f_2(\boldsymbol{x})$ can achieve its goal. The next stage is to consider the second priority level and formulate the second stage of programming:

$$\begin{aligned} \min \quad & z_2 = d_3^- + d_4^- \\ \text{s.t.} \quad & 3x_1 + 5x_2 + 22.5 - d_1^+ = 150, \\ & 7x_1 + 3x_2 - d_2^+ = 135, \\ & 6x_1 + 4x_2 + d_3^- - d_3^+ = 130, \\ & -3x_1 + 5x_2 + d_4^- - d_4^+ = 100, \\ & x_1 + x_2 \le 30, \\ & 2x_1 + 3x_2 \le 60, \\ & d_i^-, d_i^+, x_1, x_2 \ge 0. \end{aligned}$$

Solving the above programming, we obtain $x_1 = 9.2$, $x_2 = 20$, $d_3^- = 0$, and $d_4^- = 27.5$. From the results, we can see that the goals of $f_2(\mathbf{x})$ and $f_3(\mathbf{x})$ can be achieved but $f_1(\mathbf{x})$ and $f_4(\mathbf{x})$ are short of the goals by 22.5 and 27.5, respectively.

3.4 MIN–MAX (TCHEBYCHEFF) GOAL PROGRAMMING

The third model of goal programming, i.e., the min–max goal programming model, is similar to the weighted goal method, but instead of minimizing the weighted sum of the deviations from the targets, the maximum deviation of any goal from the target is minimized. The goal programming of a min–max model can be described as the following mathematical system (Flavell, 1976):

$$
\begin{aligned}
\min \quad & d \\
s.t. \quad & \frac{w_i^- d_i^-}{k_i} + \frac{w_i^+ d_i^+}{k_i} \le d, \\
& f_i(x) + d_i^- - d_i^+ = t_i, \qquad i = 1,\ldots,n, \\
& g(x) \le 0 \ \text{(or writing } \boldsymbol{Ax} \le \boldsymbol{b}\text{)}, \\
& d_i^-, d_i^+ \ge 0, \\
& x \ge \mathbf{0}.
\end{aligned} \tag{3.4}
$$

This model implies the optimization of a utility function where the maximum deviation is minimized. It provides the most balanced solution among the achievements of different goals. Thus, it is the solution of maximum equity among the achievements of the different goals.

Example 3.3 Consider the following two-objective programming problem:

$$
\begin{aligned}
\max \quad & f_1(\mathbf{x}) = 3x_1 + 5x_2 \\
\max \quad & f_2(\mathbf{x}) = 6x_1 + 4x_2 \\
s.t. \quad & x_1 + x_2 \le 30, \\
& 2x_1 + 3x_2 \le 60, \\
& x_1, x_2 \ge 0.
\end{aligned}
$$

If $f_1(\boldsymbol{x})$ is equally important as $f_2(\boldsymbol{x})$, we can formulate the following min–max goal programming:

$$
\begin{aligned}
\min\ & d \\
s.t.\ & \frac{d_1^-}{15} \le d, \\
& \frac{d_2^-}{18} \le d, \\
& 3x_1 + 5x_2 + d_1^- - d_1^+ = 150, \\
& 6x_1 + 4x_2 + d_2^- - d_2^+ = 180, \\
& x_1 + x_2 \le 30, \\
& 2x_1 + 3x_2 \le 60, \\
& d_1^- \cdot d_1^+ = 0,\ d_2^- \cdot d_2^+ = 0, \\
& d_1^-, d_2^-, x_1, x_2 \ge 0.
\end{aligned}
$$

Solving the above problem, we can obtain $d_1^- = 53.57$, $d_2^- = 64.29$, $x_1 = 10.71$, and $x_2 = 12.86$. We can then calculate $f_1(\boldsymbol{x}) = 96.43$ and $f_2(\boldsymbol{x}) = 115.71$, respectively.

3.5 FUZZY GOAL PROGRAMMING

Sometimes decision makers have difficulty determining precise targets and goals of objectives. Hence, the linguistically vague statement is more suitable for dealing with this situation and fuzzy goal programming may be considered. A fuzzy goal programming problem starts with finding x:

$$
\begin{aligned}
lexmin\ & \sum_{j=1}^{l} P_j \left[\sum_{i=1}^{n} \left(\frac{w_i^- d_i^- + w_i^+ d_i^+}{k_i} \right)^p \right]^{1/p}, \quad p \ge 1 \\
s.t.\ & f_i(\boldsymbol{x}) = \widetilde{T}_i, \quad i = 1, \ldots, n, \\
& g(x) \le \boldsymbol{0} \quad (\text{or writing } \boldsymbol{Ax} \le \boldsymbol{b}) \\
& x \ge \boldsymbol{0},
\end{aligned}
\tag{3.5}
$$

where T_i denotes the ith linguistic goal. To solve the above problem, we should first define the membership functions of these linguistic statements. For the sake of simplicity, let all membership functions be symmetrically triangular functions. Then the membership functions of objectives can be represented as:

$$
\begin{aligned}
& find\ \boldsymbol{x} \\
& s.t.\ f_i(\boldsymbol{x}) = T_i, \quad i = 1, \ldots, n, \\
& g(\boldsymbol{x}) \le \boldsymbol{0} \quad (\text{or writing } \boldsymbol{Ax} \le \boldsymbol{b}),
\end{aligned}
\tag{3.6}
$$

$$\mu_i(x)=\begin{cases}[f_i(\boldsymbol{x})-(t_i-s_i)]/s_i, & \text{if} \quad t_i-s_i\le f_i(\boldsymbol{x})\le t_i,\\ [(t_i+s_i)-f_i(\boldsymbol{x})]/s_i, & \text{if} \quad t_i\le f_i(\boldsymbol{x})\le t_i+s_i,\\ 0, & \text{otherwise,}\end{cases} \tag{3.7}$$

where s_i denotes the spread or the maximum acceptable deviations of the ith fuzzy target or goal. We next formulate the fuzzy goal programming based on the max–min operation of fuzzy functions as follows:

$$\begin{aligned} \max \quad & \alpha \\ s.t. \quad & [f_i(\boldsymbol{x})-(t_i-s_i)]/s_i\ge\alpha, \quad \text{for some } i,\\ & t_i-s_i\le f_i(\boldsymbol{x})\le t_i,\\ & [(t_i+s_i)-f_i(\boldsymbol{x})]/s_i\ge\alpha, \quad \text{for other } i,\\ & t_i\le f_i(\boldsymbol{x})\le t_i+s_i,\\ & \alpha\in[0,1], \qquad \boldsymbol{x}\ge 0.\end{aligned} \tag{3.8}$$

Let $f_i(\boldsymbol{x})/s_i=t_i/s_i-d_i^+$. Then, the first two constraints can be expressed as:

$$\alpha+d_i^+\le 1, \tag{3.9}$$

and

$$\frac{f_i(\boldsymbol{x})}{s_i}-d_i^+=\frac{t_i}{s_i} \tag{3.10}$$

Similarly, let $f_i(\boldsymbol{x})/s_i=t_i/s_i+d_i^-$. We can rewrite the next two constraints as:

$$\alpha+d_i^-\le 1, \tag{3.11}$$

and

$$\frac{f_i(\boldsymbol{x})}{s_i}+d_i^-=\frac{t_i}{s_i} \tag{3.12}$$

Then, we can transform fuzzy goal programming into the following linear goal programming model:

$$\begin{aligned} \max \quad & \alpha \\ s.t. \quad & f_i(x)+s_id_i^- - s_id_i^+=t_i,\\ & \alpha+d_i^- - d_i^+\le 1,\\ & d_i^-,d_i^+\ge 0, \quad d_i^-\cdot d_i^+=0,\\ & \alpha\in[0,1], \quad x\ge\boldsymbol{0}, \quad i=1,\ldots,n.\end{aligned} \tag{3.13}$$

Example 3.4 Consider the following fuzzy two-objective goal programming problem. To find x:

$$\begin{aligned} s.t. \quad & f_1(\boldsymbol{x}) = 3x_1 + 5x_2 = \widetilde{100} \\ & f_2(\boldsymbol{x}) = 6x_1 + 4x_2 = \widetilde{110}, \\ & x_1 + x_2 \leq 30, \\ & 2x_1 + 3x_2 \leq 60, \\ & x_1, x_2 \geq 0, \end{aligned}$$

where goal 1 is about 150 with a maximum allowable deviation of 10, goal 2 is about 180 with a maximum allowable deviation of 8, constraint 1 is about 30 with a maximum allowable deviation of 3, and constraint 2 is about 60 with a maximum allowable deviation of 5.

$$\begin{aligned} \max \quad & \alpha \\ s.t. \quad & 3x_1 + 5x_2 + 10d_1^- - 10d_1^+ = 100, \\ & 6x_1 + 4x_2 + 8d_2^- - 8d_2^+ = 110, \\ & x_1 + x_2 \leq 30, \\ & 2x_1 + 3x_2 \leq 60, \\ & \alpha + d_1^- + d_1^+ \leq 1, \\ & \alpha + d_2^- + d_2^+ \leq 1, \\ & d_i^- \cdot d_i^+ \geq 0, \quad d_i^- \cdot d_i^+ = 0, \quad i = 1, 2, \\ & x_1, x_2 \geq 0. \end{aligned}$$

Solving the above problem, we can obtain the overall satisfactory degree of $\alpha =$ 0.7222 and $x_1 = 8.33$, $x_2 = 14.44$, $d_1^- = 0.28$, and $d_2^- = 0.28$. Finally, we can calculate $f_1(x) = 97.22$ and $f_1(x) = 107.77$, respectively.

In addition, fuzzy preemptive goal programming can be formulated in a similar way. For simplicity, we assume only two-priority levels and the $f_i(\boldsymbol{x})$ goal has higher priority than the $f_j(\boldsymbol{x})$ goal. We then solve the first sub-problem using the following mathematical programming:

$$\begin{aligned} \max \quad & \alpha \\ s.t. \quad & f_i(\boldsymbol{x}) + s_i d_i^- - s_i d_i^+ = t_i, \\ & \alpha + d_i^- - d_i^+ \leq 1, \\ & d_i^-, d_i^+ \geq 0, \quad d_i^- \cdot d_i^+ = 0, \\ & \alpha \in [0,1], \quad \boldsymbol{x} \geq \boldsymbol{0}, \ i = 1, \ldots, n \end{aligned} \tag{3.14}$$

Solving the above problem, we can obtain the values of derivation variables, i.e., d_i^- and d_i^+. Next, we can consider the second priority level under the condition that achievement of the solution from the first sub-problem is satisfied. Here we assume d_i^- and d_i^+ are both 0, i.e. the $f_i(\boldsymbol{x})$ goal is fully achieved. Then, the second priority level of $f_j(\boldsymbol{x})$ can be formulated as:

$$\begin{aligned}
\max\ & \alpha \\
s.t.\ & f_j(\boldsymbol{x}) + s_j d_j^- - s_j d_j^+ = t_j, \\
& f_i(\boldsymbol{x}) = t_i, \\
& \alpha + d_j^- - d_j^+ \le 1, \\
& d_j^-, d_j^+ \ge 0, \quad d_j^- \cdot d_j^+ = 0, \\
& \alpha \in [0,1], \quad \boldsymbol{x} \ge \boldsymbol{0}, \; i = 1,\ldots n;\; j = 1,\ldots,m
\end{aligned} \tag{3.15}$$

When the priority levels are extended to r levels, we can repeat the above procedures until all priority levels are exhausted.

Example 3.5 Consider the following fuzzy two-objective goal programming problem. We start by finding x:

$$\begin{aligned}
s.t.\ & f_1(\boldsymbol{x}) = 3x_1 + 5x_2 = \widetilde{100}, \text{(second priority level)} \\
& f_2(\boldsymbol{x}) = 6x_1 + 4x_2 = \widetilde{110}, \text{(first priority level)} \\
& x_1 + x_2 \le 30, \\
& 2x_1 + 3x_2 \le 60, \\
& x_1, x_2 \ge 0,
\end{aligned}$$

where $f_2(\boldsymbol{x})$ is more important than $f_1(\boldsymbol{x})$.

$$\begin{aligned}
\text{First-priority level: } \max\ & \alpha \\
s.t.\ & 6x_1 + 4x_2 + 8d_2^- - 8d_2^+ = 110, \\
& x_1 + x_2 \le 30, \\
& 2x_1 + 3x_2 \le 60, \\
& \alpha + d_2^- + d_2^+ \le 1, \\
& d_2^- \cdot d_2^+ \ge 0, \quad d_2^- \cdot d_2^+ = 0, \\
& x_1, x_2 \ge 0.
\end{aligned}$$

At the first priority level, we obtain the first priority satisfaction $\alpha = 1$, i.e., $d_2^- = 0$ and $d_2^+ = 0$. We can then deal with the second priority level:

Second-priority level: max α

$$\begin{aligned} s.t.\quad & 6x_1 + 4x_2 = 110, \\ & 3x_1 + 5x_2 + 10d_1^- - 10d_1^+ = 100, \\ & x_1 + x_2 \le 30, \\ & 2x_1 + 3x_2 \le 60, \\ & \alpha + d_1^- + d_1^+ \le 1, \\ & d_1^- \cdot d_1^+ \ge 0, \quad d_1^- \cdot d_1^+ = 0, \\ & x_1, x_2 \ge 0. \end{aligned}$$

Solving the above problem, we obtain the overall satisfaction α = 0.7, $d_1^- = 0.3$ $x_1 = 9$, and $x_2 = 14$. Hence, based on the above information, we can calculate $f_1(x) = 97$ and $f_1(x) = 110$, respectively.

4 Compromise Solution and TOPSIS

In this chapter, the compromise solutions and TOPSIS (technique for order preference by similarity to ideal solution) for MODM are introduced. Both methods involve the concept of the *Lp*-norm and find the optimal solutions based on reference points.

4.1 COMPROMISE SOLUTION

In a multiple objective programming (MOP) problem, an ideal (or utopian) point is usually not attainable if trade-offs between objectives exist. Hence, Yu (1973) proposed the compromise solutions to determine the optimal solution closest to the ideal point among Pareto solutions based on the L_p-norm distance. Figure 4.1 depicts the concept. The L_p-norm distance between a point and an ideal point can be defined as:

$$d_p = \left\| f^* - f \right\|_p, \quad p = 1, \ldots, \infty \tag{4.1}$$

In a generalized optimal problem, the distance measured by the L_p-norm between a point and the ideal point can be presented as shown in Figure 4.2. The lower left square belongs to the maximized problems (maximize all the objective functions) and is the case covered in this chapter. From Figure 4.2, it can be seen that the shape of $p = 1$ is a square diamond, $p = 2$ is a circle, and $p = \infty$ is a square. The different shapes of the L_p-norm may result in a different result due to the optimal solution. Other kinds of the L_p-norm are less discussed because they have no concrete meanings in practice.

The procedures of the compromise solutions can be demonstrated by considering a multiple objective programming (MOP) problem as follows:

$$\begin{aligned} \max \quad & z(\boldsymbol{x}) = [z_1(\boldsymbol{x}), z_2(\boldsymbol{x}), \ldots, z_n(\boldsymbol{x})] \\ s.t. \quad & g(\boldsymbol{x}) \le \boldsymbol{b} \quad (\text{or writing } \boldsymbol{Ax} \le \boldsymbol{b}), \\ & \boldsymbol{x} \ge \boldsymbol{0}. \end{aligned} \tag{4.2}$$

The first step of the compromise solution is to determine the ideal point of each objective. This can be done by optimizing each objective as follows:

$$\begin{aligned} \max \quad & z_i(\boldsymbol{x}) \\ s.t. \quad & (\boldsymbol{x}) \le \boldsymbol{b} \\ & \boldsymbol{x} \ge \boldsymbol{0}. \end{aligned} \tag{4.3}$$

Then we can obtain the ideal point as $z^* = (z_1^*, z_2^*, \ldots, z_n^*)$. Next, we want to determine which point located on the Pareto solutions is closest to the ideal point as the optimal

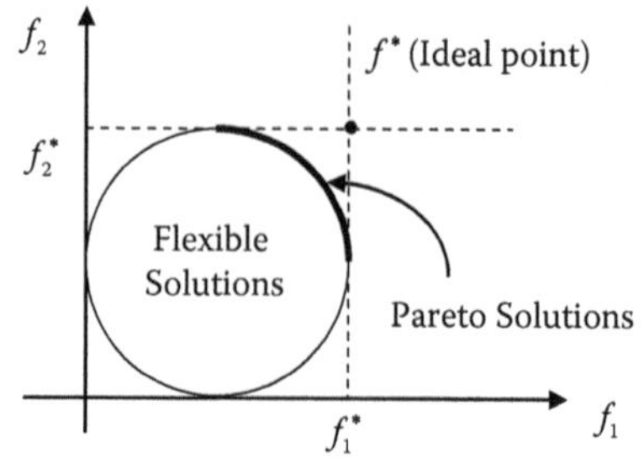

FIGURE 4.1 Concept of compromise solutions.

solution. We can use the concept of the L_p-norm to measure the distance between objective values and the ideal point and formulate a compromise solution method (Yu, 1973):

$$\min \quad d_p = \left\{ \sum_{i=1}^{n} w_i^p [z_i^*(\boldsymbol{x}) - z_i(\boldsymbol{x})]^p \right\}^{1/p}, \quad p = 1, \ldots, \infty \tag{4.4}$$

$$s.t. \quad g(\boldsymbol{x}) \le \boldsymbol{b} \quad (\text{or writing } Ax \le b$$

$$\boldsymbol{x} \ge \boldsymbol{0}.$$

where w_i denotes the importance of the ith objective. Besides using the traditional L_p-norm, Duckstein (1984) proposed the normalized L_p-norm and formulated the compromise solutions as:

$$\min \quad d_p = \left\{ \sum_{i=1}^{n} w_j^p \left[\frac{z_i^*(\boldsymbol{x}) - z_i(\boldsymbol{x})}{z_i^*(\boldsymbol{x}) - z_i^-(\boldsymbol{x})} \right]^p \right\}^{1/p}, \quad p = 1, \ldots, \infty \tag{4.5}$$

$$s.t. \quad g(\boldsymbol{x}) \le \boldsymbol{b} \ (\text{or writing } Ax \le b)$$

$$\boldsymbol{x} \ge \boldsymbol{0}.$$

where $z_i^*(\boldsymbol{x})$ and $z_i^-(\boldsymbol{x})$ denote the maximal value (or aspiration level, or positive ideal point) and the minimum value (or the worst value, or negative ideal point) of the ith goal respectively.

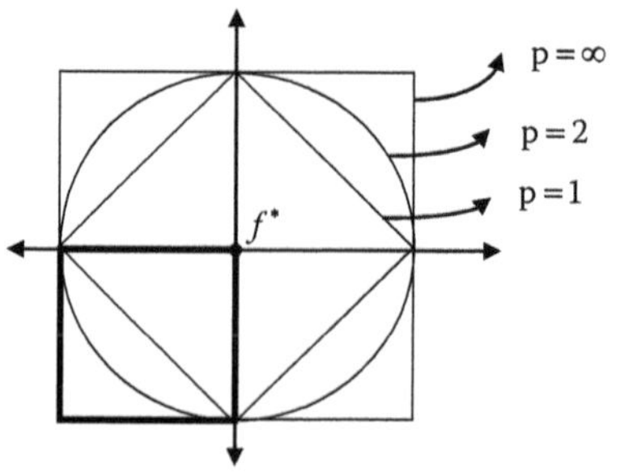

FIGURE 4.2 Concept of L_p-norm distance.

Next, we depict a two-objective case, as shown in Figure 4.3, to illustrate the concepts of the PIS and the NIS. If 'max' is better and/or 'min' is better in each objective, we can first define the PIS and the NIS, respectively, as:

Vector $\boldsymbol{f}^*$ = PIS
 = {max $f_i(x)$, $\forall i$, and/or min $f_j(x)$, $\forall j$; or setting the aspiration level of each objective}

and

Vector $\boldsymbol{f}^-$ = NIS
 = {min f_i (x), $\forall_i$, and/or max $f_j(x)$, $\forall j$; or setting the worst value of each objective}

Example 4.1 Consider a two-objective programming problem as follows:

$$\begin{aligned} \max \quad & -3x_1 + 5x_2 \\ \max \quad & 7x_1 - 4x_2 \\ s.t. \quad & 2x_1 + 3x_2 \le 30, \\ & 5x_1 + 3x_2 \le 45, \\ & 2x_1 + x_2 \ge 6, \\ & x_1, x_2 \ge 0. \end{aligned}$$

The first step of the compromise solution is to determine an ideal point. This can be done by optimizing each objective separately. Hence, the ideal point of the above problem can be calculated as $z^* = (50,63)$ and $z^- = (-27,-40)$. Then, if we set $p = 2$, we can formulate the following compromise solution programming:

$$\min \quad d_{p=2} = 0.5 \times \left\{ \left[\frac{50-(-3x_1+5x_2)}{50-(-27)} \right]^2 + \left[\frac{63-(7x_1-4x_2)}{63-(-40)} \right]^2 \right\}^{1/2}$$

$$\begin{aligned} s.t. \quad & 2x_1 + 3x_2 \le 30, \\ & 5x_1 + 3x_2 \le 45, \\ & 2x_1 + x_2 \ge 6, \\ & x_1, x_2 \ge 0. \end{aligned}$$

Finally, we can obtain $x_1 = 5.25$, $x_2 = 6.25$, and $d_{p=2} = 0.6696$. The objective values are $f_1(\boldsymbol{x}) = 15.48$ and $f_2(\boldsymbol{x}) = 11.78$, respectively. On the other hand, if we set $p = \infty$, we can also formulate the min–max compromise solution programming:

$$\begin{aligned} & \min \ v \\ s.t. \quad & \left[\frac{50-(-3x_1+5x_2)}{50-(-27)} \right] \le v, \\ & \left[\frac{63-(7x_1-4x_2)}{63-(-40)} \right] \le v, \\ & 2x_1 + 3x_2 \le 30, \end{aligned}$$

$$5x_1 + 3x_2 \le 45,$$

$$2x_1 + x_2 \ge 6,$$

$$x_1, x_2 \ge 0.$$

We obtain $x_1 = 5.43$, $x_2 = 5.96$ and $v = 0.4741$. The corresponding objective values are $f_1(\boldsymbol{x}) = 13.49$ and $f_2(\boldsymbol{x}) = 14.17$.

Comparing the results of the compromise solutions with $p = 2$ and $p = \infty$, we can see that the optimal solution is different due to the different definition of distance. Usually, $p = 1$, $p = 2$ and $p = \infty$ are widely used in dealing with MOP problems. Other distances are less used because they lack practical meanings.

4.2 TOPSIS FOR MODM

In addition to compromise solutions, the technique for order preference by similarity to ideal solution (TOPSIS; Huang and Yoon, 1981) is another multiple objective programming (MOP) method using the concept of L_p-norm. TOPSIS considers both the positive ideal solution (PIS) and the negative ideal solution (NIS) to determine the optimal solution of a MODM problem that should be the closest to the PIS and the farthest from the NIS.

Next, we depict a two-objective case, as shown in Figure 4.3, to illustrate the concepts of the PIS and the NIS. We first define the PIS and the NIS, respectively, as:

$$f^* = \text{PIS} = \{\max / f_j(x), \forall j, \text{ and/or} \min f_i(x), \forall i\}$$

and

$$f^- = \text{NIS} = \{\min f_j(x), \forall j, \text{ and/or} \max f_i(x), \forall i\}$$

where $i \in \boldsymbol{J}$ *and* $j \in \boldsymbol{I}$. $\boldsymbol{I}$ is a set of benefit attribute, and $\boldsymbol{J}$ is a set of cost attribute. A optimal solution should belong to the Pareto solutions and be closest of NIS and farthest to NIS. Surely, we can use L_p-norm to measure the distance between objective values, the PIS and the NIS as follows:

$$d_p^{pis}(\boldsymbol{x}) = \left\{ \sum_{i \in I} w_i^p \left[\frac{f_i^* - f_i(\boldsymbol{x})}{f_i^* - f_i^-} \right]^p + \sum_{j \in J} w_j^p \left[\frac{f_j(\boldsymbol{x}) - f_j^*}{f_j^- - f_j^*} \right]^p \right\}^{1/p} \tag{4.6}$$

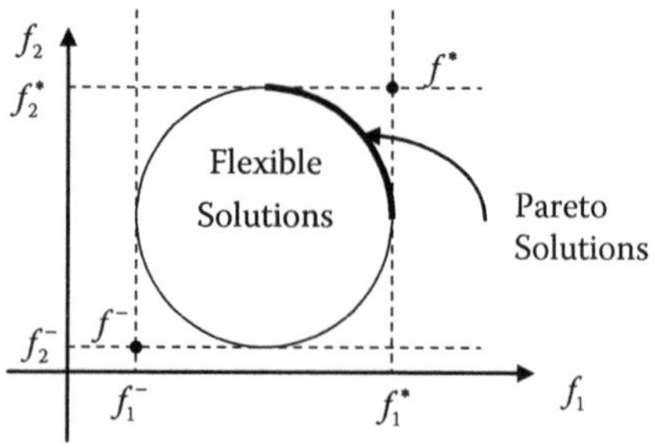

FIGURE 4.3 Concepts of PIS and NIS.

and

$$d_p^{nis}(\boldsymbol{x}) = \left\{ \sum_{i \in I} w_i^p \left[\frac{f_i(\boldsymbol{x}) - f_i^-}{f_i^* - f_i^-} \right]^p + \sum_{j \in J} w_j^p \left[\frac{f_j^- - f_j(\boldsymbol{x})}{f_j^- - f_j^*} \right]^p \right\}^{1/p} \tag{4.7}$$

We can then transform the concept of TOPSIS to solve the following two-objective programming:

$$\begin{aligned} \max \quad & d_p^{nis}(\boldsymbol{x}) \\ \min \quad & d_p^{pis}(\boldsymbol{x}) \\ s.t. \quad & \boldsymbol{Ax} \le \boldsymbol{b}, \\ & x \ge 0 \end{aligned} \tag{4.8}$$

or the following fractional single-objective programming:

$$\begin{aligned} \min \quad & \frac{d_p^{pis}(\boldsymbol{x})}{d_p^{nis}(\boldsymbol{x})} \\ s.t. \quad & \boldsymbol{Ax} \le \boldsymbol{b}, \\ & \boldsymbol{x} \ge \boldsymbol{0}. \end{aligned} \tag{4.9}$$

Example 4.2 Let us reconsider the problem of Example 4.1 as follows:

$$\begin{aligned} \max \quad & f_1(\boldsymbol{x}) = -3x_1 + 5x_2 \\ \max \quad & f_2(\boldsymbol{x}) = 7x_1 - 4x_2 \\ s.t. \quad & 2x_1 + 3x_2 \le 30, \\ & 5x_1 + 3x_2 \le 45, \\ & 2x_1 + x_2 \ge 6, \\ & x_1, x_2 \ge 0. \end{aligned}$$

By solving the above objective separately, we obtain PIS = (50,63) and NIS = (–27,–40). We then set p = 2 and formulate TOPSIS for the above problem by calculating the following fractional programming problem:

$$\min \quad \frac{d_{p=2}^{pis}(\boldsymbol{x}) = 0.5 \times \left[\left(\frac{50-(-3x_1+5x_2)}{77} \right)^2 + \left(\frac{63-(7x_1-4x_2)}{103} \frac{63-(7x_1-4x_2)}{103} \right)^2 \right]^{1/2}}{d_{p=2}^{nis}(\boldsymbol{x}) = 0.5 \times \left[\left(\frac{(-3x_1+5x_2)+27}{77} \right)^2 + \left(\frac{(7x_1-4x_2)+40}{103} \right)^2 \right]^{1/2}}$$

$$\begin{aligned} s.t. \quad & 2x_1 + 3x_2 \le 30, \\ & 5x_1 + 3x_2 \le 45, \\ & 2x_1 + x_2 \ge 6, \\ & x_1, x_2 \ge 0. \end{aligned}$$

Finally, we can obtain the optimal solution of the above problem as $x_1 = 5.00$ and $x_2 = 6.67$. Then, we can derive $f_1(\boldsymbol{x}) = 18.35$ and $f_2(\boldsymbol{x}) = 8.32$.

4.3 FUZZY COMPROMISE SOLUTION AND TOPSIS

In this section, the compromise solutions and TOPSIS are extended to fuzzy environments. All parameters of the models are considered fuzzy numbers and we deal with the following situation:

$$\begin{aligned} \max \quad & \tilde{z}(\boldsymbol{x}) = [\tilde{z}_1(\boldsymbol{x}), \tilde{z}_2(\boldsymbol{x}), \ldots, \tilde{z}_n(\boldsymbol{x})] \\ s.t. \quad & \tilde{g}(\boldsymbol{x}) \le \tilde{\boldsymbol{b}} \text{ (or writing } \tilde{\boldsymbol{A}}x \le \tilde{\boldsymbol{b}}), \\ & \boldsymbol{x} \ge \boldsymbol{0}. \end{aligned} \tag{4.10}$$

To solve the above fuzzy MOP, we first introduce the concept of possibilistic distribution of fuzzy numbers. Let $\boldsymbol{x}_\beta^\alpha$ be a solution of Equation (4.10), where $\alpha \in [0,1]$ denotes the level of possibility and $\beta \in [0,1]$ denotes the grade of compromise to which the solution satisfies all of the fuzzy goals. Next, let $\tilde{m}_\alpha$ be the α-cut of a triangular fuzzy number $\tilde{m}$, defined as

$$\tilde{m}_\alpha = \{m \in \text{supp}(\tilde{m}) \mid \mu_{\tilde{m}}(m) \ge \alpha\}, \tag{4.11}$$

where $\text{supp}(\tilde{m})$ is the support of $\tilde{m}$. If we define $\tilde{m}_\alpha^l$ and $\tilde{m}_\alpha^u$ as the lower and upper bounds of the α-cut of $\tilde{m}$, respectively, we can obtain

$$\tilde{m}_\alpha^l \le \tilde{m}_\alpha \le \tilde{m}_\alpha^u. \tag{4.12}$$

For a specific value of α-cut, we can replace fuzzy objectives by the upper bound and the lower bound such that

$$\tilde{z}_i(\boldsymbol{x})_\alpha^u = \sum_{j=1}^{m} (\tilde{c}_{ij})_\alpha^u x_j, \qquad i = 1,\ldots,n. \tag{4.13}$$

If objectives are to be minimized, we can rewrite Equation (4.13) as

$$\tilde{z}_i(\boldsymbol{x})_\alpha^l = \sum_{j=1}^{m} (\tilde{c}_{ij})_\alpha^l x_j, \qquad i = 1,\ldots,n. \tag{4.14}$$

In addition, we can use the above technique to deal with fuzzy constraints. Therefore, for constraints $\tilde{g}(\boldsymbol{x}) \le \tilde{\boldsymbol{b}}$ or $\tilde{g}(\boldsymbol{x}) \ge \tilde{\boldsymbol{b}}$, we can formulate

$$\sum_{j=1}^{m} (\tilde{a}_{ij})_\alpha^l x_j \le (\tilde{b}_k)_\alpha^u, \qquad k = 1,\ldots,r \tag{4.15}$$

or

$$\sum_{j=1}^{m} (\tilde{a}_{ij})_{\alpha}^{u} x_j \le (\tilde{b}_k)_{\alpha}^{l}, \qquad k = 1,\ldots,r. \tag{4.16}$$

Then, the original fuzzy MOP problems can be rewritten as the following crisp MOP problems:

$$\begin{aligned} \max \quad & \tilde{z}_i(\boldsymbol{x})_{\alpha}^{l} = \sum_{j=1}^{m} (\tilde{c}_{ij})_{\alpha}^{l} x_j, \qquad i = 1,\ldots,n \\ s.t. \quad & \sum_{j=1}^{m} (\tilde{a}_{ij})_{\alpha}^{l} x_j \le (\tilde{b}_k)_{\alpha}^{u}, \qquad k = 1,\ldots,r, \\ & \boldsymbol{x} \ge \boldsymbol{0}. \end{aligned} \tag{4.17}$$

Solving the above objective function independently, we can obtain the positive ideal point z_i^* and negative ideal point z_i^-, respectively. We then formulate the following linear programming to solve the grade of compromise of the solution:

$$\begin{aligned} \min \quad & \beta \\ s.t. \quad & \frac{[\tilde{z}_i^* - \sum_{j=1}^{m} (\tilde{c}_{ij})_{\alpha}^{u} x_j]}{(\tilde{z}_i^* - \tilde{z}_i^-)} \le \beta, \\ & \sum_{j=1}^{m} (\tilde{a}_{ij})_{\alpha}^{l} x_j \le (\tilde{b}_k)_{\alpha}^{u}, \qquad k = 1,\ldots,r, \\ & \boldsymbol{x} \ge \boldsymbol{0}. \end{aligned} \tag{4.18}$$

For the above fuzzy compromise solution problems in which the value of α is decreased gradually, the value of β increases steadily. Therefore, the best solution is obtained at $\alpha = \beta$ while objectives are equally important.

Example 4.3 Let us reconsider the following fuzzy two-objective mathematical problem:

$$\begin{aligned} \max \quad & \tilde{f}_1(\boldsymbol{x}) = -3x_1 + \tilde{5}x_2 \\ \max \quad & \tilde{f}_2(\boldsymbol{x}) = \tilde{7}x_1 - 4x_2 \\ s.t. \quad & 2x_1 + \tilde{3}x_2 \le \widetilde{30}, \\ & \tilde{5}x_1 + 3x_2 \le \widetilde{45}, \\ & \tilde{2}x_1 + x_2 \ge \tilde{6}, \\ & x_1, x_2 \ge 0. \end{aligned}$$

Note that $\tilde{5} = (4,5,6)$, $\tilde{7} = (6,7,8)$, $\tilde{3} = (2,3,4)$, $\widetilde{30} = (25,30,35)$, $\widetilde{45} = (40,45,50)$, $\tilde{2} = (1,2,3)$ and $\tilde{6} = (5,6,7)$. Here, we assume two objectives are equally important. Then, we can transform the above fuzzy numbers into the possibilistic form and formulate the following deterministic linear programming:

$$\begin{aligned}
\max \quad & \tilde{f}_1(\boldsymbol{x}) = -3x_1 + (6-\alpha)x_2 \\
\max \quad & \tilde{f}_2(\boldsymbol{x}) = (8-\alpha)x_1 - 4x_2 \\
s.t. \quad & 2x_1 + (2+\alpha)x_2 \le (35-5\alpha), \\
& (4+\alpha)x_1 + 3x_2 \le (50-5\alpha), \\
& (3-\alpha)x_1 + x_2 \ge (8+\alpha), \\
& (x_1, x_2 \ge 0.
\end{aligned}$$

Next, we should set a specific α-cut and derive the positive and negative ideal points of the two objectives by optimizing each objective independently. Then, we can obtain the degree of the compromise coefficient β. For example, if we set α-cut = 0.5, we can obtain the positive ideal points as (71.5, 79.17) and the negative ideal points as (–31.68, –52). We then derive the compromise coefficient by solving $\max \beta$:

$$\begin{aligned}
& \max \beta \\
s.t. \quad & \frac{[z_1^* + 3x_1 + (6-\alpha)x_2]}{(z_1^* - z_1^-)} \le \beta, \\
& \frac{[z_2^* - (8-\alpha)x_1 - 4x_2]}{(z_2^* - z_2^-)} \le \beta, \\
& 2x_1 + (2+\alpha)x_2 \le (35-5\alpha), \\
& (4+\alpha)x_1 + 3x_2 \le (50-5\alpha), \\
& (3-\alpha)x_1 + x_2 \ge (8+\alpha), \\
& x_1, x_2 \ge 0
\end{aligned}$$

and derive β = 0.4855. Table 4.1 shows the different α-cuts and corresponding compromise coefficients. The table indicates the optimal solution at α = β = 0.4860 with x_1 = 5.85 and x_2 = 7.11. The corresponding goals are $f_1(\boldsymbol{x}) = 21.67$ and $f_2(\boldsymbol{x}) = 15.49$.

Conversely, we can derive fuzzy TOPSIS for fuzzy MOP using the similar method above. First, we derive the PIS and the NIS of each objective, respectively, according to Equation (4.17). Then we can choose the following mathematical programming to obtain the result of fuzzy TOPSIS:

$$\begin{aligned}
& \min \beta \\
& s.t.\ \tilde{d}_p^{pis}(\boldsymbol{x}) \ge \beta, \\
& \tilde{d}_p^{nis}(\boldsymbol{x}) \le 1-\beta \\
& \sum_{j=1}^{m} (\tilde{a}_{ij})_\alpha^l x_j \le (\tilde{b}_k)_\alpha^u, \quad k = 1,\ldots,r, \\
& \boldsymbol{x} \ge \boldsymbol{0}.
\end{aligned} \tag{4.19}$$

TABLE 4.1
Optimal Solutions for Various α-Cuts

α–cut	f_1^*	f_1^-	f_2^*	f_2^-	β	x_1	x_2	$f_1(x)$	$f_2(x)$
1.0	50.00	–27.00	63.00	–40.00	0.4741	5.43	5.96	13.49	14.17
0.9	53.64	–27.86	65.93	–42.07	0.4759	5.52	6.16	14.85	14.53
0.8	57.57	–28.75	69.00	–44.29	0.4779	5.60	6.37	16.32	14.86
0.7	61.83	–29.68	72.22	–46.67	0.4811	5.69	6.58	17.80	15.23
0.6	66.46	–30.65	75.61	–49.23	0.4828	5.76	6.83	19.58	15.34
0.5	71.50	–31.68	79.17	–52.00	0.4855	5.84	7.08	21.40	15.48
0.4860*	72.23	–31.81	79.67	–52.40	0.4860*	5.85	7.11	21.67	15.49
0.4	77.00	–32.73	82.91	–55.00	0.4886	5.91	7.34	23.38	15.52
0.3	83.02	–33.84	86.85	–58.26	0.4920	5.96	7.62	25.53	15.46
0.2	86.61	–35.00	91.00	–61.82	0.4895	6.08	7.82	27.08	16.19
0.1	89.29	–36.22	95.38	–65.71	0.4849	6.23	7.99	28.43	17.27
0	92.00	–37.50	100.00	–66.67	0.4850	6.44	8.08	29.19	19.16

Note: 0.4860* denotes the optimal solution at $\alpha = \beta = 0.4860$ with $x_1 = 5.85$ and $x_2 = 7.11$.

The optimal solution of fuzzy TOPSIS at $\alpha = \beta$. To illustrate fuzzy TOPSIS, we can reformulate Example 4.3 to solve the following mathematical programming:

$$\min \beta$$

$$s.t. \frac{[z_1^* + 3x_1 - (6-\alpha)x_2]}{(z_1^* - z_1^-)} \le \beta,$$

$$\frac{[z_2^* - (8-\alpha)x_1 + 4x_2]}{(z_2^* - z_2^-)} \le \beta,$$

$$\frac{[-3x_1 + (6-\alpha)x_2 - z_1^-]}{(z_1^* - z_1^-)} \le 1-\beta,$$

$$\frac{[(8-\alpha)x_1 - 4x_2 - z_2^-]}{(z_2^* - z_2^-)} \le 1-\beta,$$

$$2x_1 + (2+\alpha)x_2 \le (35-5\alpha),$$

$$(4+\alpha)x_1 + 3x_2 \le (50-5\alpha),$$

$$(3-\alpha)x_1 + x_2 \ge (8+\alpha),$$

$$x_1, x_2 \ge 0$$

If we set $\alpha = 0.5$, we can derive the PIS = (71.50, 79.17) and the NIS = (–31.68, –52). Then the compromise coefficient $\beta = 0.4855$. In addition, we can obtain $x_1 = 5.84$ and $x_2 = 7.08$. The corresponding goals are $f_1(\boldsymbol{x}) = 21.40$ and $f_2(\boldsymbol{x}) = 15.48$.

5 De Novo Programming and Changeable Parameters

When dealing with MODM problems, we usually confront a situation in which it is almost impossible to optimize all criteria in a given system. This property requires so-called trade-offs—we cannot increase the level of satisfaction for a criterion without decreasing the level for another criterion. Zeleny (1981, 1986) proposed De Novo programming to re-design or re-shape given systems to achieve aspirations or desired levels. He suggested that trade-offs are properties of an inadequately designed system and thus can be eliminated by re-designing a better, preferably optimal, system.

5.1 DE NOVO PROGRAMMING

The usefulness of De Novo programming can be illustrated by the following MODM problem (Zeleny, 1982). Assume a factory makes two products (suits and dresses) in quantities x and y. Each of them requires five resources (nylon, velvet, silver thread, silk, and golden thread) according to technologically determined requirements. The unit prices of these resources are shown in Table 5.1.

Two objectives, namely profit (f_1) and quality (f_2), are considered by the company. We can formulate the following two-objective mathematical programming:

$$\begin{aligned}
\max \quad & f_1 = 400x_1 + 300x_2 \\
\max \quad & f_2 = 6x_1 + 8x_2 \\
s.t. \quad & 4x_1 \le 20, \\
& 2x_1 + 6x_2 \le 24, \\
& 12x_1 + 4x_2 \le 60, \\
& 3x_2 \le 0.5, \\
& 4x_1 + 4x_2 \le 26, \\
& x_1, x_2 \ge 0.
\end{aligned}$$

where f_1 and f_2 denote the profit and quality objectives, respectively. Let the two objectives be equally important. Then, if we employ the compromise solutions and set $p = 2$, we can obtain the optimal solution as $x_1 = 3.9837$, $x_2 = 2.5163$,

TABLE 5.1
Resource Allocation in Zeleny's Example

		Technological Coefficients		
Unit Price	**Resource**	$x = 1$	$y = 1$	**No. of Units**
30	Nylon	4	0	20
40	Velvet	2	6	24
9.5	Silver thread	12	4	60
20	Silk	0	3	10.5
10	Golden thread	4	4	26

$f_1 = 2348.37$, and $f_2 = 44.03$. However, the solution may be unsatisfactory for decision makers because of the inappropriate resource portfolio.

Therefore, Zeleny (1995) proposed the concept of optimal portfolio of resources based on a design integrating system resources to eliminate trade-offs in a new system. The original idea of de novo programming was that productive resources should not be engaged individually because they are not independent.

For example, when the design budget of a new optimal system exceeds the total available budget, Zeleny (1995) suggested an optimum path ratio to contract the new design budget to the size of the available budget along an optimal path. Later, additional concepts such as fuzzy coefficients (Li and Lee, 1990), optimum-path ratios (Shi, 1995), and 0–1 programming (Kim et al., 1993), were proposed to cover more complicated situations. By releasing the constraint of fixed resources, de novo programming attempts to eliminate or minimize limitations to achieve a desired solution.

Zeleny (2005) provided eight major optimality concepts according to dual classification: single versus multiple criteria versus the extent of the *given*, ranging from *all but* to *none except*, as shown in Table 5.2. The classification clearly distinguishes de novo programming from traditional optimal methods. Instead of optimizing objective functions for fixed resources, de novo programming seeks to resolve optimal design problems with changeable resources. Traditionally, resource allocation

TABLE 5.2
Eight Concepts of Optimality

Criteria Given	Single	Multiple
Criteria and alternatives	Traditional optimality	Multi-criteria decision making
Criteria only	Optimal design (de novo programming)	Optimal design (de novo programming)
Alternatives only	Optimal valuation (limited equilibrium)	Optimal valuation (limited equilibrium)
Value complex only	Cognitive equilibrium (matching)	Cognitive equilibrium (matching)

problems (Hackman and Platzman, 1990) can be considered by maximizing the following multi-objective knapsack problem:

$$\begin{aligned} \max \quad & z = \boldsymbol{Cx} \\ s.t. \quad & \boldsymbol{Ax} \le \boldsymbol{b}, \\ & \boldsymbol{x} \ge \boldsymbol{0}. \end{aligned} \tag{5.1}$$

where matrix C and vector x denote given resource parameters, matrix A denotes the technological coefficient, and b denotes the maximum limited resource portfolio. Because the components of b are determined in advance, an ideal point usually is not attainable for the properties of trade-offs among multiple criteria. Therefore, the key to optimizing objective functions depends on the appropriate resource parameters and portfolio. In practice, however, it may be hard to achieve aspiration levels due to inappropriate resource allocation.

In addition, although it is rational to allocate resources using the above equation in a hierarchical system, when resource allocation problems are subject to market-based systems, the factor of unit price must be considered and the traditional methods are no longer suitable. To achieve optimal resource allocation, de novo programming is proposed. The procedures of de novo programming can be described as follows:

1. Find the aspiration level vector ($\boldsymbol{z}^u$) by solving each objective function of a system separately as

$$\begin{aligned} \max \quad & z_k^u = \boldsymbol{c}_k \boldsymbol{x}, \ k = 1, \ldots, m \\ s.t. \quad & V\boldsymbol{x} \le B, \\ & \boldsymbol{x} \ge \boldsymbol{0}. \end{aligned} \tag{5.2}$$

   where $\boldsymbol{z}^u = [z_1^u \quad z_2^u \quad \cdots \quad z_m^u]$ denotes the aspiration level vector, $V = p$A denotes the unit cost vector, p is the resource's unit price vector, and B denotes the total budget.
2. Identify the minimum budget B^* and its corresponding resource allocation ($\boldsymbol{x}^*$ and $\boldsymbol{b}^* = \boldsymbol{Ax}^*$) with the aspiration level, such as

$$\begin{aligned} \min \quad & B^* = V\boldsymbol{x}^* \\ s.t. \quad & \boldsymbol{Cx} = z^*, \\ & \boldsymbol{x} \ge \boldsymbol{0}. \end{aligned} \tag{5.3}$$

3. Use the optimum-path ratio (r) to obtain the final solution (z, x and b).

$$z = r \times z^*, \tag{5.4}$$

$$\boldsymbol{x} = r \times \boldsymbol{x}^* \tag{5.5}$$

and

$$\boldsymbol{b} = r \times \boldsymbol{b}^*, \tag{5.6}$$

where

$$r = B / B^*. \tag{5.7}$$

Example 5.1

We can give a numerical example to demonstrate de novo programming procedures. If a company wants to optimize the objectives of profit (f_1) and customer satisfaction (f_2) with limited resources, we can formulate the following programming:

$$\begin{aligned} \max \quad & f_1 = 12x_1 + 25x_2 \\ \max \quad & f_2 = 5x_1 + 2x_2 \\ s.t. \quad & 3x_1 + 5x_2 \le b_1, \\ & 6x_1 + 4x_2 \le b_2, \\ & x_1, x_2 \ge 0, \end{aligned}$$

where profit and customer satisfaction are equally important. For the sake of simplicity, the unit price of b_1 is equal to 0.8, the unit price of b_2 is 0.7, and the company budget equals 300. We can reformulate the above mathematical programming as:

$$\begin{aligned} \max \quad & f_1 = 12x_1 + 25x_2 \\ \max \quad & f_2 = 5x_1 + 2x_2 \\ s.t. \quad & 0.8 \times (3x_1 + 5x_2) + 0.7 \times (6x_1 + 4x_2) \le 300, \\ & x_1, x_2 \ge 0. \end{aligned}$$

To optimize two objectives separately, we can obtain $z_1^u = 1102.940$ when the decision variable $(x_1, x_2) = (0, 44.11765)$ and $z_2^u = 227.273$ when the decision variable $(x_1, x_2) = (45.45455, 0)$, respectively. We can then calculate the minimum budget B^* and the corresponding resource allocation ($\boldsymbol{x}^*$ and $\boldsymbol{b}^*$) as $B^* = 414.8091$, $\boldsymbol{x}^* = [34.41530 \quad 27.59826]'$, and $\boldsymbol{b}^* = [241.2372 \quad 316.88484]'$. Next, we can derive the optimum-path ratio (r) to obtain the final solution of resource allocation:

$$\begin{aligned} \boldsymbol{z} &= [797.646 \quad 164.364]', \\ \boldsymbol{x} &= [24.889 \quad 19.959]' \\ \boldsymbol{b} &= [174.463 \quad 229.171]', \end{aligned}$$

where r = 0.7232. Based on the calculated results, we should buy $b_1 = 174.463$ and $b_2 = 229.171$ under the total budget B = 300 to produce $x_1 = 24.889$

and $x_2 = 19.959$ for obtaining profit $f_1 = 797.646$ and customer satisfaction $f_2 = 164.364$.

In addition, readers can reconsider the beginning example to solve the de novo programming problem and obtain the optimal portfolio as $x = 4.03$ and $y = 2.54$, $b_1 = 16.12$, $b_2 = 23.3$, $b_3 = 58.52$, $b_4 = 7.62$, and $b_5 = 26.28$ under the total budget $B = 300$ to generate $f_1 = 2375$ and $f_2 = 44.5$.

5.2 DE NOVO PROGRAMMING BY GENETIC ALGORITHMS

In the previous section, we assumed the two objectives of profit and customer satisfaction were equal. However, the importance levels of two objectives may vary because of different preferences of decision makers. If we reconsider the steps of de novo programming, we can see that only one solution of the efficient frontier is derived. Although Shi (1995) provided six types of optimum ratios to consider other possibilities of resource allocation, decision makers like to consider efficient whole solutions and then pick one based on utility. Therefore, we can reconsider Example 5.1 and depict its decision and objective spaces as shown in Figure 5.1.

We can see that the solution of de novo programming proposed by Zenley is only one point of the efficient frontier. Although Shi (1995) provided six types of optimum ratios for decision makers to consider in other situations, they are also the special points of the efficient frontier. Hence, it is effective to derive all efficient solutions via de novo programming.

Here we adopt genetic algorithms to derive possible solutions of Example 5.1. The parameters of genetic algorithms are arranged as follows: populations = 100, generations = 1000, crossover rate = 0.8, mutation rate = 0.01, and crossover type = uniform. We picked five solutions of the result and calculated the corresponding resouce portfolio and objective functions, as shown in Table 5.3.

From the results of Table 5.3, we can consider other conditions when determining the solution of de novo programming For example, if a decision maker hopes company profit will exceed 900, Ratio 5 should be the optimization. On the other hand, if he or she wants customer satisfaction to be larger than 180, Ratio 1 should be the choice. However, Zenley's method cannot meet additional conditions or many practical applications.

Genetic algorithms also provide operational convenience when calculating de novo programming. That is, genetic algorithms do not need to derive the optimal

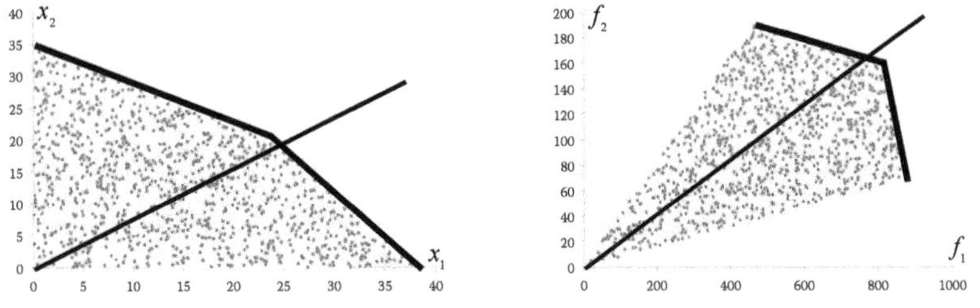

FIGURE 5.1 Decision space and objective space of Example 5.1.

TABLE 5.3
Various Ratios of De Novo Programming Derived by Genetic Algorithms

Type	x_1	x_2	b_1	b_2	Budget	f_1	f_2
Ratio 1	33.021	12.023	159.179	246.217	300	696.833	189.150
Ratio 2	27.869	17.067	168.944	235.484	300	761.114	173.480
Ratio 3	23.084	21.701	177.757	225.308	300	819.531	158.822
Ratio 4	16.686	27.918	189.648	211.789	300	898.182	139.267
Ratio 5	13.514	30.968	195.381	204.956	300	936.364	129.506
Zenley	24.889	19.959	174.463	229.171	300	797.646	164.364

solution of each objective function separately and set up only related parameters to obtain all possible solutions simultaneously.

5.3 DE NOVO PROGRAMMING BY COMPROMISE SOLUTION

Genetic algorithms provide an efficient method to derive Pareto solutions of de novo programming when a decision maker's preference is uncertain. However, these Pareto solutions may be cumbersome if the preference of the decision maker is precise. In this situation, conventional MODM methods can be used to deal with de novo programming.

In this section, we adopt the compromise solutions proposed by Yu (1973) and Zeleny (1972,1973,1975), to derive the solution of de novo programming when the importance of objective functions are known. The concepts of the compromise solutions are described below.

Assume the outcome space of a trade-off of two objectives (f_1 and f_2) is presented as shown in Figure 5.2. Generally, when each objective is characterized as "more is better," y^* should be a unique and desired target or called the ideal point. However, y^* is usually not attainable due to the trade-off among objectives.

Therefore, the problem of calculating the ideal point is transformed into finding a point that is attainable and closest to the ideal. This problem triggers another issue:

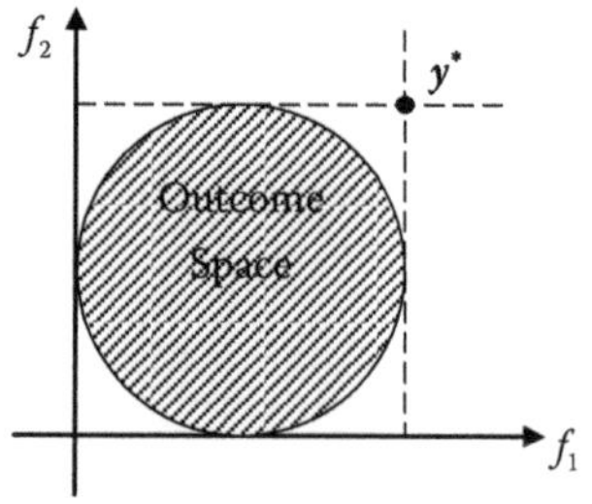

FIGURE 5.2 Outcome space of bi-objective programming problem.

how to define the distance between two vectors. Generally, the distance from vector $\boldsymbol{y}$ to $\boldsymbol{y}^*$ can be defined as the following L_p-norm:

$$r(\boldsymbol{y};p) = \|\boldsymbol{y} - \boldsymbol{y}^*\|_p = \left(\sum_i |y_i - y_i|^p \right)^{1/p} \tag{5.8}$$

where $p \geq 1$. The $r(y;p)$ provides the measurement of distance (or regret) from $\boldsymbol{y}$ to $\boldsymbol{y}^*$ according to the L_p-norm. Then, the compromise solutions can be formulated as:

$$\begin{aligned} &\min r(\boldsymbol{y};p) = \|\boldsymbol{y} - \boldsymbol{y}^*\|_p \\ &s.t.\ \boldsymbol{X} = \{\boldsymbol{x} \in \boldsymbol{X} \mid g_k(\boldsymbol{x}) \leq 0, \quad k = 1,\ldots,m\}. \end{aligned} \tag{5.9}$$

Let us reconsider Example 5.1 by using compromise solutions. First, we should calculate the ideal point $\boldsymbol{y}^*$ by optimizing each objective function separately as $\boldsymbol{y}^*$ = (1102.940, 227.273). Second, if the objective profit f_1 is twice as important as the objective customer satisfaction f_2, the compromise solution of Example 5.1 can be formulated as:

$$\begin{aligned} \min \quad & r(\boldsymbol{y};\infty) = v \\ s.t. \quad & \frac{2}{3} \times \frac{1102.940 - (12x_1 + 25x_2)}{1102.940} \leq v, \\ & \frac{1}{3} \times \frac{227.273 - (5x_1 + 2x_2)}{227.273} \leq v, \\ & 0.8 \times (3x_1 + 5x_2) + 0.7 \times (6x_1 + 4x_2) \leq 300, \\ & x_1, x_2 \geq 0. \end{aligned}$$

$p = \infty$. By calculating the above mathematical programming, we obtain $x_1 = 24.890$, $x_2 = 19.960$, and $v = 0.277$, where v can be viewed as the measurement of regret. When the value of v increases, the more regret or trade-off among objective functions occurs. Then we derive the corresponding resource portfolio as $b_1 = 174.470$, $b_2 = 229.180$, $B = 300$, $f_1 = 797.680$, and $f_2 = 164.370$.

The compromise solutions provide a better way than searching all Pareto solutions to determine the final resource portfolio if the preference of a decision maker is clear. In addition, it is more efficient and flexible than Zeleny's or Shi's methods to derive an optimal solution. Although the compromise solution method is used here to solve de novo programming, other MCDM methods such as goal programming can also be used similarly.

5.4 EXTENSIONS OF DE NOVO PROGRAMMING

Several extensions of de novo programming have been proposed. Zeleny (1989) considered one kind of extension of a de novo programming formulation to approximate the real concerns of free-market producers:

$$\begin{aligned} \max \quad & \sum_j c_j(x_j)x_j - \left(\sum_{i\in I_1} p_i b_i\right)^{\pi_1} - \cdots - \left(\sum_{i\in I_r} p_i b_i\right)^{\pi_r} \\ s.t. \quad & \sum_j a_{ij}x_j - b_i \le 0, \quad i \in \boldsymbol{I}, \\ & \left(\sum_{i\in \boldsymbol{I}} p_i b_i\right)^{\beta} \le B, \end{aligned} \tag{5.10}$$

where

$$I = I_1 \cup \ldots \cup I_r, \quad I_s \cup I_{s+1} = 0, \quad 0 < \pi_s < 1, \quad s = 1,\ldots,r, \quad \beta \ge 1$$

and

$$c_j(x_j) = \begin{cases} c_j^1, & x_j \le x_j^1 \\ c_j^2, & x_j < x_j \le x_j^2 \\ \vdots & \\ c_j^{k_j}, & x_j^{k_j-1} \le x_j, \end{cases}$$

where

$$c_j^h \ge c_j^{h+1}, \quad h = 1,\ldots,k_j.$$

In addition, Li and Lee (1993) proposed de novo programming with fuzzy parameters:

$$\begin{aligned} \max \quad & \tilde{z}_k^u = \tilde{\boldsymbol{c}}_k \boldsymbol{x} \quad k = 1,\ldots,m \\ s.t. \quad & \tilde{\boldsymbol{A}}\boldsymbol{x} \le \tilde{\boldsymbol{b}}, \\ & \tilde{\boldsymbol{p}}'\boldsymbol{b} \le \tilde{B}, \\ & \boldsymbol{x} \ge \boldsymbol{0}. \end{aligned} \tag{5.11}$$

To solve the above fuzzy mathematical programming, we can consider the concept of the membership function and reformulate Equation (5.11):

$$\begin{aligned} \max \quad & (\tilde{z}_k)_\alpha = \mu_{\tilde{c}_k}^{-1}[\alpha]\boldsymbol{x}, \quad k = 1,\ldots,m \\ s.t. \quad & \mu_{\tilde{p}'}^{-1}[\alpha]\mu_{\tilde{A}}^{-1}[\alpha]\boldsymbol{x} \le \mu_{\tilde{B}}^{-1}[\alpha], \\ & \alpha \in [0,1], \\ & \boldsymbol{x} \ge \boldsymbol{0}, \end{aligned} \tag{5.12}$$

where $[\alpha]$ denotes the α*–cut* operator.

Example 5.2 To demonstrate de novo programming with fuzzy parameters, we modify Example 5.1:

$$\begin{aligned} \max \quad & \tilde{f}_1 = \tilde{c}_{11}x_1 + \tilde{c}_{12}x_2 \\ \max \quad & \tilde{f}_2 = \tilde{c}_{21}x_1 + \tilde{c}_{22}x_2 \\ s.t. \quad & \tilde{a}_{11}x_1 + \tilde{a}_{12}x_2 \le \tilde{b}_1, \\ & \tilde{a}_{21}x_1 + \tilde{a}_{22}x_2 \le \tilde{b}_2, \\ & x_1, x_2 \ge 0, \end{aligned}$$

where $\tilde{c}_{11} = (11,11,13)$, $\tilde{c}_{12} = (24,24,26)$, $\tilde{c}_{21} = (4,4,6)$, $\tilde{c}_{22} = (1,1,3)$, $\tilde{a}_{11} = (2,4,4)$, $\tilde{a}_{12} = (4,6,6)$, $\tilde{a}_{21} = (5,7,7)$, $\tilde{a}_{22} = (3,5,5)$, $\tilde{p}_1 = (0.7,0.9,0.9)$, $\tilde{p}_2 = (0.6,0.8,0.8)$, and $B = 300$. Here, the total budget is a crisp value since company budgets are usually determined in advance and fixed. Assume $\tilde{c}_{ij}$ is monotonically decreasing and $\tilde{a}_{ij}$ and $\tilde{p}_{ij}$ are monotonically increasing. The membership function of each fuzzy parameter can be defined, respectively, as:

$$\begin{aligned} &\mu_{\tilde{c}_{11}}^{-1}[\alpha] = 13 - 2\alpha, && \mu_{\tilde{c}_{12}}^{-1}[\alpha] = 26 - 2\alpha, \\ &\mu_{\tilde{c}_{21}}^{-1}[\alpha] = 6 - 2\alpha, && \mu_{\tilde{c}_{22}}^{-1}[\alpha] = 3 - 2\alpha, \\ &\mu_{\tilde{a}_{11}}^{-1}[\alpha] = 2 + 2\alpha, && \mu_{\tilde{a}_{12}}^{-1}[\alpha] = 4 + 2\alpha, \\ &\mu_{\tilde{a}_{21}}^{-1}[\alpha] = 5 + 2\alpha, && \mu_{\tilde{a}_{22}}^{-1}[\alpha] = 3 + 2\alpha, \\ &\mu_{\tilde{p}_1}^{-1}[\alpha] = 0.7 + 0.2\alpha, && \mu_{\tilde{p}_2}^{-1}[\alpha] = 0.6 + 0.2\alpha. \end{aligned}$$

We then transform the above fuzzy two-objective mathematical programming into the following crisp two-objective mathematical programming:

$$\begin{aligned} \max \quad & f_1 = c_{11}x_1 + c_{12}x_2 \\ \max \quad & f_2 = c_{21}x_1 + c_{22}x_2 \\ s.t. \quad & p_1(a_{11}x_1 + a_{12}x_2) + p_2(a_{21}x_1 + a_{22}x_2) \le B, \\ & c_{ij} \in \tilde{c}_{ij}[\alpha], \quad a_{ij} \in \tilde{a}_{ij}[\alpha], \quad p_{ij} \in \tilde{p}_{ij}[\alpha], \\ & x_1, x_2 \ge 0, \end{aligned}$$

where $[\alpha]$ denotes the α–*cut* operator.

We assume the objective profit (f_1) is twice as important as the objective customer satisfaction f_2. Then we use the compromise solutions and set $p = \infty$ and $[\alpha] = 0.5$ to derive two-objective fuzzy de novo programming so that we can compare the result with the previous example. The final solution can be obtained as $x_1 = 20.633$, $x_2 = 24.091$, $v = 0.023$, $b_1 = 182.356$, $b_2 = 220.164$, $B = 300$, $f_1 = 849.882$, and $f_2 = 151.348$.

Although the compromise solution method is used to deal with the above fuzzy de novo programming, other kinds of MODM methods such as goal programming and multi-objective genetic algorithms can be used easily to derive optimal solutions.

5.5 MOP WITH CHANGEABLE PARAMETERS

Although previous models extended traditional mathematical programming to deal with more practical problems, they cannot satisfy the purpose of this paper. Figure 5.3 highlights the purpose of this paper.

The original idea of de novo programming is to reallocate production resources so that system trade-offs can be eliminated and the ideal point achieved. However, a question arises: What if the ideal point still cannot be satisfied by a decision maker? While Chiang Lin et al. (2007) proposed another model to resolve the problem, their model still cannot ensure that the level desired by a decision maker can be achieved. This might be true if adding the effects of y and z still cannot achieve the desired point.

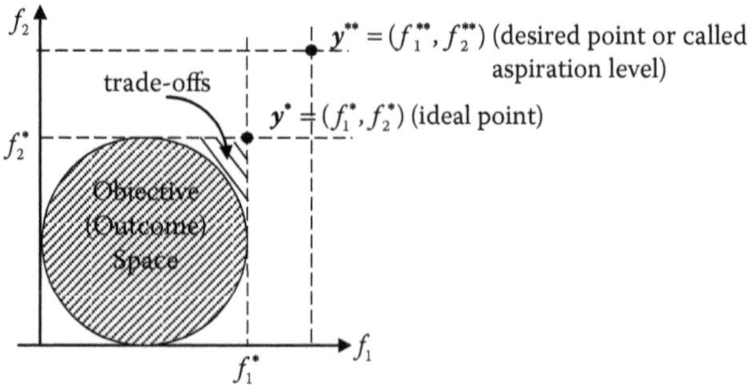

FIGURE 5.3 Basic concept of desired point.

In addition, what if a company cannot consider the effects of y or z? Maybe the above questions should be transformed into another question: If a decision maker wants to reach a desired point, how can the system be adjusted or redesigned to achieve that point? A decision makers may prefer to know how to achieve the desired point (aspiration level) via a redesigned system instead of optimizing an existing system.

It is obvious that if the parameters of a system remain constant, the best solution is the ideal point if no trade-off among objective functions happens. However, the desired point (aspiration level) that is better than the ideal point can never be achieved. Hence, if we want to achieve the desired point, we must upgrade the parameters of a system. Usually, a company may improve its objective or technological coefficients by importing new equipment and methods, expanding its fixed resources, promoting human resources by innovating technology, increasing the total budget, or both, or implementing other ideas.

We now develop possible models to redesign or reshape a system to achieve the desired point (aspiration level) according to the concept of changeable space, including decision space and objective (outcome) space. Assume a company has n objectives to be achieved and m products are produced. We can incorporate the concepts of financing decisions into MOP and formulate the following model:

Model 5.1: MOP with Changeable Budgets

$$\begin{aligned} &\min \quad \widehat{B} \\ &s.t. \quad \sum_{j=1}^{m} c_{ij}x_{ij} \geq f_i^{**}(\boldsymbol{x}), \qquad i=1,\ldots,n, \\ &\qquad \boldsymbol{p}'\boldsymbol{A}\boldsymbol{x} \leq B+\widehat{B}, \\ &\qquad < \text{extra conditions for } \widehat{B} > \\ &\qquad \boldsymbol{x} \geq \boldsymbol{0}, \end{aligned} \tag{5.13}$$

where c_{ij} denotes the jth coefficient of the ith objective function, $f_i^{**}(\boldsymbol{x})$ denotes the desired value of the ith objective, $\boldsymbol{p}$ denotes the unit price vector of resources, B is the original budget, and $\widehat{B}$ denotes the extra budget obtained from financing decisions.

Example 5.3

Let us reconsider the beginning example of producing suits and dresses. If the objective functions and constraints are constant, we can obtain the optimal solution from de novo programming as $f_1 = 2375$ and $f_2 = 44.5$. However, the decision maker feels unsatisfied with the results and hopes to increase f_1 (profit) from 2375 to 2600 and f_2 (quality) from 44.5 to 60. One way to achieve the desired solution is to borrow money from capital markets based on financing decisions. Table 5.4 lists relevant information.

TABLE 5.4
Information Table for Example 5.3

Unit Price	Resource	Technological Coefficients		No. of Units
		$x = 1$	$y = 1$	
30	Nylon	4	0	b_1
40	Velvet	2	6	b_2
9.5	Silver thread	12	4	b_3
20	Silk	0	3	b_4
10	Golden thread	4	4	b_5

The problem of Example 5.3 is to derive the minimum extra budget that can achieve the desired point and determine the corresponding resource allocation. Next, we formulate the following linear programming scheme to solve the problem of Example 5.3:

$$\begin{aligned}
\min\ & \widehat{B} \\
s.t.\ & 400x_1 + 300x_2 \geq 2600, \\
& 6x_1 + 8x_2 \geq 60, \\
& 30 \times 4x_1 + 40 \times (2x_1 + 6x_2) + 9.5 \times (12x_1 + 4x_2) + 20 \times 3x_2 \\
& +10 \times (4x_1 + 4x_2) \leq 2600 + \widehat{B}, \\
& x_1, x_2 \geq 0.
\end{aligned}$$

Solving the above problem, we can obtain the extra budget need $\widehat{B} = 376$ and production factors $x_1 = 2$ and $x_2 = 6$. The corresponding resource allocation can be calculated as $b_1 = 8$, $b_2 = 40$, $b_3 = 48$, $b_4 = 18$, and $b_5 = 32$. The corresponding profit and quality indices equal 2600 and 60, respectively.

Comparing the results from de novo programming and the proposed models, we note that the proposed method can find a way to achieve the desired point that cannot be reached with de novo programming. It is clear that the only way to proceed beyond the ideal point is to utilize outside help. Therefore, if the system hopes to achieve the desired point, it needs an additional \$376 (or other monetary units). However, if the extra budget need is equal to zero, the original system is sufficient to achieve the desired point and the proposed model reduces to de novo programming.

Besides borrowing money from capital markets, a company can also achieve its desired goal by improving the objective coefficients of a system, e.g., by utilizing economics of scale, electronic commerce, and total quality management (TQM) and eliminating third parties. In this situation, a company should consider the unit-improving cost of each objective coefficient and determine the optimal budget allocation between improving costs and improving production resources. It can then develop a relevant MOP model with changeable objective coefficients.

Model 5.2: MOP with Changeable Objective Coefficients

$$
\begin{aligned}
&\min \quad \widehat{B} \\
&s.t. \quad \sum_{j=1}^{m}(c_{ij}+\widehat{c}_{ij})x_{ij} \geq f_i^{**}(\boldsymbol{x}), \quad i=1,\dots,n, \\
&\boldsymbol{p}'\boldsymbol{A}\boldsymbol{x}+\sum_{i=1}^{n}\sum_{j=1}^{m} p_{ij}^{c}\widehat{c}_{ij} \leq B+\widehat{B}, \qquad (5.14) \\
&< \text{extra conditions for } p_{ij}^{c} \text{ and } \widehat{c}_{ij} > \\
&\boldsymbol{x} \geq \boldsymbol{0},
\end{aligned}
$$

where p_{ij}^{c} denotes the unit upgrade cost with respect to the jth product coefficient of the ith objective function and $\widehat{c}_{ij}$ is the jth upgrade product coefficient of the ith objective function.

Example 5.4

We again utilize the previous example of producing suits and dresses. If the company cannot borrow money from capital markets but still hopes to increase f_1 (profit) from 2375 to 2600 and f_2 (quality) from 44.5 to 60, another method is to improve its objective coefficients through strategies or technologies. Therefore, we assume the unit-improving costs of the objective coefficients are \$0.200, \$0.289, \$2.225, and \$2.487, respectively, as shown in Table 5.5.

We can formulate the following mathematical programming for achieving the desired points by improving objective coefficients:

$$
\begin{aligned}
&\min \quad \widehat{B} \\
&s.t. \quad (400+\widehat{c}_{11})x_1+(300+\widehat{c}_{12})x_2 \geq 2600, \\
&(6+\widehat{c}_{21})x_1+(8+\widehat{c}_{22})x_2 \geq 60, \\
&30\times 4x_1+40\times(2x_1+6x_2)+9.5\times(12x_1+4x_2)+20\times 3x_2 \\
&+10\times(4x_1+4x_2)+(0.200\widehat{c}_{11}+0.289\widehat{c}_{12}+2.225\widehat{c}_{21}+2.487\widehat{c}_{22}) \leq 2600+\widehat{B} \\
&x_1, x_2 \geq 0.
\end{aligned}
$$

Solving the above problem, we determine the extra budget as $\widehat{B}=0$. This result means that no extra budget is needed for achieving the desired point. Then, we can derive $x_1=4.43$, $x_2=2.70$, $\widehat{c}_{11}=3.51$, $\widehat{c}_{12}=0.18$, $\widehat{c}_{21}=1.44$, and $\widehat{c}_{22}=2.00$. In addition, the corresponding resource allocation can be assigned as $b_1=17.72$, $b_2=25.06$, $b_3=17.72$, $b_4=8.10$, $b_5=28.52$, profit $=2600$, and quality index $=60$.

TABLE 5.5
Information Table for Example 5.4

Objective Coefficients				Technological Coefficients		
$x = 1$	$y = 1$	Unit Price	Resource	$x = 1$	$y = 1$	No. of Units
400 ($0.200)	300 ($0.289)	30	Nylon	4	0	b_1
6 ($2.225)	8 ($2.487)	40	Velvet	2	6	b_2
		9.5	Silver thread	12	4	b_3
		20	Silk	0	3	b_4
		10	Golden thread	4	4	b_5

We should highlight that $\hat{B} \neq 0$ means that we still cannot achieve the desired point via improving objective coefficients. Therefore, it indicates an extra budget allocation is needed for the system to achieve the desired point. Otherwise, we should consider another possibility.

The last situation discussed here is that a company may expand its outcome space by upgrading technology coefficients of a system. For example, it can pursue the upgrade by adopting business process reengineering (BPR), new information technologies, or enterprise resource management (ERP) to increase production efficiency. Hence, we can use the above description to formulate the following model:

MODEL 5.3: MOP WITH CHANGEABLE TECHNOLOGICAL COEFFICIENTS

$$
\begin{aligned}
&\min \quad \hat{B} \\
&\text{s.t.} \quad \sum_{j=1}^{m} c_{ij} x_{ij} \geq f_i^{**}(x), \quad i = 1, \ldots, n, \\
&\qquad p'(A - \hat{A})x + \sum_{k=1}^{r} \sum_{j=1}^{m} p_{kj}^{a} \hat{a}_{kj} \leq B + \hat{B}, \\
&\qquad < \text{extra condition for } \boldsymbol{p_{ij}^{a}} \text{ and } \boldsymbol{\hat{a}_{ij}} > \\
&\qquad \boldsymbol{x} \geq \boldsymbol{0},
\end{aligned}
\tag{5.15}
$$

where $\hat{A} = [\hat{a}]_{kj}$ is the upgrading technological coefficient matrix and p_{kj}^{a} is the unit upgrading cost with respect to the jth technology coefficient of the kth constraint.

Example 5.5

We can follow the previous example of producing suits and dresses to address the following problem. If the company still cannot achieve the desired point by adding extra budget or improving objective coefficients, one more technique is updating

TABLE 5.6
Information Table for Example 5.5

Objective Coefficients				Technological Coefficients		
x = 1	y = 1	Unit Price	Resource	x = 1	y = 1	No. of Units
400	300	30	Nylon	4 ($0.5)	0	b_1
6	8	40	Velvet	2 ($0.5)	6 ($0.27)	b_2
		9.5	Silver thread	12 ($0.27)	4 ($0.26)	b_3
		20	Silk	0	3 ($0.25)	b_4
		10	Golden thread	4 ($0.25)	4 ($0.25)	b_5

the technological coefficients of the system. We assume that each unit updating cost of a technological coefficient can be defined and presented as shown in Table 5.6.

Incorporating the information of the unit updating cost of the technological coefficients, we can formulate the following mathematical programming model:

$$\begin{aligned}
\min\quad & \hat{B} \\
s.t.\quad & 400x_1 + 300x_2 \geq 2600, \\
& 6x_1 + 8x_2 \geq 60, \\
& 30 \times (4 - \hat{a}_{11})x_1 + 40 \times ((2 - \hat{a}_{21})x_1 + (6 - \hat{a}_{22})x_2) + 9.5 \times ((12 - \hat{a}_{31})x_1 \\
& +(4 - \hat{a}_{32})x_2) + 20 \times (3 - \hat{a}_{42})x_2 + 10 \times ((4 - \hat{a}_{51})x_1 + (4 - \hat{a}_{52})x_2) \\
& +0.5\hat{a}_{11} + 0.5\hat{a}_{21} + 0.27\hat{a}_{22} + 12\hat{a}_{31} + 4\hat{a}_{32} + 3\hat{a}_{42} + 4\hat{a}_{51} + 4\hat{a}_{52} \leq 2600 + \hat{B} \\
& x_1, x_2 \geq 0.
\end{aligned}$$

The $\hat{B} = 0$ extra budget result means that no extra budget is needed for achieving the desired point. Then we can derive $x_1 = 2.42$, $x_2 = 5.69$, $\hat{a}_{11} = 2.03$, $\hat{a}_{21} = 1.27$, $\hat{a}_{22} = 0.30$, $\hat{a}_{31} = 0.27$, $\hat{a}_{32} = 0.25$, $\hat{a}_{42} = 0.26$, $\hat{a}_{51} = 0.25$, and $\hat{a}_{52} = 0.26$. The corresponding resource allocation can be assigned as $b_1 = 4.76$, $b_2 = 34.20$, $b_3 = 49.72$, $b_4 = 15.59$, $b_5 = 30.36$, profit = 2600, and quality index = 60.

Based on the previous models, we can use one of three methods to try to achieve the desired point. In practice, the three situations may exist simultaneously. Therefore, a more general model of changeable parameters can be considered to incorporate the previous three situations:

$$\begin{aligned}
\min\quad & \hat{B} \\
s.t.\quad & \sum_{j=1}^{m} (c_{ij} + \hat{c}_{ij})x_{ij} \geq f_i^{**}(\boldsymbol{x}), \qquad i = 1,\ldots,n, \\
& \boldsymbol{p}'(\boldsymbol{A} - \hat{\boldsymbol{A}})\boldsymbol{x} + \sum_{i=1}^{n}\sum_{j=1}^{m} p_{ij}^{c}\hat{c}_{ij} + \sum_{i=1}^{n}\sum_{j=1}^{m} p_{ij}^{a}\hat{a}_{ij} \leq B + \hat{B}, \\
& < \text{extra conditions for } \hat{B},\ p_{ij}^{c},\ \hat{c}_{ij},\ p_{ij}^{a} \text{ and } \hat{a}_{ij} > \\
& \boldsymbol{x} \geq \boldsymbol{0},
\end{aligned} \tag{5.16}$$

If $\widehat{B}=0$, the desired point can be achieved without increasing the budget. The decision maker can improve objective coefficients when $\widehat{c}_{ij}\neq 0$, update technological coefficients when $\widehat{a}_{ij}\neq 0$, or both results are present ($\widehat{c}_{ij}\neq 0$ and $\widehat{a}_{ij}\neq 0$) to achieve the desired point. On the other hand, if $\widehat{B}\neq 0$, the original system cannot be achieved by updating objective or technological coefficients. The only alternative is to increase the budget by borrowing money from capital markets.

According to the above example, we can illustrate how to expand the changeable spaces for achieving desired point (aspiration level), as shown in Figure 5.4.

The decision spaces of the example are expanded gradually through resource reallocation or changeable parameters. Therefore, the corresponding objective spaces

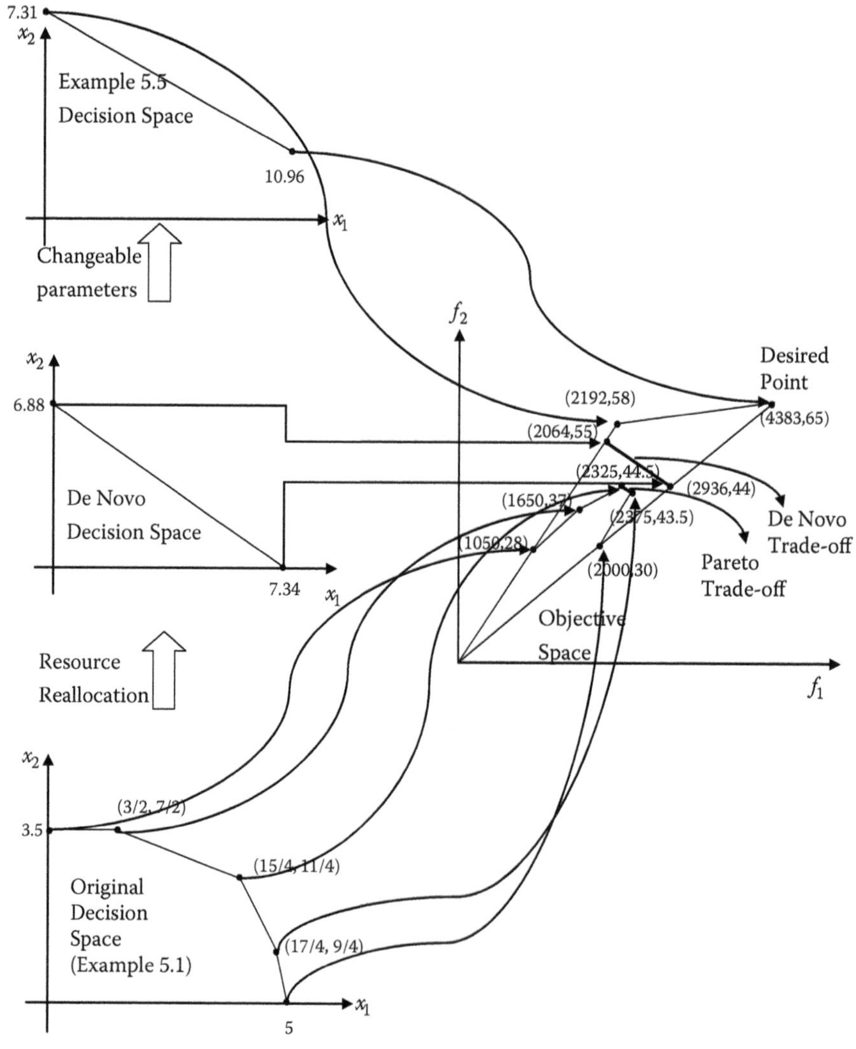

FIGURE 5.4 Changeable spaces for achieving desired point.

are changeable and create the possibility of achieving better solutions. As we know, the original decision space will cause a Pareto solution due to the trade-off between objectives. Although de novo programming enables the possibility of resource reallocation to obtain a better result, it still causes de novo trade-off because of the fixation of system parameters. Therefore, the only way to achieve the desired point is to expand the decision space via changeable parameters to reach the desired point without trade-off between objectives.

The concept of habitual domains (HDs) has been proposed to support the flexible or changeable parameters of decision spaces. Habitual domains (Yu, 1980, 1984, 1985, 1990, 1991) consist of the set of human thinking, judging, responding, experience, and knowledge. Therefore, it is clear that HDs play a key role in affecting human behavior. To improve the quality of decision making, people should review two major possibilities: (1) how to polish existing HDs and (2) how to expand existing HDs.

Many papers contain proposals for efficiently expanding HDs via a method known as competence set analysis (Yu and Zhang, 1989, 1990, 1992; Li and Yu, 1994). These methods can be used to significantly improve the quality of decision making. The four elements within a habitual domain are:

1. The potential domain (PD) is a collection of ideas and actions that can potentially be activated.
2. The actual domain (AD) is a set of ideas and actions that are actually activated.
3. The activation probabilities (AP) are the probabilities that ideas and actions in PD also belong to AD.
4. The reachable domain (RD) is a set of ideas and actions that can be attained from a given set in an HD.

Since decision processes depend on the evolution of HDs, an expanding HD can result in effective decision and preferred solutions. Hence, the expanding HD can be considered as a changeable parameter of a system.

Next, we compare the differences between goal programming and the proposed method as follows. Goal programming was proposed by Charnes and Cooper (1961) to deal with linear MOP problems. The idea behind goal programming is to seek a solution nearest the ideal point. Thus, a decision maker should first assign a target or goal to each objective and then minimize the distance between targets and objectives (usually we use L_p-norm to define the distance between the targets and objectives) to find the solution.

A number of differences between goal programming and the proposed methods can be described. First, the purpose of goal programming is to find a solution based on Pareto solutions that is closest to the ideal point. However, the purpose of the proposed models is to find out a way to reach the desired point by possible changes. Hence, the optimal solution of goal programming is usually worse than the ideal point because of trade-offs among objective functions. On the other hand, the proposed models can achieve a desired point, which is better than the ideal point. In sum, goal programming optimizes objective functions with a system and is also known as inside optimization. However, the proposed models minimize extra budget with a flexible system; this is known as outside optimization.

Second, goal programming optimizes the objective functions of a system with fixed parameters. That is, goal programming determines what to do to obtain the optimal solution of a system. On the other hand, the proposed models tolerate parameters of a system are changeable so that the desired point can be achieved through adjustments. Hence, goal programming is more suitable for dealing with "what to do" problems and the proposed models try to answer "how to do" problems.

Finally, the essence of goal programming is to determine optimal solutions to MOPs. Therefore, like other traditional MOP methods such as compromise solutions, goal programming is an optimization method. However, the essence of the proposed models seeks to change or modify an original system so that a desired point can be achieved. Hence, the proposed method can be regarded as an optimal system design method rather than an optimization method.

We highlighted the differences between de novo programming and the proposed models. De novo programming incorporates unit price data for resources into traditional mathematical programming to design a better system. The proposed method considers resource unit prices and relaxes other parameters such as objective and technological coefficients to achieve changeability. The optimal solution of de novo programming is the ideal point. The proposed method seeks ways to achieve the desired point.

Finally, we highlight the philosophical difference between the proposed method and other MOP methods, including de novo programming, from the perspective of formal analysis (Tzeng and Huang, 2011). All traditional MOP models can be considered normative models that focus on the problems that decision makers ideally encounter. However, the proposed method should be regarded as a prescriptive model that considers the methods that decision makers should apply to improve their decisions.

6 Multi-Stage Programming

Multi-stage programming can be considered a special network problem that finds the shortest path joining two points in a given network. However, unlike other shortest path problems that focus on finding the shortest path from the source node to any other node in a network, multi-stage programming seeks the shortest path from the source to each sequential stage. In this chapter, we will introduce dynamic programming, which is widely used for multi-stage network problems and the related application of this programming technique.

6.1 DYNAMIC PROGRAMMING

The concept of dynamic programming comes from the principle of optimality (Bellman, 1952, 1953) as a solution for sequential optimization problems. That is, an optimal solution has the property that whatever the initial state and the initial decisions are, the remaining decisions must constitute an optimal solution with regard to the state resulting from the first decision. The network problems of multi-stage programming are shown in Figure 6.1.

To model the illustrated problem, we first define the notations. Let the distance between node i and node j be c_{ij} and $f_t(j)$ be the node j in stage t. If we want to find the shortest distance from node 0 to node 9, we can begin by finding $f_4(6)$, $f_4(6)$, $f_4(6)$, and $f_4(6)$, i.e., finding $\min\{c_{69}, c_{79}, c_{89}\}$.

Next, we can find $f_3(4)$ and $f_3(5)$. Using $f_3(4)$ for an example, we apply the following equation:

$$f_3(4) = \min_j\{c_{4,j} + f_4(j)\}, \qquad \forall j = 6,7,8. \tag{6.1}$$

Then, we can use the procedure until $f_0(0)$ results to obtain the shortest distance from node 0 to node 9. This method is called dynamic programming with backward recursion. However, if we start from node 0 to find the shortest distance, the process is called forward recursion. We can generalize a k-stage problem as the following dynamic programming equation:

$$f_t(j) = \min_j\{c_{ij} + f_t(j)\}, \qquad \forall t = 1,\ldots,k, \tag{6.2}$$

where c_{ij} denotes the distance between node i and node j and $f_t(j)$ is the node j in stage t.

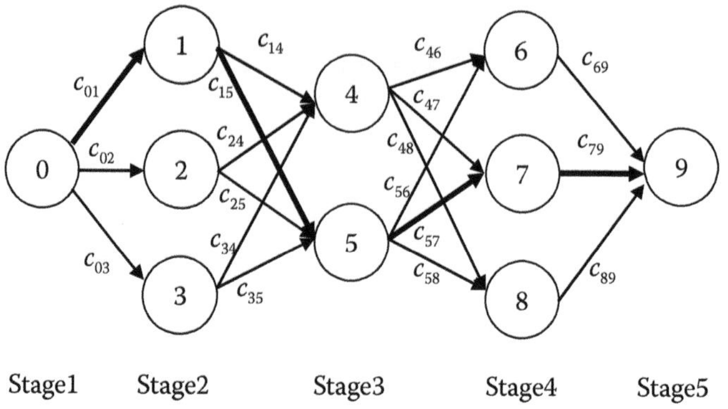

FIGURE 6.1 Example of a multi-stage programming problem.

It is easy to understand that if a multi-stage problem is too complicated, dynamic programming is far superior for achieving explicit enumeration. We can summarize the characteristics of a multi-stage problem as follows (Winston, 1994):

1. A multi-stage problem can be divided into stages with a decision required at each stage.
2. Each stage has a number of associated stages.
3. The decision chosen at any stage describes how the current stage is transformed into the next stage.
4. Given the current stage, the optimal decision for each of the remaining stages must not depend on previously reached stages or previously chosen decisions (principle of optimality).

Example 6.1

We can reconsider the above five-stage programming problem as the following shortest route problem. Assume you live in Taipei and plan to see a friend in Shanghai and want to take this opportunity to travel in China. If your funds are limited, you must determine the most economical way to enjoy each day of your trip. Figure 6.2 displays the possible cities to visit and the corresponding costs for each day.

We will employ backward recursion to determine the optimal path of the trip. We can use dynamic programming to deal with this five-stage programming problem. First, we calculate the trip costs from Day 3 to Day 4 separately, as shown in Table 6.1. Next, we consider the trip costs from Day 2 to Day 4 as shown in Table 6.2. The solution of Table 6.2 can be explained as follows. The trip cost from Jinjiang to Ningbo is equal to 7. Hence, if we decide to arrive at Jinjiang in Day 2 and at Ningbo in Day 3, we should cost 12 units to Shanghai after Day 1.

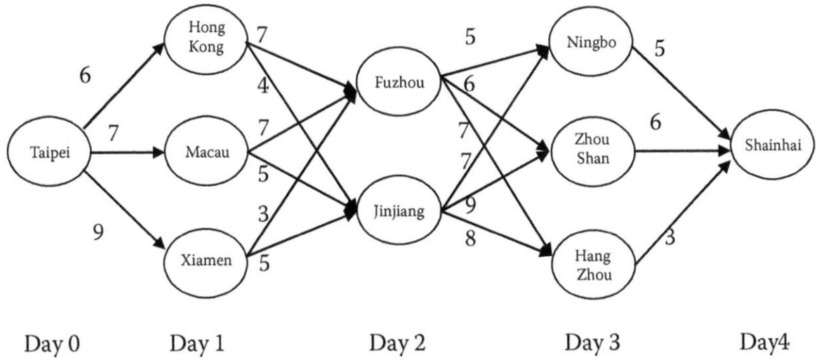

FIGURE 6.2 Multi-stage problem for trip planning.

However, if we decide to stay in Zhoushan or Hangzhou on Day 3, we must cost 12 or 10 units to Shanghai after Day 1. If we stay in Jinjiang on Day 2, we should go to Hangzhou to minimize the trip cost on Day 3. We can then calculate the trip costs from Day 1 to Day 4 in a similar way, as shown in Table 6.3.

The explanation of Table 6.3 is similar to that for Table 6.2. That is, if we stay in Hong Kong and/or Macau, we should go to Jinjiang on Day 2. Otherwise, we should arrive in Fuzhou to minimize trip cost. The total cost from Day 0 to Day 4 is relatively easy to calculate by adding the cost from Taipei to Hong Kong, Macau, and Xiamen and the optimal solution derived from Table 6.3, as shown in Table 6.4.

TABLE 6.1
Trip Costs for Day 3 to Day 4

Stage 4 ($c_{i,j}$)	**Day 4**	**Optimal Solution**	
Day 3	Shanghai	$f_4(\cdot)$	x_5^*
Ningbo	5	5	Shanghai
Zhoushan	6	6	Shanghai
Hangzhou	3	3	Shanghai

TABLE 6.2
Trip Costs for Day 2 to Day 4

Stage 3 ($c_{i,j} + f_3(j)$)	**Day 3**	**Day 3**	**Day 3**	**Optimal Solution**	
Day 2	Ningbo	Zhoushan	Hangzhou	$f_3(\cdot)$	x_4^*
Fuzhou	5 + 5 = 10	6 + 6 = 12	7 + 3 = 10	10	Ningbo/Hangzhou
Jinjiang	7 + 5 = 12	9 + 6 = 12	8 + 3 = 10	10	Hangzhou

TABLE 6.3
Trip Costs for Day 1 to Day 4

Stage 2 ($c_{i,j} + f_3(j)$)	Day 2	Day 2	Optimal Solution	
Day 1	Fuzhou	Jinjiang	$f_2(\cdot)$	x_3^*
Hong Kong	7 + 10 = 17	4 + 10 = 14	14	Jinjiang
Macau	7 + 10 = 17	5 + 10 = 15	15	Jinjiang
Xiamen	3 + 10 = 13	5 + 15 = 13	13	Fuzhou

TABLE 6.4
Trip Costs for Day 0 to Day 4

Stage 1 ($c_{i,j} + f_3(j)$)	Day 1	Day 1	Day 1	Optimal Solution	
Day 0	Hong Kong	Macau	Xiamen	$f_2(\cdot)$	x_2^*
Taipei (x_1^*)	6 + 14 = 20	7 + 15 = 22	9 + 13 = 22	20	Hong Kong

From the above results, we can determine the optimal path of the 4-day trip as Taipei → Hong Kong → Jinjiang → Hangzhou → Shanghai at a total cost of 20 units. Readers can use the forward recursion approach to derive the same solution.

The applications of multi-stage programming include multi-stage process planning (MPP) problems (Jensen and Barnes, 1980; Sancho, 1986), cargo-loading or knapsack problems (McLeod, 1983), job shop scheduling (Shah, 1998), and operations issues. In addition to deterministic dynamic programming, if the path from the former stage to the latter stage depends on some probability, the dynamic programming process is called probabilistic dynamic programming. Readers can refer to Bertsekas (1987), Cooper and Cooper (1981), and Smith (1991) to compare the differences between deterministic and probabilistic dynamic programming.

6.2 APPLICATION OF MULTI-STAGE PROBLEM: COMPETENCE SETS

The concept of the competence set was proposed by Yu (1990a and b) to resolve a specific decision problem by acquiring the necessary ideas, information, skills, and knowledge. Competence set analysis involves identification of the true competence set and the decision maker's competence set and following the efficient expansion path to good decisions.

Among these issues, the method of expanding the existing competence set optimally is highlighted. Several methods such as the minimum spanning tree (Yu and Zhang, 1992), the mathematical programming method (Shi and Yu, 1996), and the deduction graphs (Grefenstette, 1991), have been proposed to obtain the optimal path. The optimal expansion path from the existing competence set to the true competence set can be described as follows.

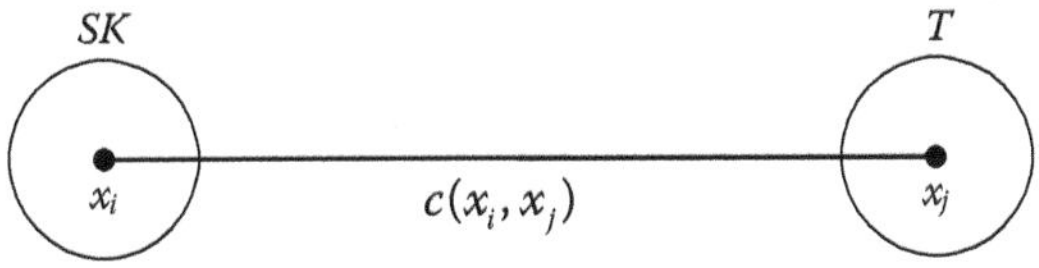

FIGURE 6.3 Cost functions of competence sets.

Let $HD = SK \cup T$, where HD (habitual domain) includes all the relevant skills needed to solve a particular problem, SK denotes the already acquired competence set and T denotes the true required competence set. Therefore, the optimal expansion path can be obtained by minimizing the following equation:

$$\min\{c(x_i, x_j), \quad \text{where} \quad x_i \in SK \quad \text{and} \quad x_j \in T\}, \tag{6.3}$$

where $c(x_i, x_j)$ denotes the cost and/or time of acquiring x_j from x_i. Figure 6.3 represents the corresponding graph.

We can use an example to illustrate the optimal expansion path using the minimum spanning tree. Let the $SK = \{a\}$, $T = \{a, b, c, d\}$. The cost function is shown in the following matrix:

Cost	*a*	*B*	*c*	*d*
A	0	2	6	8
B	8	0	1	4
C	8	2	0	1
D	1	2	3	0

In order to determine the first step of the expansion path, we must consider the cost information as follows:

Process	*b*	*c*	*d*
C (a, process)	2	6	8

The first step is $a \rightarrow b$. Next, consider the following cost to determine the second step:

Process	*C*	*d*
c($\{a \rightarrow b\}$, process)	1	4

Therefore, the optimal second expansion path is $b \rightarrow c$, and the optimal expansion path is $a \rightarrow b \rightarrow c \rightarrow d$.

The optimal expansion path of competence sets can be viewed as a special case of the multi-stage programming problem. Figure 6.4 illustrates the concept.

The main difference between the problems in Sections 6.1 and 6.2 is that traditional multi-stage programming seeks an optimal node in each stage to minimize the cost. On the other hand, competence sets search the optimal path to link all necessary nodes (including inner and intra stages) in a multi-stage situation.

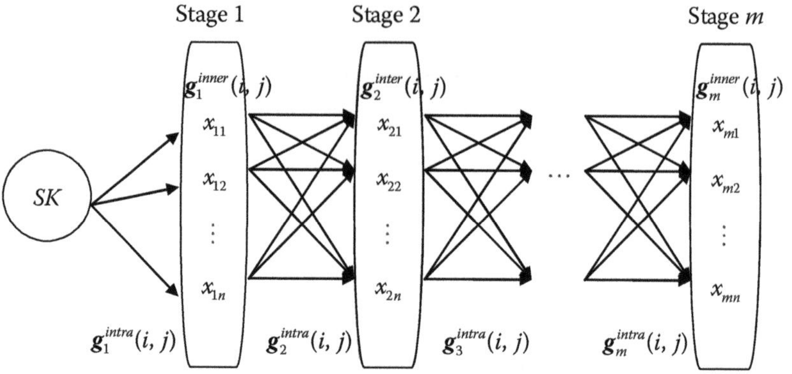

FIGURE 6.4 Concept of competence set.

We can employ the routing method (Shi and Yu, 1996) to select the optimal expansion path based on the following mathematical programming model:

$$\min z = \sum c_{ij} x_{ij} \tag{6.4}$$

$$s.t. \sum_{i=0}^{n} x_{ij} = 1, \qquad j = 1, 2, \ldots, n,$$

$$u_i - u_j + (n+1)x_{ij} \le n, \qquad 1 \le i, j \le n, \qquad i \ne j, \quad \forall_{ij},$$

$$u_i, u_j \in \{0, 1, \ldots, n\}.$$

where $c_{ij} = c(x_i, x_j)$ denotes the cost of acquiring x_j from x_i and u_i denotes the subsidiary variable.

Although many scholars extend competence sets to consider further situations such as asymmetric acquiring cost (Shi and Yu, 1996) and group decisions (Li and Yu, 1994), these papers address only one-stage, one-objective situations. However, we can extend competence sets to consider multi-stage, multi-objective problems or fuzzy environments for more complicated applications.

6.3 FUZZY MULTI-STAGE MULTI-OBJECTIVE COMPETENCE SETS

In order to formulate the fuzzy multi-objective competence set, the fuzzy mathematical programming model is employed here. The fuzzy programming problem (Carlsson and Korhonen, 1986) can be represented as follows:

$$\max \quad \tilde{z} = \sum_i \tilde{c}_i \tilde{x}_i \tag{6.5}$$

$$s.t. \quad \tilde{X} = \{(x, \mu(x)) \mid (\tilde{\boldsymbol{A}}\boldsymbol{x})_i \le \tilde{\boldsymbol{b}}_i, \forall i, x \ge 0, \mu(x) \in [0,1]\}$$

where $\mu(x)$ denotes the membership function of x. By setting the adequate membership function and α-cut, we can transform Equation (6.5) into the following equation to derive the optimal solution of the fuzzy programming problem.

$$\max \quad \tilde{z} = \sum_{j=1}^{n} \mu_{\tilde{c}_j}^{-1}(\alpha)x_j \tag{6.6}$$

$$\text{s.t.} \quad \sum_{j=1}^{n} \mu_{\tilde{a}_{ij}}^{-1}(\alpha)x_j \le \sum_{j=1}^{n} \mu_{\tilde{b}_j}^{-1}(\alpha), \qquad \forall i = 1,\ldots,m,$$

$$x_j \ge 0, \qquad \forall j = 1,\ldots,n.$$

Now, based on the concepts above, we can formulate the optimal fuzzy multi-criteria expansion process as the following mathematical programming model:

$$\text{min/max} \quad \tilde{z}_1 = \sum \mu_{\tilde{c}_{1ij}}^{-1}(\alpha)x_{1ij} \tag{6.7}$$

$$\text{min/max} \quad \tilde{z}_2 = \sum \mu_{\tilde{c}_{2ij}}^{-1}(\alpha)x_{2ij}$$

$$\vdots$$

$$\text{min/max} \quad \tilde{z}_m = \sum \mu_{\tilde{c}_{mij}}^{-1}(\alpha)x_{mij}$$

$$\text{s.t.} \quad \sum_{i=0}^{n} x_{ij} = 1, \qquad j = 1,2,\ldots,n,$$

$$u_i - u_j + (n+1)x_{ij} \le n, \qquad 1 \le i,j \le n, \qquad i \ne j,$$

$$\forall x_{ij}, u_i \in \{0,1,\ldots n\}.$$

where $c_{ij} = c(x_i, x_j)$ denotes the cost of acquiring x_j from x_i and u_i denotes the subsidiary variable. As mentioned previously, because some criteria are intangible and conflict with each other, Pareto solutions can be derived by using MOEA. Then, decision-makers can select the final optimal expansion process based on their preferences.

Example 6.2

We will demonstrate a fuzzy two-objective (i.e., cost and benefit) expansion of competence sets. Let $SK = \{x_0\}$, $T \backslash SK = \{x_1, x_2, x_3, x_4, x_5, x_6, x_7\}$. The fuzzy cost and the fuzzy benefit functions that represent interval values are shown in Tables 6.5 and 6.6. Note that the M symbol denotes the infeasible route and will be treated as a minimum number in our fuzzy mathematical programming model. In addition, the membership of the cost and benefit functions is assumed to be triangular in form.

TABLE 6.5
Cost Function of Fuzzy Competence Set

Cost	x_0	x_1	x_2	x_3	x_4	x_5	x_6	x_7
x_0	M	(4.9, 5.9)	(6.1, 7.1)	(2.7, 3.7)	(2.6, 3.6)	(3.2, 4.2)	(5.5, 6.5)	(3.4, 4.4)
x_1	M	M	(3.9, 4.9)	(4.7, 5.7)	(3.9, 4.9)	(4.0, 5.0)	(4.3, 5.3)	(3.5, 4.5)
x_2	M	(4.4, 5.4)	M	(3.7, 4.7)	(6.4, 7.4)	(5.7, 6.7)	(5.9, 6.9)	(4.5, 5.5)
x_3	M	(6.3, 7.3)	(2.7, 3.7)	M	(5.8, 6.8)	(6.4, 7.4)	(6.4, 7.4)	(3.9, 4.9)
x_4	M	(5.5, 7.5)	(3.9, 4.9)	(5.6, 6.6)	M	(3.3, 4.3)	(2.8, 3.8)	(2.8, 3.8)
x_5	M	(4.4, 5.4)	(4.5, 5.5)	(4.0, 5.0)	(6.4, 7.4)	M	(6.3, 7.3)	(2.8, 3.8)
x_6	M	(4.0, 5.0)	(2.9, 3.9)	(4.9, 5.9)	(6.1, 7.1)	(6.0, 7.0)	M	(4.7, 5.7)
x_7	M	(6.3, 7.3)	(2.6, 3.6)	(4.1, 5.1)	(6.0, 7.0)	(3.1, 4.1)	(2.8, 3.8)	M

By using Equation (6.7) and letting the α-cut equal 0.8 (other results that set α *at* 0.2 and 0.5 are shown in the appendix at the end of this chapter), we can formulate the optimal fuzzy multi-criteria expansion model based on the data from Tables 6.5 and 6.6. To obtain Pareto solutions, we next set the adequate parameters of MOEA as shown in Table 6.7. After generating and calculating the optimal

TABLE 6.6
Benefit Function of Fuzzy Competence Set

Benefit	x_0	x_1	x_2	x_3	x_4	x_5	x_6	x_7
x_0	M	(4.5, 5.5)	(4.6, 5.6)	(5.1, 6.1)	(2.8, 3.8)	(2.9, 3.9)	(4.9, 5.9)	(3.6, 4.6)
x_1	M	M	(5.3, 6.3)	(2.9, 3.9)	(3.6, 4.6)	(2.8, 3.8)	(3.5, 4.5)	(4.9, 5.9)
x_2	M	(5.8, 6.8)	M	(3.7, 4.7)	(4.9, 5.9)	(5.6, 6.6)	(5.9, 6.9)	(3.5, 4.5)
x_3	M	(5.2, 6.2)	(5.8, 6.8)	M	(6.5, 7.5)	(3.8, 4.8)	(5.2, 6.2)	(5.0, 6.0)
x_4	M	(3.2, 4.2)	(5.9, 6.9)	(4.0, 5.0)	M	(3.4, 4.4)	(3.5, 4.5)	(5.7, 6.7)
x_5	M	(4.7, 5.7)	(6.0, 7.0)	(6.4, 7.4)	(4.6, 5.6)	M	(6.4, 7.4)	(4.8, 5.8)
x_6	M	(5.4, 6.4)	(2.8, 3.8)	(5.6, 6.6)	(5.7, 6.7)	(6.0, 7.0)	M	(2.6, 3.6)
x_7	M	(6.4, 7.4)	(3.0, 4.0)	(4.9, 5.9)	(5.5, 6.5)	(4.3, 5.3)	(2.7, 3.7)	M

TABLE 6.7
Parameter Settings of MOEA

Parameter	Value
Population size	100
Selection strategy	Tournament
Maximum number of generations	1000
Crossover rate	0.9
Mutation rate	0.01

TABLE 6.8
Pareto Solutions of Competence Set

α = 0.8	Optimal Expansion Process							Cost	Benefit
Model 1	0→1	0→3	1→2	1→4	1→6	3→7	6→5	31.0	38.6
Model 2	0→1	0→4	0→5	0→7	4→2	5→3	5→6	30.0	38.1
Model 3	0→1	0→3	0→4	1→6	4→2	4→7	7→5	25.7	37.4
Model 4	0→1	0→3	0→4	1→6	3→2	4→7	7→4	24.5	37.3
Model 5	0→1	0→3	0→4	0→5	0→7	3→4	7→6	24.0	33.0
Model 6	0→3	0→4	0→5	0→7	5→1	7→2	7→6	23.4	30.4

generations, we obtain six optimal expansion processes, i.e., Pareto solutions, as shown in Table 6.8. For example, Model 1 depicts the optimal expansion process as shown in Figure 6.5 to obtain the optimal costs equal to 31.0 and the optimal benefits equal to 38.6.

On the basis of the results, a decision maker can select one of the six paths based on his or her preferences or subjective judgments to determine the final optimal expansion process.

Competence set analysis has been used for many applications such as learning sequences for decision makers (Hu et al., 2002) and for consumer decision problems (Chen, 2001 and 2002). However, these papers consider only situations involving one criterion and the crisp function. In practice, decision makers usually determine the optimal expansion process based on multiple criteria that may conflict with each other. Therefore, Pareto solutions should be derived to determine the final expansion process based on the decision maker's preferences. In addition, due to uncertainties and subjective judgment, the concept of fuzzy sets should be incorporated into competence set analysis.

In this chapter, the fuzzy multi-criteria expansion model is proposed to deal with the above problems. In order to obtain Pareto solutions efficiently and correctly, MOEA is employed. A numerical example is used to demonstrate the

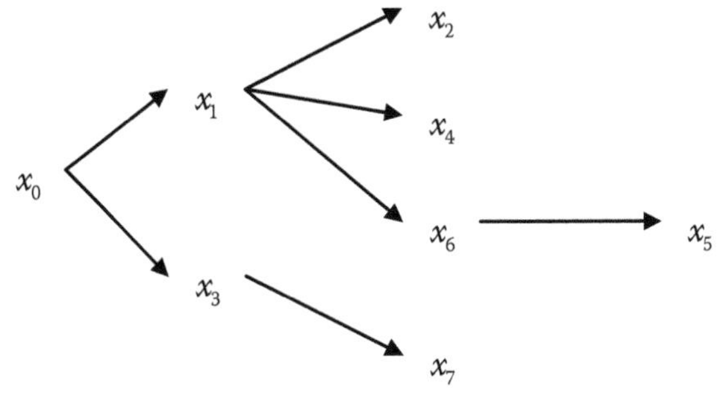

FIGURE 6.5 Optimal expansion process for Model 1.

proposed method. On the basis of the simulated results, we can obtain six non-dominated solutions. For those who are risk averse, Model 6 may be the optimal expansion process. However, Model 1 may be the optimal expansion process for a risk taker.

APPENDIX

By setting $\alpha = 0.2$ and $\alpha = 0.5$, we can obtain two other Pareto solutions as shown in Tables A.1 and A.2.

TABLE A.1
Pareto Solutions of Competence Set $\alpha = 0.2$

$\alpha = 0.2$	Optimal Expansion Process							Cost	Benefit
Model 1	0 → 3	0 → 5	0 → 7	4 → 2	5 → 6	7 → 1	7 → 4	37.7	37.2
Model 2	0 → 3	0 → 7	1 → 4	5 → 2	5 → 6	6 → 1	7 → 5	33.5	35.8
Model 3	0 → 3	0 → 6	0 → 7	6 → 1	7 → 2	7 → 4	7 → 5	32.9	33.2
Model 4	0 → 3	0 → 7	4 → 6	7 → 1	7 → 2	7 → 4	7 → 5	32.5	32.8
Model 5	0 → 3	0 → 7	1 → 4	5 → 6	6 → 1	6 → 2	7 → 5	31.9	32.6
Model 6	0 → 3	0 → 7	2 → 1	4 → 5	7 → 2	7 → 4	7 → 5	30.6	32.2
Model 7	0 → 3	0 → 7	6 → 1	7 → 2	7 → 4	7 → 5	7 → 6	30.2	31.0

TABLE A.2
Pareto Solutions of Competence Set $\alpha = 0.5$

$\alpha = 0.5$	Optimal Expansion Process							Cost	Benefit
Model 1	0 → 3	0 → 6	1 → 7	2 → 1	3 → 2	3 → 4	3 → 5	34.5	40.3
Model 2	0 → 3	0 → 6	0 → 7	2 → 1	3 → 2	3 → 4	3 → 5	34.4	39.0
Model 3	0 → 1	0 → 3	0 → 5	1 → 2	4 → 7	5 → 4	5 → 6	34.0	38.0
Model 4	0 → 3	0 → 7	2 → 1	3 → 2	3 → 4	3 → 5	4 → 6	31.7	37.6
Model 5	0 → 1	0 → 3	0 → 5	1 → 2	1 → 4	4 → 7	5 → 6	31.5	37.0
Model 6	0 → 1	0 → 3	0 → 5	1 → 2	1 → 4	2 → 6	4 → 7	31.1	36.5
Model 7	0 → 3	0 → 6	2 → 1	3 → 2	4 → 5	4 → 6	7 → 4	28.8	36.2
Model 8	0 → 1	0 → 3	0 → 4	4 → 2	4 → 5	4 → 7	7 → 6	26.5	33.6
Model 9	0 → 1	0 → 3	0 → 4	4 → 5	4 → 7	7 → 2	7 → 6	25.2	30.7

7 Multi-Level Multi-Objective Programming

Multi-level programming is a special kind of mathematical programming that has different objectives with respect to various levels. In addition, one level of the hierarchy may have its objective function determined by decision variables controlled at other levels. Multi-level multi-objective programming can be considered an extension of multi-level programming because each level has multi-objective functions. In this chapter, we introduce bi-level programming as the beginning step of multi-level multi-objective programming.

7.1 BI-LEVEL PROGRAMMING

The original formulation for bi-level programming can be traced to Bracken and McGill (1973). Then, Candler and Norton (1977) used the bi-level and multi-level programming designations in their technical report. Although their algorithm for solving multi-level programming is incorrect (it can deal only with convex sets and the higher levels are non-convex sets), the basic premise of their problem has received much attention. Later, the problem of bi-level or multi-level programming was widely used for dealing with special issues of the game theory (Stackelberg, 1952). This static game has fixed leaders and a continuous decision space may be defined to encompass multi-level optimization problems. In addition, other contributors for multi-level programming were made by Aiyoshi and Shimizu (1981a and b, 1982), Bard (1982), Bard and Falk (1982), Bialas and Karwan (1978, 1982), and Candler and Norton (1977a and b, 1981).

The structure of bi-level or multi-level programming facilitates the formulation of the problems that involve hierarchical decision processes. Traditionally, bi-level programming is defined as:

$$\begin{aligned} \max_{x,y} \quad & F(x,y) \\ s.t. \quad & g(x,y) \le 0, \end{aligned} \tag{7.1}$$

where y for each value of x is the solution of the lower level problem:

$$\begin{aligned} \max_{y} \quad & f(x,y) \\ s.t. \quad & h(x,y) \le 0 \end{aligned} \tag{7.2}$$

where $x \in \Re^{nx}$ is the upper level variable, $y \in \Re^{ny}$ is the lower level variable, $F, f : \Re^{nx+ny} \to \Re$, $g : \Re^{nx+ny} \to \Re^{nu}$, and $h : \Re^{nx+ny} \to \Re^{nl}$. Note that $g(x,y) \le 0$ is the upper level constraint, $h(x,y) \le 0$ is the lower level constraint, $F(x,y)$ is the upper level objective function, and $h(x,y)$ is the lower level objective function.

In addition, the relaxed problem associated with bi-level programming problems can be described as:

$$\begin{aligned} &\max_{x,y} \quad F(x,y) \\ &s.t. \quad g(x,y) \le 0, \quad h(x,y) \le 0 \end{aligned} \tag{7.3}$$

The result of Equation (7.3) is the lower bound for the optimal value of the bi-level programming problem. Next, we give some important definitions and notions of bi-level programming as follows:

1. The constraint region:

$$\Omega = \{(x,y) : g(x,y) \le 0, \qquad h(x,y) \le 0\} \tag{7.4}$$

2. For each upper variable, the lower level feasible set:

$$\Omega(x) = \{y\text{: } h(x,y) \le 0\}. \tag{7.5}$$

3. For each upper variable, the lower level reaction set:

$$M(x) = \{y\text{: } y \in argmin\{f(x,y)\text{: } y \in \Omega(x)\}\}. \tag{7.6}$$

4. For each upper variable and any value of the lower variable in $M(x)$, the lower level optimal value:

$$v(x) = f(x,y). \tag{7.7}$$

5. The induced region:

$$IR = \{(x,y)\text{: } (x,y) \in \Omega, \;\; y \in M(x)\}. \tag{7.8}$$

The induced region is the feasible set of bi-level programming and usually non-convex. If $f(x,y)$ and $h(x,y)$ are convex functions in y for all values of x, the bi-level programming is convex. Surely convex bi-level programming has received most of the attention because the lower level problem can be replaced by its Karush-Kuhn-Tucker (KKT) conditions to obtain equivalent one-level mathematical programming under an appropriate constraint qualification.

However, as Dempe et al. (2006) pointed out, the presence of many lower level Lagrange multipliers and an abstract term involving co-derivatives makes the procedure difficult to apply in practice. Hence, Fliege and Vicente (2006) proposed a mapping concept in which a bi-level programming problem can be converted to an equivalent four-objective optimization problem with a special cone dominance concept.

7.2 MULTIPLE LEVEL PROGRAMMING

Although bi-level programming is a special case of multiple level programming when the level is equal to two, the complexity of multiple level programming increases significantly when the number of levels exceeds two (Blair, 1992). Let us illustrate a case for the application of multiple level programming.

Assume a firm plans a multiple level resource allocation program. First, an upper-level decision maker determines the specific tolerance levels and a lower-level decision maker optimizes the maximum objective values under the upper-level constraints. If a final solution does not exist, the upper-level decision maker negotiates with the lower-level decision maker to determine a level that both can accept.

The procedures of multiple level resource allocation problems are illustrated in Figure 7.1. Based on the above concepts, multiple level resource allocation problems can be considered to maximize the following knapsack equation:

$$
\begin{aligned}
&\text{Level 1:} && \max \quad z_1 = \boldsymbol{c}_{11}\boldsymbol{x}_1 + \boldsymbol{c}_{12}\boldsymbol{x}_2 + \cdots + \boldsymbol{c}_{1p}\boldsymbol{x}_p \\
&\text{Level 2:} && \max \quad z_2 = \boldsymbol{c}_{21}\boldsymbol{x}_1 + \boldsymbol{c}_{22}\boldsymbol{x}_2 + \cdots + \boldsymbol{c}_{2p}\boldsymbol{x}_p \\
& && \quad \vdots \\
&\text{Level } m\text{:} && \max \quad z_m = \boldsymbol{c}_{m1}\boldsymbol{x}_1 + \boldsymbol{c}_{m2}\boldsymbol{x}_2 + \cdots + \boldsymbol{c}_{mp}\boldsymbol{x}_p \qquad (7.9) \\
& && \quad s.t. \quad \boldsymbol{a}_{11}\boldsymbol{x}_1 \le \boldsymbol{b}_1, \\
& && \qquad\qquad \boldsymbol{a}_{21}\boldsymbol{x}_2 \le \boldsymbol{b}_2, \\
& && \qquad\qquad \vdots \\
& && \qquad\qquad \boldsymbol{a}_{m1}\boldsymbol{x}_m \le \boldsymbol{b}_m, \\
& && \qquad\qquad \boldsymbol{x} \ge \boldsymbol{0},
\end{aligned}
$$

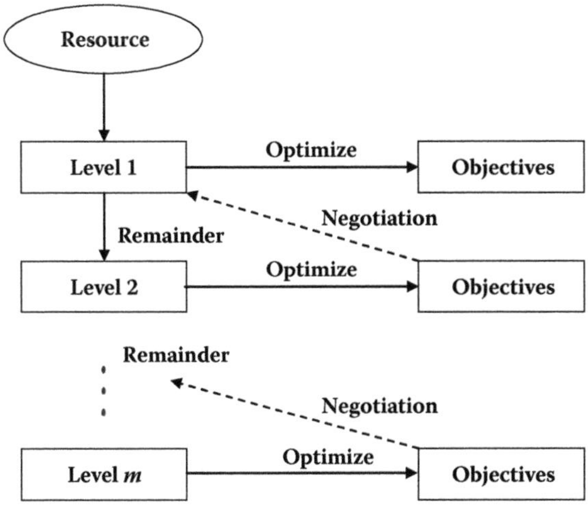

FIGURE 7.1 Multiple level resource problems.

where $\boldsymbol{c}_{ij}$ and $\boldsymbol{x}_i$ denote given resource parameters at the ith level, which usually are represented as technological coefficients and products, respectively, and $\boldsymbol{b}_i$ denotes the maximum limited resource portfolios on the ith level.

Similar to bi-level programming, one of the important characteristics of multiple-level programming is that a decision maker at a certain level of hierarchy has his or her own objective function and may affect (or be affected by) other levels. That means that a decision maker can introduce the instruments to adjust the objective functions of each level. Note that these instruments may include the allocation and use of resources at lower levels and the advantages obtained from other levels (Gaur and Arora, 2008).

7.3 FUZZY PROGRAMMING FOR MULTI-LEVEL MULTI-OBJECTIVE PROGRAMMING

In previous situations involving bi-level or multi-level programming, each level considers only one objective function. In practice, multi-objective functions may appear in a level. Hence, if we relax the condition of the objective function, we can formulate multi-level multi-objective programming. When levels are equal to two, multi-level multi-objective programming is reduced to bi-level multi-objective programming.

For a vector maximization p-level hierarchical system that has a maximization type objective function, we can formulate the following programming:

$$
\begin{aligned}
&\max_{x_1} \quad f_1(\boldsymbol{x}) = \boldsymbol{c}_{11}\boldsymbol{x}_1 + \boldsymbol{c}_{12}\boldsymbol{x}_2 + \cdots + \boldsymbol{c}_{1P}\boldsymbol{x}_P \\
&\max_{x_2} \quad f_2(\boldsymbol{x}) = \boldsymbol{c}_{21}\boldsymbol{x}_1 + \boldsymbol{c}_{22}\boldsymbol{x}_2 + \cdots + \boldsymbol{c}_{2P}\boldsymbol{x}_P \\
&\qquad\qquad \vdots \\
&\max_{x_P} \quad f_p(\boldsymbol{x}) = \boldsymbol{c}_{p1}\boldsymbol{x}_1 + \boldsymbol{c}_{p2}\boldsymbol{x}_2 + \cdots + \boldsymbol{c}_{pP}\boldsymbol{x}_P \\
&\text{s.t.} \quad \boldsymbol{A}_{i1}\boldsymbol{x}_1 + \boldsymbol{A}_{i2}\boldsymbol{x}_2 + \cdots + \boldsymbol{A}_{iP}\boldsymbol{x}_P \le \boldsymbol{b}_i, \qquad i = 1,2,\ldots,m, \\
&\qquad \boldsymbol{x}_1 \ge \boldsymbol{0}, \boldsymbol{x}_2 \ge \boldsymbol{0}, \ldots, \boldsymbol{x}_P \ge \boldsymbol{0}.
\end{aligned}
\tag{7.10}
$$

Note that we assume the system above includes one decision maker on each level, n decision variables, and m constraints. Let $\boldsymbol{x} = \boldsymbol{x}_1 \cup \boldsymbol{x}_2 \cup \ldots \cup \boldsymbol{x}_P$ and $n = n_1 + n_2 + \cdots + n_p$ where decision vector $\boldsymbol{x}_P$, $p = 1,2,\ldots,P$ is under the control of the pth level decision maker and has n_p decision variables. The decision maker on the pth level ($p = 1,2,\ldots,P$) individually solves the maximization problem:

$$
\begin{aligned}
&\max_{x_P} \quad f_p(\boldsymbol{x}) = \boldsymbol{c}_{p1}\boldsymbol{x}_1 + \boldsymbol{c}_{p2}\boldsymbol{x}_2 + \cdots + \boldsymbol{c}_{pP}\boldsymbol{x}_P \\
&\text{s.t.} \quad \boldsymbol{A}_{i1}\boldsymbol{x}_1 + \boldsymbol{A}_{i2}\boldsymbol{x}_2 + \cdots + \boldsymbol{A}_{iP}\boldsymbol{x}_P \le \boldsymbol{b}_i, \qquad i = 1,2,\ldots,m, \\
&\qquad \boldsymbol{x}_1 \ge \boldsymbol{0}, \boldsymbol{x}_2 \ge \boldsymbol{0}, \ldots, \boldsymbol{x}_P \ge \boldsymbol{0}.
\end{aligned}
\tag{7.11}
$$

Solving the above programming, we calculate the individual optimal solution x_p^o, $p = 1,2,\ldots,P$.

Sinha (2003) proposed a fuzzy approach for dealing with multi-level programming problems. First we set the minimum acceptable degree of satisfaction for objective function f_p, β_p, i.e., $\mu[f_p(x)] \geq \beta_p$. Then we calculate the membership function for any maximization type objective function as:

$$\mu[f_p(x)] = \begin{cases} 0, & f_p(x) \leq f_p^U \\ [f_p(x) - f_p^L]/[f_p^U - f_p^L], & f_p^L \leq f_p(x) \leq f_p^U \\ 1, & f_p(x) \leq f_p^L \end{cases} \tag{7.12}$$

where f_p^U and f_p^L are the upper and lower bounds of the pth level programming, respectively.

Next, we set α_1 as the minimum acceptable degree of satisfaction for x_1, i.e., $\mu_{x_1}(x_1) \geq \alpha_1$, and incorporate the concept of the tolerance interval that enlarges the flexible solution of the lower level. Consider the bi-level situation for example. The first level decision maker can set up the negative tolerance value e_{1-} on x_1 and the positive tolerance value e_{1+} on x_1. Then the first-level decision maker can assign the membership function for x_1 as:

$$\mu_{x_1}(x_1) = \begin{cases} [x_1 - (x_1^o - e_{1-})]/e_{1-}, & (x_1^o - e_{1-}) \leq x_1 \leq x_1^o, \\ [(x_1^o + e_{1+}) - x_1]/e_{1+}, & x_1^o \leq x_1 \leq (x_1^o - e_{1-}), \\ 0, & \text{otherwise}, \end{cases} \tag{7.13}$$

where x_1^o denotes the optimal solution of the first level. For simplicity, we assume the membership function in this chapter is linear. However, other types of membership functions such as trapezoidal, exponential, or logarithmic can be used in a similar way. We then use the above concept to set the minimum acceptable degree of satisfaction for objective f_2 and the corresponding membership function. Finally, we try to maximize $\beta_1, \alpha_1, \beta_2$ simultaneously. Let $\lambda = \min\{\beta_1, \alpha_1, \beta_2\}$. We can solve the second-level auxiliary problem as formulating the following programming:

$$\begin{aligned} \max \quad & \lambda \\ s.t. \quad & A_{i1}x_1 + A_{i2}x_2 + \cdots + A_{iP}x_P \leq b_i, \quad i = 1,\ldots,m, \\ & \mu_{x_1}(x_1) \geq \lambda i, \quad \mu_{f_1}[f_1(x)] \geq \lambda, \quad \mu_{f_2}[f_2(x)] \geq \lambda, \\ & x_1 \geq 0, x_2 \geq 0, \ldots, x_P \geq 0, \quad \lambda \in [0,1], \end{aligned} \tag{7.14}$$

where i is a column vector, which has $n_1 P$ components of value 1. Next, we consider the following different multi-level programming as follows.

Example 7.1 Consider a bi-level programming problem as follows:

$$
\begin{aligned}
\max_{x_1, x_2} \quad & f_1 = 4x_1 + 3x_2 + 2x_3 + x_4 \\
\max_{x_3, x_4} \quad & f_2 = x_1 + 2x_2 + 3x_3 + 4x_4 \\
s.t. \quad & x_1 + x_2 + x_3 + x_4 \le 10, \\
& x_1 + x_4 \le 5, \\
& x_2 + x_3 \le 4, \\
& x_1, x_2, x_3, x_4 \ge 0.
\end{aligned}
$$

Solving the above objective individually, we obtain the optimal solution $f_1 = 32$ at (5,4,0,0) and $f_2 = 32$ at (0,0,4,5), respectively. Then we can calculate $f_1^U = 32$; $f_1^L = 13$ and $f_2^U = 32$; $f_2^L = 13$ from the pay-off matrix. Let the first-level decision maker decide $x_1 = 5$ with 2 negative and 2 positive tolerances and $x_2 = 4$ with 1.5 negative and 1.5 positive tolerance. Finally, we can solve the second-level problem as:

$$
\begin{aligned}
\max \quad & \lambda \\
s.t. \quad & x_1 + x_2 + x_3 + x_4 \le 10 \\
& x_1 + x_4 \le 5, \\
& x_2 + x_3 \le 4, \\
& (f_1 - 13)/(32 - 13) \ge \lambda, \\
& (f_2 - 13)/(32 - 13) \ge \lambda, \\
& [x_1 - (5 - 2)]/2 \ge \lambda; \quad [(5 + 2) - x_1]/2 \ge \lambda, \\
& [x_2 - (5 - 1.5)]/1.5 \ge \lambda; \quad [(5 + 1.5) - x_2]/2 \ge \lambda, \\
& x_1, x_2, x_3, x_4 \ge 0
\end{aligned}
$$

Then, we can obtain the satisfactory solution as $x_1 = 3.57$, $x_2 = 2.92$, $x_3 = 1.08$, $x_4 = 1.43$, and $\lambda = 0.283$. We can calculate the two objective values as $f_1 = 26.63$ and $f_2 = 18.37$.

We now consider the extension from bi-level programming to multi-level programming. We can define λ as the minimum of all the minimum acceptable levels of satisfaction of the pth-level system. The pth-level programming problems can be presented as:

$$
\begin{aligned}
\max \quad & \lambda \\
s.t. \quad & \boldsymbol{A}_{i1}\boldsymbol{x}_1 + \boldsymbol{A}_{i2}\boldsymbol{x}_2 + \cdots + \boldsymbol{A}_{iP}\boldsymbol{x}_P \le \boldsymbol{b}_i, \quad i = 1, \ldots, m, \\
& \mu_{x_1}(\boldsymbol{x}_1) \ge \lambda \boldsymbol{I}, \quad \mu_{x_2}(\boldsymbol{x}_1) \ge \lambda \boldsymbol{I}, \ldots, \quad \mu_{x_{(P-1)}}(\boldsymbol{x}_{(P-1)}) \ge \lambda \boldsymbol{I}, \\
& \mu_{f_1}[f_1(\boldsymbol{x})] \ge \lambda, \quad \mu_{f_2}[f_2(\boldsymbol{x})] \ge \lambda, \ldots, \quad \mu_{f_p}[f_p(\boldsymbol{x})] \ge \lambda, \\
& \boldsymbol{x}_1 \ge \boldsymbol{0}, \boldsymbol{x}_2 \ge \boldsymbol{0}, \ldots, \boldsymbol{x}_P \ge \boldsymbol{0}, \lambda \in [0, 1].
\end{aligned}
$$

Example 7.2 Let us extend the previous example as a multi-level programming sequence as follows:

$$\max_{x_1} \quad f_1 = 4x_1 + 3x_2 + 2x_3 + x_4$$

$$\max_{x_4} \quad f_2 = x_1 + 2x_2 + 3x_3 + 4x_4$$

$$\max_{x_2,x_3} \quad f_2 = x_1 + 4x_2 + 3x_3 + x_4$$

$$\begin{aligned} s.t. \quad & x_1 + x_2 + x_3 + x_4 \le 10, \\ & x_1 + x_4 \le 5, \\ & x_2 + x_3 \le 4, \\ & x_1, x_2, x_3, x_4 \ge 0. \end{aligned}$$

The optimal solutions of the objectives are $f_1 = 32$ at (5,4,0,0); $f_2 = 32$ at (0,0,4,5); and $f_3 = 21$ at (0,4,0,5). Obviously, there is no satisfactory solution. Let us consider the second-level situation. From the pay-off matrix, we obtain $f_1^U = 32$; $f_1^L = 13$, and $f_2^U = 32$; $f_2^L = 13$. If we consider $x_1 = 5$ with 2 negative and 2 positive tolerances, we can solve the satisfactory solution as $x_1 = 3.8$, $x_2 = 0$, $x_3 = 4$, $x_4 = 1.2$, and $\lambda = 0.4$.

From the satisfactory solution of the second level, we can re-calculate $f_1^U = 32$; $f_1^L = 24.4$, $f_2^U = 32$; $f_2^L = 20.6$, and $f_3^U = 21$; $f_3^L = 17$. Next, if we decide $x_1 = 3.8$ with 1 negative and 1 positive tolerance and $x_4 = 1.2$ with 1 negative and 1 positive tolerance, the satisfactory solution of the problem is $x_1 = 2.8$, $x_2 = 3$, $x_3 = 1$, $x_4 = 2.2$, and $\lambda = 0$. The corresponding objectives are $f_1 = 24.4$, $f_2 = 20.6$, and $f_3 = 20$. Next, we consider multi-level multi-objective programming problems as follows.

Example 7.3 Let us consider the following bi-level bi-objective programming:

$$\max_{x_1,x_2} \quad f_{11} = 4x_1 + 3x_2 + 2x_3 + x_4$$

$$\max_{x_1,x_2} \quad f_{12} = 2x_1 + 4x_2 + x_3 + 3x_4$$

$$\max_{x_3,x_4} \quad f_{21} = x_1 + 2x_2 + 3x_3 + 4x_4$$

$$\max_{x_3,x_4} \quad f_{22} = 3x_1 + x_2 + 4x_3 + 2x_4$$

$$\begin{aligned} s.t. \quad & x_1 + x_2 + x_3 + x_4 \le 10, \\ & x_1 + x_4 \le 5, \\ & x_2 + x_3 \le 4, \\ & x_1, x_2, x_3, x_4 \ge 0. \end{aligned}$$

The optimal solution of each objective can be calculated as $f_{11} = 32$ at (5,4,0,0); $f_{12} = 31$ at (0,4,0,5); $f_{21} = 32$ at (0,0,4,5); and $f_{22} = 31$ at (5,0,4,0).

Next, from the pay-off matrix, we obtain $f_{11}^U = 32$; $f_{11}^L = 13$, $f_{12}^U = 31$; $f_{12}^L = 14$, $f_{21}^U = 32$; $f_{21}^L = 13$ and $f_{22}^U = 31$; and $f_{22}^L = 26$. Then, the first-level decision maker sets $x_1 = 5$ with 1 negative and 1 positive tolerance and $x_2 = 4$ with 2.5 negative and 2.5 positive tolerances. We obtain the final optimal solution as $x_1 = 5$, $x_2 = 1.6$, $x_3 = 2.4$, $x_4 = 0$, and $\lambda = 0.04$. We can calculate the objective values as $f_{11} = 29.6$, $f_{12} = 18.8$, $f_{21} = 15.4$, and $f_{22} = 26.2$. Multi-level multi-objective problems can be easily considered as extensions of Example 7.3.

Many methods have been proposed to solve multi-level multi-objective programming. For example, Baky (2010) used the fuzzy goal programming approach; Wang et al. (1996) adopted the decomposition method; Yano (2007) employed interactive fuzzy decision making. Readers can also consider the nonlinear and fuzzy situations of multi-level multi-objective programming. Since solving multi-level multi-objective programming is normally a complex process, globally optimal solutions are achieved rarely. Hence, evolutionary algorithms are very suitable for dealing with complex problems.

8 Data Envelopment Analysis

Traditionally, data envelopment analysis (DEA) and regression-based methods such as deterministic and stochastic models were widely used to measure the technical efficiency of decision-making units (DMUs). The main difference between DEA and regression-based methods is that DEA is a non-parametric approach while regression-based methods are parametric. Several papers compared DEA with regression-based methods with respect to efficiency, flexibility, robust, assumptions, and sample sizes (Cooper and Tone, 1997; Ruggiero, 1998; Chen, 2002).

The abandonment of DEA for measuring technical efficiency has been suggested due to the disadvantages of sensitivity to outliers and failure to reveal measurement errors (Schmidt, 1985; Greene, 1993). However, the most critical problem of using regression-based methods is mis-specification (Giannakas et al., 2003; Gonzalez and Castro, 2001). It is necessary to specify a particular production function (e.g., Cobb-Douglas or translog form) before measuring the frontiers of DMUs, and different production functions may yield different results. However, it is hard to specify a correct production function in advance because of the complex relations of input and output variables.

This chapter attempts to provide a flexible and robust method for finding the production function automatically so that the linear and nonlinear relations between input and output variables can be considered. Section 8.1 introduces the concepts of DEA and regression-based frontier models for evaluating the technical efficiency of DMUs.

8.1 TRADITIONAL DEA

DEA is a mathematical programming technique that can calculate the relative efficiencies of DMUs according to multiple inputs and outputs. Thus, when facing topics that involve investigating the efficiency of converting multiple inputs into multiple outputs, DEA can be an appropriate and useful technique. Furthermore, DEA makes it possible to benchmark the best practice DMU and provides estimates of the potential improvements for DMUs that are considered inefficient.

DEA has been applied extensively in the managerial and economics fields to solve multi-criterion problems. Weber (1996) and Liu et al. (2000) applied DEA in evaluation of suppliers for an individual product. DEA has been applied to evaluate the performance of private sector facilities such as banks and power plants (Berg et al., 1991; Golany et al., 1994). Instead of investigating the operation efficiency of corporations in various industries, DEA is also considered a good technique to provide the performance indicators when outputs are not defined clearly, for example, the productivity of public sectors such as universities, hospitals, and government institutions (Bedard, 1985).

The origin of DEA can be traced to the simplex algorithm used to estimate production possibility frontiers and access technical efficiency proposed by Dantzig (1951). Farrell (1957) developed a measure of technical efficiency calculated from sample data. Charnes et al. (1978) reintroduced and developed the mathematical method as DEA. Some comprehensive reviews are provided by Boussofiane et al. (1991), Seiford (1996), and Charnes et al. (1989).

The first DEA model discussed here is the CCR (Charnes, Cooper and Rhodes) model introduced in 1978. The input-oriented CCR form of a DEA mathematical programming model is:

$$\max h_0 = \frac{\sum_{r=1}^{t} u_r y_{rj0}}{\sum_{i=1}^{m} v_i x_{ij0}} \tag{8.1}$$

$$\begin{aligned} s.t. \quad & \frac{\sum_{r=1}^{t} u_r y_{rj}}{\sum_{i=1}^{m} v_i x_{ij}} \le 1, \quad j = 1,\ldots,n, \\ & u_r \ge \varepsilon > \mathbf{0}, \quad r = 1,\ldots,t, \\ & v_i \ge \varepsilon > \mathbf{0}, \quad i = 1,\ldots,m, \end{aligned}$$

where u_r is the weight of output r, v_i is the weight of input i, y_{rj} is the output r of DMU j, x_{ij} is the amount of input i of DMU j, t is the number of outputs, m is the number of inputs, n is the number of DMUs, and ε is a small positive number (in general, ε is taking $\varepsilon = 10^{-6}$.

To maximize the efficiency score of a DMU j_0, the objective function, we choose a set of weights for all inputs and outputs. The constraint set ensures that the efficiency scores of all DMUs will not exceed 1.0. The last two constraint sets ensure that all inputs and outputs are included in the model, that is, no weights are set at 0. A score of 1 represents an efficient DMU j_0; other values are considered inefficient.

Equation (8.1) is the ratio form of DEA that has an infinite number of solutions. To find a solution, the formula can be converted into a linear programming problem by moving the denominator in the first constraint set in Equation (8.1) and setting the denominator to 1:

$$\max h_0 = \sum_{r=1}^{t} u_r y_{rj0} \tag{8.2}$$

$$\begin{aligned} \textbf{s.t.} \quad & \sum_{i=1}^{t} v_i x_{ij0} = 1, \\ & \sum_{r=1}^{t} u_r y_{rj} - \sum_{i=1}^{m} v_i x_{ij} \le 0, \quad j = 1,\ldots,n, \\ & u_r \ge \varepsilon > 0, \quad r = 1,\ldots,t, \\ & v_i \ge \varepsilon > 0, \quad i = 1,\ldots,m \end{aligned}$$

where u_r is the weight of output r, v_i is the weight of input i, y_{rj} is the amount of output r of DMU j, x_{ij} is the amount of input i of DMU j, t is the number of outputs, m is the number of inputs, n is the number of DMUs, and ε is a small positive number. The dual model of Equation (8.2) is:

$$\max \quad z_0 = \varepsilon\left[\sum_{i=1}^{m} s_i^+ + \sum_{r=1}^{t} s_r^-\right] \tag{8.3}$$

$$s.t. \quad z_0 x_{ij0} - \sum_{j=1}^{n} x_{ij}\lambda_i - s_i^- = 0, \qquad i = 1,\ldots,m,$$

$$\sum_{j=1}^{n} y_{rj}\lambda_j - s_r^+ = y_{rj0}, \qquad r = 1,\ldots,t,$$

$$\lambda_j, s_i^+, s_r^- \geq 0,$$

where z_0, s_i^-, s_r^+ are the dual variables. Based on Equation (8.3), a DMU j_0 is efficient in the dual optimal solution, that is, when $z_0 = 1$ and $s_i^- = s_r^+ = 0$ for all i and r. Conversely, for an inefficient DMU, appropriate adjustments of the inputs and outputs can be proposed according to the difference in performance to the efficient level.

Example 8.1

Assume five DMUs are to be evaluated by their input and output information shown in Table 8.1. We can use the input-oriented CCR to obtain the results of each DMU as shown in Table 8.2. The results from the CCR model indicate that the DMUs B and C are both efficient. However, DMU B outperforms DMU C in the criterion of time as a benchmark for another DMU. Hence, we can consider DMU B as the best followed by DMU C, DMU E, DMU D, and DMU A. We next utilize the BCC model to reconsider the above problem and obtain the efficient scores of DMUs shown in Table 8.3. The results from the BCC model show that DMU B Is the best unit; DMUs C and D perform equally; and DMUs A and E are inefficient.

Several kinds of DEA models have been proposed more recently. For example, interval DEA models (Cooper et al., 1999) and fuzzy DEA models (Kuo et al., 2006; Guo and Tanaka, 2001; Entani et al., 2002) are proposed to deal with imprecise input and output data. Network DEA models (Fare and Grosskopf, 1996, 2000) are used for evaluating network divisions of DMUs and multi-objective programming approaches for DEA (Yu et al., 2004; Chiang and Tzeng, 2000). Here, we introduce network DEA models based on their popularity in recent years and discuss multi-objective programming for DEA because of its relevance.

TABLE 8.1
Input and Output Data for Example 8.1

	Input		Output	
DMU	x_1	x_2	y_1	y_2
A	16	10	4	2
B	10	8	6	6
C	12	14	8	4
D	18	16	10	8
E	14	14	8	6

TABLE 8.2
CCR Model for Example 8.1

DMU	Score	Benchmark (Lambda)	Times as Benchmark for Other DMU
A	0.533333	B (0.666667)	0
B	1	B (1.000000)	3
C	1	C (1.000000)	2
D	0.901639	B (1.229508); C (0.327869)	0
E	0.904762	B (0.666667); C (0.500000)	0

TABLE 8.3
BCC Model for Example 8.1

DMU	Score	Benchmark (Lambda)	Times as Benchmark for Other DMU
A	0.8	B (1.000000)	0
B	1	B (1.000000)	2
C	1	C (1.000000)	1
D	1	D (1.000000)	1
E	0.952381	B (0.333333); C (0.333333); D (0.333333)	0

8.2 NETWORK DEA

Network DEA was proposed by Fare and Grosskopf (1996, 2000) as a general form of the traditional DEA model. In traditional DEA, a DMU is the basic unit of analysis, and we can only evaluate the performance of each DMU from input and output values. Therefore, the processes of DMUs that transform inputs to outputs, are viewed as a black box. However, network DEA considers sub-units or sub-technologies of DMUs as basic units of DEA. Thus, we can determine network efficiency by calculating the interrelated production frontiers.

We now describe the notations of network. Let $\boldsymbol{x} \in \Re_+^n$ be the inputs of DMUs; ${}_0^i\boldsymbol{x}$ is the network exogenous vector and indicates the input vector from source 0 used in activity i; P^j represents the jth sub-technologies; ${}_j^k\boldsymbol{y}$ is the output produced by P^j; and $P^j(\cdot)$ denotes the jth output set. Figure 8.1 illustrates the network DEA concept. Based on the figure, the network model can be represented as:

$$\begin{aligned} P(x) = \{ & \left({}_1^4y + {}_2^4y + {}_3^4y\right) : \\ & \left({}_1^4y + {}_1^3y\right) \in P^1\left({}_0^1x\right) \\ & \left({}_1^4y + {}_1^3y\right) \in P^1\left({}_0^2x\right) \\ & {}_3^4y \in P^1\left({}_0^3x, {}_1^3y + {}_2^3y\right) \\ & {}_0^1x + {}_0^2x + {}_0^3x \le x \} \end{aligned} \tag{8.4}$$

The network DEA model can utilize radial measures of efficiency such as the CCR (Charnes et al., 1978) or BBC (Banker et al., 1984) models as the basic DEA methodology and the production possibility set.

Tone and Tsutsui (2009) proposed the slack-based measure (SBM) approach for evaluating the efficiency of DMUs. The major difference between the SBM approach and previous method is that SBM is a non-radial method and is suitable for measuring efficiencies when inputs and outputs may change non-proportionally. Therefore, this

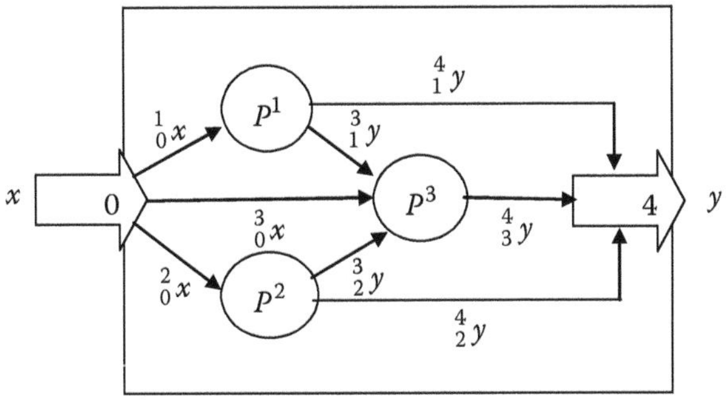

FIGURE 8.1 Example of network DEA.

method can decompose the overall efficiency into divisional units. In addition, the SBM approach can rank the divisions (Cooper et al., 2007; Tsutsui and Goto, 2009).

In this section, we introduce the SBM approach (Tone and Tsutsui, 2009). Assume n DMUs ($j = 1,\ldots,n$) consist of K divisions ($k = 1,\ldots,K$). Let m_k and r_k be the numbers of inputs and outputs to Division k, respectively. A link leads from Division k to Division h by (k,h) and the set of links by L. Then, the input resources to DMU$_k$ are defined as $\{\boldsymbol{x}_j^k \in \Re_+^{m_k}, j = 1,\ldots,n; k = 1,\ldots,K\}$; the output products from DMU$_k$ are defined as $\{\boldsymbol{y}_j^k \in \Re_+^{r_k}, j = 1,\ldots,n; k = 1,\ldots,K\}$; and the linking intermediate products from Division k to Division h are defined as $\{\boldsymbol{z}_j^{(k,h)} \in \Re_+^{t(k,h)}, j = 1,\ldots,n; (k,h) \in L\}$ where $t_{(k,h)}$ is the number of items in Link (k,h). The production possibility set $\{\boldsymbol{x}^k, \boldsymbol{y}^k, \boldsymbol{z}^{(k,h)}\}$ can be defined by:

$$\boldsymbol{x}^k \geq \sum_{j=1}^{n} \boldsymbol{x}_j^k \lambda_j^k, \quad k = 1,\ldots,K, \tag{8.5}$$

$$\boldsymbol{y}^k \geq \sum_{j=1}^{n} \boldsymbol{y}_j^k \lambda_j^k, \quad k = 1,\ldots,K, \tag{8.6}$$

$$\boldsymbol{z}^{(k,h)} \geq \sum_{j=1}^{n} \boldsymbol{z}_j^{(k,h)} \lambda_j^k, \tag{8.7}$$

$$\boldsymbol{z}^{(k,h)} \geq \sum_{j=1}^{n} \boldsymbol{z}_j^{(k,h)} \lambda_j^h, \tag{8.8}$$

$$\sum_{j=1}^{n} \lambda_j^k = 1, \quad \lambda_j^k \geq 0, \tag{8.9}$$

where $\lambda^k \in \Re_+^n$ is the intensity vector corresponding to the kth division. Note that if we delete the Equation (8.9) from the above model, we can deal with the CRS instead of VRS case. DMU$_o$ ($o = 1,\ldots,n$) can be represented by

$$\begin{aligned} \boldsymbol{x}_o^k &\geq \boldsymbol{x}^k \lambda^k + \boldsymbol{s}^{k-}, \quad k = 1,\ldots,K, \\ y_o^k &\geq y^k \boldsymbol{\lambda}^k - \boldsymbol{s}^{k+}, \quad k = 1,\ldots,K, \\ \boldsymbol{e}^k \lambda^k &= 1, \quad k = 1,\ldots,K, \ \lambda^k, \boldsymbol{s}^{k-}, \boldsymbol{s}^{k+} \geq \boldsymbol{0} \end{aligned} \tag{8.10}$$

where $\boldsymbol{x}^k = (\boldsymbol{x}_1^k,\ldots,\boldsymbol{x}_n^k) \in \Re^{m_k \times n}$, $\boldsymbol{y}^k = (\boldsymbol{y}_1^k,\ldots,\boldsymbol{y}_n^k) \in \Re^{r_k \times n}$, and $\boldsymbol{s}^{k-}(\boldsymbol{s}^{k+})$ are the input (output) slack vectors. Note that the SBM approach claims all observed data are positive. If negative or zero data are considered, we should transform them to positive data first (Tone, 2004). In order to define the linking constraints, two possible cases can be considered:

Free link value case — The linking activities are freely determined while maintaining continuity between input and output:

$$z^{(k,h)}\lambda^h = z^{(k,h)}\lambda^k, \forall (k,h), \tag{8.11}$$

where

$$z^{(k,h)} = (z_1^{(k,h)}, \ldots, z_n^{(k,h)}) \in \Re^{t(k,h)\times n}. \tag{8.12}$$

Fixed link value case — The linking activities are kept unchanged:

$$z_o^{(k,h)} = z^{(k,h)}\lambda^h, \forall (k,h), \tag{8.13}$$

$$z_o^{(k,h)} = z^{(k,h)}\lambda^k, \forall (k,h).$$

The first case indicates that the link flow may increase or decrease in the optimal solution of the linear programming. On the other hand, the latter case indicates that intermediate products are beyond the control of DMUs. Finally, we can define the efficiency scores of each DMU corresponding to the selected orientation, i.e., input, output, or non-oriented.

8.2.1 Input-Oriented Efficiency

The input-oriented efficiency of DMU_o can be calculated as:

$$\theta_o^* = \min_{\lambda^k, s^{k-}} \sum_{k=1}^{K} w^k \left[1 - \frac{1}{m_k} \left(\sum_{i=1}^{m_k} \frac{s_i^{k-}}{x_{io}^k} \right) \right] \tag{8.14}$$

$$s.t.\ (8.10) \text{ and } (8.11) \text{ or } (8.13),$$

where w^k is the importance of the kth division such that $\sum_{k=1}^{K} w^k = 1$. Hence, if $\theta_o^* = 1$, the DMU_o is called overall input-efficient. Furthermore, we can define the input-oriented divisional efficiency score as:

$$\theta_k = 1 - \frac{1}{m_k} \left(\sum_{i=1}^{m_k} \frac{s_i^{k-*}}{x_{io}^k} \right), \quad \forall k = 1, \ldots, K \tag{8.15}$$

where s_i^{k-*} is the optimal input slack s_i^{k-*} of Equation (8.14). The DMU_o is called input-efficient for the kth division if $\theta_k = 1$. In addition, the overall input-efficient score is the weighted arithmetic mean of the divisional scores.

8.2.2 Output-Oriented Efficiency

The output-oriented efficiency of DMU_o can be calculated as solving the following programming:

$$1/\tau_o^* = \max_{\lambda^k, s^{k-}} \sum_{k=1}^{K} w^k \left[1 + \frac{1}{r_k} \left(\sum_{r=1}^{r_k} \frac{s_r^{k+}}{y_{ro}^k} \right) \right] \tag{8.16}$$

$$s.t. \text{ (8.10) and (8.11) or (8.13)}$$

where w^k is the importance of the kth division such that $\sum_{k=1}^{K} w^k = 1$. Therefore, if $\tau_o^* = 1$, the DMU_o is called overall output efficient. We can also define the output-oriented divisional efficiency score for the kth division as:

$$\tau_k = \frac{1}{1 + r_k \left(\sum_{r=1}^{k} \frac{s_r^{k+*}}{y_{ro}^k} \right)}, \quad \forall k = 1, \ldots, K \tag{8.17}$$

where s_r^{k+*} is the optimal output slack s_i^{k-*} of Equation (8.16). The output-oriented overall efficiency score is the weighted harmonic mean of the divisional scores.

8.2.3 Non-Oriented Efficiency

The non-oriented efficiency of DMU_o can be defined as the following fractional programming:

$$\rho_o^* = \min_{\lambda^k, s^{k-}, s^{k+}} \frac{\sum_{k=1}^{K} w^k w^k \left[1 - \frac{1}{m_k} \left(\sum_{i=1}^{m_k} \frac{s_i^{k-}}{x_{io}^k} \right) \right]}{\sum_{k=1}^{K} w^k \left[1 + \frac{1}{r_k} \left(\sum_{r=1}^{r_k} \frac{s_r^{k+}}{y_{ro}^k} \right) \right]} \tag{8.18}$$

$$s.t. \quad \text{(8.10) and (8.11) or (8.13),}$$

where w^k is the importance of the kth division such that $\sum_{k=1}^{K} w^k = 1$. When $\rho_o^* = 1$, the DMU_o is called overall efficient. Then, the non-oriented divisional efficiency score for the kth division can be defined as:

$$\rho_k = \frac{1 - \frac{1}{m_k} \left(\sum_{i=1}^{m_k} \frac{s_i^{k-*}}{x_{io}^k} \right)}{1 + r_k \left(\sum_{r=1}^{r_k} \frac{s_r^{k+*}}{y_{ro}^k} \right)}, \quad \forall k = 1, \ldots, K \tag{8.19}$$

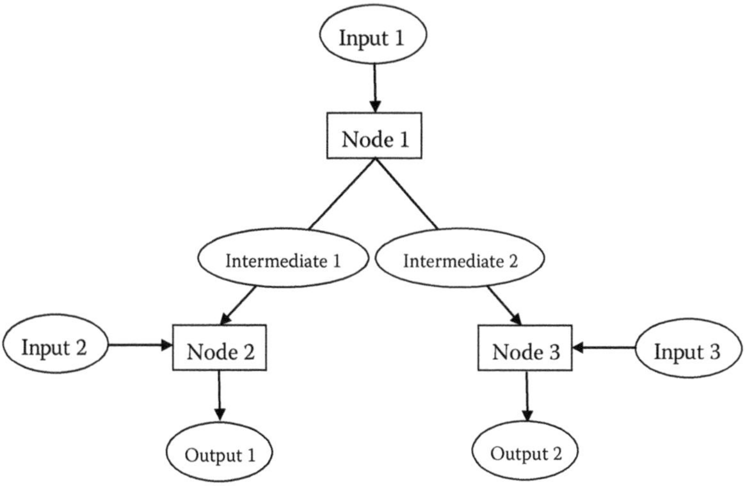

FIGURE 8.2 Network DEA model.

where s_i^{k-*} and s_r^{k+*} are the optimal input and output slacks of Equation (8.18). The other properties of network SBM models are discussed in Tone and Tsutsui (2009).

Example 8.2

Consider a network DEA model as shown in Figure 8.2. Assume the data of the model are listed in Table 8.4. That is, there are three nodes (or divisions) with two intermediates in the system. In addition, we assume all divisions are equally important. We want use the network SBM model to evaluate the DMUs.

First, we employ the input-oriented SBM model with CRS and VRS to evaluate the efficiency scores of the DMUs. Then, we can obtain the results as shown in Tables 8.5 and 8.6.

TABLE 8.4
Network DEA Data

	Node 1	Node 2		Node 3		Intermediates	
DMU	Input 1	Input 2	Output 1	Input 3	Output 2	Intermediate 1	Intermediate 2
A	5	4	7	6	9	6	6
B	3	3	5	9	6	8	9
C	10	4	8	7	6	6	3
D	7	3	3	7	7	7	8
E	6	7	6	8	6	7	9
F	8	5	6	8	4	2	6
G	3	2	6	2	5	4	5

TABLE 8.5
Results of Input-Oriented Network DEA with CRS

	Overall	Divisional Score			Benchmark (Lambda)			Times as Benchmark		
DMU	Score	Node 1	Node 2	Node 3	Node 1	Node 2	Node 3	Node 1	Node 2	Node 3
A	0.674	0.600	0.821	0.600	B (1.000)	B (0.714) G (0.571)	G (1.800)	0	0	2
B	0.489	0.444	0.577	0.444	B (0.444)	B (0.048) G (0.794)	A (0.667)	6	3	0
C	0.403	0.200	0.667	0.343	B (0.667)	G (1.333)	G (1.200)	0	0	0
D	0.476	0.222	0.538	0.667	B (0.519)	B (0.460) G (0.116)	A (0.778)	0	0	0
E	0.324	0.333	0.340	0.300	B (0.667)	B (0.286); G (0.762)	G (1.200)	0	0	0
F	0.271	0.187	0.400	0.225	B (0.500)	G (1.000)	G (0.900)	1	0	0
G	0.941	0.822	1.000	1.000	B (0.467) F (0.133)	G (1.000)	G (1.000)	0	6	4

TABLE 8.6
Results of Input-Oriented Network DEA with VRS

DMU	Overall Score	Divisional Score			Benchmark (Lambda)			Times as Benchmark		
		Node 1	Node 2	Node 3	Node 1	Node 2	Node 3	Node 1	Node 2	Node 3
A	0.783	0.600	0.750	1.000	B (0.250) G (0.750)	C (0.500) G (0.500)	A (1.000)	0	0	3
B	0.674	1.000	0.688	0.333	B (0.063) G(0.938)	B (0.063) G (0.938)	A (0.250) G (0.750)	4	2	0
C	0.677	0.388	1.000	0.643	B (0.438) C (0.125) G (0.438)	C (1.000)	D (0.500) G (0.500)	0	2	0
D	0.569	0.429	0.708	0.571	B (0.125) G (0.875)	B (0.125) G (0.875)	A (0.500) G (0.500)	0	0	1
E	0.392	0.500	0.300	0.375	B (0.063) G (0.938)	B (0.050) C (0.025) G (0.925)	A (0.250) G (0.750)	0	0	0
F	0.342	0.375	0.400	0.250	G (1.000)	G (1.000)	G (1.000)	0	0	0
G	1.000	1.000	1.000	1.000	G (1.000)	G (1.000)	G (1.000)	6	5	5

The explanation of tables is similar to that for traditional DEA models. However, we can understand the performance of each division via the network DEA models. Readers can use output-oriented or non-oriented network models to compare the results with Example 8.2.

8.3 FUZZY MULTI-OBJECTIVE PROGRAMMING (FMOP) TO DEA

FMOP to DEA provides a unitary weight (μ^*, ω^*) for all DMUs evaluated by an equal standard (Chiang and Tzeng 2000, 2003). By this approach, we can obtain the efficiency rating of each DMU more fairly. Moreover, all DMUs can be treated simultaneously, which makes the method effective for handling large numbers of DMUs.

8.3.1 Model 1

$$\max z_1 = \frac{\sum_{r=1}^{s} \mu_r y_{r1}}{\sum_{i=1}^{m} \omega_i x_{i1}} \tag{8.20}$$

$$\max z_2 = \frac{\sum_{r=1}^{s} \mu_r y_{r2}}{\sum_{i=1}^{m} \omega_i x_{i2}}$$

$$\vdots$$

$$\max z_n = \frac{\sum_{r=1}^{s} \mu_r y_{rn}}{\sum_{i=1}^{m} \omega_i x_{in}}$$

$$s.t. \quad \frac{\sum_{r=1}^{s} \mu_r y_{rk}}{\sum_{i=1}^{m} \omega_i x_{ik}} \le 1, \quad k = 1, 2, \ldots, n$$

$$\mu_r, \omega_i \ge \varepsilon > 0$$

where x_{ik} denotes the value of the ith input for the kth DMU, y_{jk} is the value of jth output for the kth DMU, and ε is a small positive number called the non-Archimedean quantity that denotes the unitary weight for all DMUs in fuzzy multiple objective programming to DEA.

By considering the efficiencies of all DMUs, we can establish a multiple objective linear programming (MOLP) model. This model can be solved by the fuzzy multiple objective linear programming (FMOLP) approach, as proposed by Zimmermann (1978). FMOP to DEA adopts this approach to obtain common weights that can maximize all DMU efficiencies. FMOLP utilizes membership function transfers of multiple objective functions into a single objective function. The membership function is as follows:

$$\mu_k(z_k) = \begin{cases} 0, & z_k \le z_k^L \\ \dfrac{z_k - z_k^L}{z_k^R - z_k^L}, & z_k^L \le z_k \le z_k^R \\ 1, & z_k \ge z_k^R \end{cases} \tag{8.21}$$

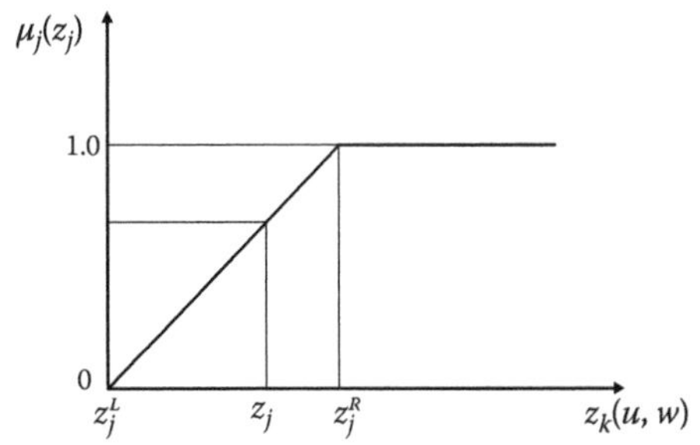

FIGURE 8.3 Linear membership function of z_k.

where z_k^L and z_k^R are the negative ideal solution and the positive ideal solution, respectively, for the value of the objective function z_k, such that the degree of membership function is [0, 1]. The geometric view of the linear membership function is shown in Figure 8.3.

The degree of membership function of z_k in $\mu(z_k)$ refers to the achievement level of the efficiency ratio for DMU_k. The problem of obtaining the maximum decision is to choose (μ^*, ω^*), such that:

$$\max_{\mu,\omega} \min_k \{\mu_k(z_k) \mid k = 1, 2, \ldots, n\} \tag{8.22}$$

$$\text{s.t.} \quad \frac{\sum_{r=1}^{s} \mu_r y_{rk}}{\sum_{i=1}^{m} \omega_i x_{ik}} \le 1$$

$$\mu(z_k) \ge \alpha$$

$$\mu_r, \omega_i \ge \varepsilon > 0, \forall r, i.$$

Let the achievement level of the objective functions for Model 1 be at a larger level:

$$\alpha = \frac{z_k - z_k^L}{z_k^R - z_k^L} \tag{8.23}$$

Equation (8.24), via variable transformation, has transformed $z_j = \alpha \cdot z_j^R + (1-\alpha) \cdot z_j^L$ where z_j is a convex combination of z_j^L and z_j^R; Equation (8.22) can be rewritten as Equation (8.24). According to the concept of multiple objective linear programming, we can determine a weight that satisfies all DMU restrictions. (μ^*, ω^*) is the common weight of all DMUs evaluated on a consistent standard of ranking.

$$\max_{\mu,\omega} \min_j \left\{ Z_j = \frac{\sum_{r=1}^{s} \mu_r y_{rj}}{\sum_{i=1}^{m} \omega_i x_{ij}} \right\} \tag{8.24}$$

$$\text{s.t.} \quad \frac{\sum_{r=1}^{s} \mu_r y_{rk}}{\sum_{i=1}^{m} \omega_i x_{ik}} \le 1, \qquad k = 1, 2, \ldots, n$$

$$\frac{\sum_{r=1}^{s} \mu_r y_{rj}}{\sum_{i=1}^{m} \omega_i x_{ij}} \ge \alpha \cdot z_j^R + (1-\alpha) \cdot z_j^L, \quad j = 1, 2, \ldots, n$$

By rewriting Model 1, Model 2 can be obtained. Model 2 is a nonlinear programming that may be solved by the bisection method, as proposed by Sakawa and Yumine (1983). By employing Model 2, a common weight (μ^*, ω^*) can be determined for all DMUs by directly ranking them.

8.3.2 Model 2

$$max \quad \alpha \tag{8.25}$$

$$s.t. \quad \sum_{r=1}^{s} \mu_r y_{rk} - \sum_{i=1}^{m} \omega_i x_{ik} \leq 0, \quad k = 1, 2, ..., n,$$

$$\sum_{r=1}^{s} \mu_r y_{rj} - \alpha \sum_{i=1}^{m} \omega_i x_{ij} \geq 0, \quad j = 1, 2, ..., n,$$

$$\mu_r,\ \omega_i \geq \varepsilon > 0$$

Next, we give an example of supplier rating to illustrate the procedures of FOMP to DEA.

Example 8.3

Assume that the efficiency levels of five firms designated A, B, C, D, and E, are to be evaluated according to their input and output data as shown in Table 8.7. In addition, assume that the target values of ideals are the best values for each attribute across the five firms. Adding a set of target values can be viewed as an ideal alternative for management, as these values contain exactly what management requires. First, we use raw data (without information on the ideal) to calculate the efficiency of each DMU as shown in Table 8.8. The results in Table 8.8 indicate that firm A is efficient and the best DMU. The rank of all five firms is A > D > C > B > E. On the other hand, if we consider the ideal as the benchmark of the DMUs, we can obtain the efficiency of each DMU as shown in Table 8.9.

Table 8.9 demonstrates that firm A is still the best DMU even though it is not efficient. The DMU ranking is A > C > B > D > E. Comparing with the results of

TABLE 8.7
Raw Data for Example 8.3

	Ideal	A	B	C	D	E
Input	52	52	82	64	100	84
Output 1	97	57.5	77.25	62.5	97	57
Output 2	76	70	76	76	76	58
Output 3	95	90	95	80	70	60
Output 4	100	70	70	70	100	40

Tables 8.8 and 8.9, we can see that the introduction of an ideal may change the rank of the DMUs and affect the final decision making.

Next, we propose the improving directions to allow firms to achieve efficiency as shown in Table 8.10. The results shown in Table 8.10 can be explained as follows. Firms D and E must decrease their Inputs to achieve a better rank. The ideal point will become adjustable after upper management needs are met. Therefore, adding an ideal list can indicate areas in which the firm should improve.

TABLE 8.8
Efficiency of DMU without Ideal Information

DMU	Efficiency	Rank
A	1	1
B	0.85	4
C	0.88	3
D	0.87	2
E	0.61	5

TABLE 8.9
Efficiency of DMU with Ideal Information

DMU	Efficiency	Rank
Ideal	**1**	**Benchmark**
A	0.92	1
B	0.63	3
C	0.81	2
D	0.52	4
E	0.47	5

TABLE 8.10
Improving Directions for Firms

	Ideal	A	B	C	D	E
Input	52	M	D	D	D	D
Output 1	97	I	I	I	M	I
Output 2	76	I	M	M	M	I
Output 3	95	I	M	I	I	I
Output 4	100	I	I	I	M	I

I = increase. D = decrease. M = maintain.

Section II

Applications of Multi-Objective Decision Making

9 Motivation and Resource Allocation for Strategic Alliances through the De Novo Perspective

In the past two decades, strategic alliances have become important business issues. One aspect of these alliances is their formation. Although many theories and models such as strategic perspective, organization learning, and other explanations have been proposed to explain alliance formation, their perspectives are usually limited (Tsang, 1998; Borys and Jemison, 1989). Additionally, the problem of resource allocation is another crucial issue and previous papers seem to discuss only why strategic alliances should be formed but fail to cover what to do next. In contrast, this chapter proposes a holistic perspective to explain the formation of strategic alliances and provides a method for optimal resource allocation among alliances.

In recent years, mainstream research can be summarized as the use of transaction cost theory and a resource-based view to explain the formation of strategic alliances. Of these two, transaction cost theory focuses on cost aspects (including transaction and product costs), whereas the resource-based view emphasizes the combination of resources among alliances. In our view, both theories represent reasons to form strategic alliances and so should be considered together.

Neither transaction cost theory nor the resource-based view will provide a method to resolve the problem of resource allocation. Traditional mathematical systems such as linear programming and dynamic programming are valid tools for providing optimal solutions in operations research areas. However, when these tools are used to allocate the combined resources of alliances, synergies cannot be explained or displayed. The reason lies in the assumption of additivity. The assumption of resource independence does not allow a synergy effect, and so is not suitable for strategic alliances.

In this chapter, transaction cost theory and the resource-based view are combined into what we call the de novo perspective and used to explain the formation of strategic alliances. In addition, the problem of optimal resource allocation between alliances is proposed using de novo programming. We present a numerical example to demonstrate the criteria of strategic alliances and assign optimal resource allocation.

Based on numerical results, we show that the motivation for strategic alliances is determined by both transaction cost and firm resources. However, whether firms enter into alliances depends on the presence of necessary and sufficient conditions. When the necessary condition is satisfied, a firm has the motivation to form strategic alliances, but only when sufficient conditions are met will a firm enter into alliances. In addition, the results also show the optimal resource allocation of firms' resources.

9.1 MOTIVATIONS FOR STRATEGIC ALLIANCES

A strategic alliance may be defined as a cooperative arrangement between two or more independent firms that exchange or share resources to gain a competitive advantage. Many studies discuss the formation of strategic alliances using various theories and models such as transaction cost theory (Anderson and Gatignon, 1986), the perspective of strategy (Porter, 1980; Hagedoorn, 1993), resource dependence theory (Preffer and Nowak, 1976; Preffer and Salancik, 1978), organizational knowledge and learning (Nelson and Winter, 1982; Kogut, 1988). and the resource-based view (Barney, 1991; Das and Teng, 2000). However, previous studies have not approached the problem from a holistic perspective.

To summarize these theories, the perspective of the strategy is seeking appropriate alliances that can improve a firm's competitive position or increase its competitive advantage. In contrast, in the resource dependence theory, the motivations for strategic alliances focus on a firm's lack of valuable resources. Organizational knowledge and learning focus on the desire of a firm to acquire or other firms' organizational knowledge.

Recently, transaction cost theory and the resource-based view have been used to explain the formation of strategic alliances and comparison of both theories were proposed (Chan and Chen, 2003; Tsang, 1998). This chapter discusses both transaction cost theory and the resource-based view along with perspectives, techniques, and examples.

9.1.1 Transaction Cost Theory

Based on an economics approach, transaction cost theory was proposed by Coase (1937) to explain the decisions made by firms regarding marketing and hierarchy issues. In brief, when the transaction cost of an exchange is high, the form of internalization will predominate and vice versa. However, one restriction is that transaction cost theory explains only extreme conditions. This limitation was extended by Williamson to explain strategic alliances (Williamson, 1975, 1985, 1991a and b).

This extension can describe how transaction cost theory uses transaction cost (determining or enforcing contract cost) and production cost (through internal coordination or managing) to determine markets or hierarchies. However, when the optimal total cost involves neither markets nor hierarchy, strategic alliances should be the best choice (Gulati, 1998; Ramanathan et al., 1997).

Williamson suggests that transaction costs should include the direct costs of managing relationships and the possible impact costs of making inferior governance decisions. These concepts can be described as bounded rationality, opportunism, asset specificity, and uncertainty (Rindfleisch and Heide, 1997; Parkhe, 1993; Dyer and Singh, 1998).

Although transaction cost theory provides a useful explanation for the formation of strategic alliances, it has a major weakness in that the analysis focuses on single-party cost minimization rather than global cost minimization (Zajac and Olsen, 1993). Furthermore, it does not assign a significant role to partner firms' resources in theorizing (Das and Teng, 2000) that led to the emergence of the resource-based view.

9.1.2 Resource-Based View

A resource-based view (Barney, 1991; Grant, 1991; Wernerfelt, 1984; Barney et al., 2001; Barney, 2001) proposes another perspective on strategic alliances, and states that the valuable resources that firms do not own serve as motives for strategic alliances. The literature provides many classifications of such resources (Barney, 1991; Das and Teng, 2000; Grant, 1991; Miller and Shamsie, 1996) but resources are generally classified as tangible (e.g., financial and technological) and intangible (e.g., knowledge-based and managerial).

Additionally, heterogeneity is the reason why firms are distinctive and acts as the basis of the resource-based view (Penrose, 1959). In order to acquire competitive advantage and the ability to respond quickly to a dynamic environment, a firm should consider how to construct and extend limited resources to develop a capability to gain a sustainable competitive advantage (Teece et al., 1997).

The three ways for a firm to build and extend its resources are through hierarchy, markets, or alliances. However, the assumption of heterogeneity across firms causes the cost of hierarchy and markets to be high. In the resource-based view, firms seek complementary resources to create synergies and acquire sustainable competitive advantages (Harrison et al., 1991; Lockett and Thompson, 2001). When the degree of heterogeneity among firms increases, the higher probability of forming alliances creates rents (Barney, 1991). In short, by way of strategic alliances, firms can gain their partners' complementary resources to enhance or reshape their internal processing to create synergies and competitive advantages within the market (Nohria and Garcia-Pont, 1991; Porter and Fuller, 1986).

Although the resource-based view imposes a reliable perspective on a firm's resources, some notable questions must be answered to explain the formation of strategic alliances. For example, what are the criteria for forming alliances when a firm lacks desired complementary resources? Obviously, not every firm with or without complementary resources enters alliances in the real world. In addition, a firm may even acquire a partner firm's resources. What should the firm do after entering an alliance? Firms cannot gain anything unless they use their newly acquired resources well. In other words, the optimization of resource allocation is the key to whether a firm can create synergies and achieve a competitive advantage.

9.2 PROBLEMS OF RESOURCE ALLOCATION

Based on the above discussions, we know that resources play a central role in the formation of strategic alliances. However, neither transaction cost theory nor the resource-based view provides a method to solve the problem of resource allocation.

In operations research, optimal resource allocation has been a popular issue and one of the notable methods for seeking a solution is mathematic programming.

Mathematical programming is a technique that distributes limited resources to competing activities in an optimal way. Of the several mathematic programming techniques, linear programming is the most popular. Linear programming was developed by Kantorovich and Koopmans (1976), and the general matrix formulation of linear programming can be described as follows.

$$\begin{aligned} \max \quad & \boldsymbol{Cx} \\ s.t. \quad & \boldsymbol{Ax} \leq \boldsymbol{b}, \\ & \boldsymbol{x} \geq \boldsymbol{0} \end{aligned} \tag{9.1}$$

where both $\boldsymbol{C} = \boldsymbol{C}_{q\times n}$ and $\boldsymbol{A} = \boldsymbol{A}_{m\times n}$ are matrices, $\boldsymbol{b} = (b_1,...,b_m)^T \in R^m$, and $\boldsymbol{x}$ = $(x_1,\ldots,x_j,\ldots,x_n)^T \in R^n$. Let the $\boldsymbol{k}^{\text{th}}$ row of $\boldsymbol{C}$ be denoted by $\boldsymbol{c}^k = (c_1^k,\ldots,c_j^k,\ldots,c_n^k) \in \boldsymbol{R}^n$, so that $\boldsymbol{c}^k\boldsymbol{x}$, $k = 1,\ldots,q$, is the kth criterion or objective function. There are several ways to solve this question, such as the simplex method or the interior point algorithm, and the solutions indicate the optimal way to distribute limited resources.

Although mathematical programming provides a way to resolve the problem of resource allocation, the basic assumption of additivity seems irrational when we extend this method to managing the resource of an alliance. This is because additivity presumes that all productive elements are independent and the total effects equal the summation of each individual effect. The most critical problem arises because this assumption makes it impossible for firms to create synergies.

The famous case typifying the problem of element independence is the emergence of mass customization. Traditionally, a firm has two ways to gain profit. One is to reduce unit cost by economic scaling while earning the same unit revenue. The other way is to increase unit revenue by customization at higher unit cost. When element independence exists, it is impossible to reduce unit cost and increase unit revenue simultaneously. However since the concept of mass customization has been proposed and is used in practice, the restriction of element independence should be released.

Based on Figure 9.1, we assume that an alliance in a market has two members (A and B) and π_A, and π_B denote the profits of A and B, respectively. The goal of both firms is to optimize profit maximization and the feasible solutions have circular shapes. Usually, compromise solutions are the best decisions in traditional mathematic programming and they fall into the $\widehat{AB}$ formula. Options with points A, B, and C (C is the ideal point) are unavailable because of the assumption of additivity.

Additivity (the combination of alliance resources) allows only 1 + 1 = 2 solutions rather than 1 + 1 > 2 answers. However, synergies are usually the results of strategic alliances. In other words, if a firm has resource constraints that cannot be changed, the traditional methodology of mathematical programming is rational and available (Babic and Pavic, 1996). However, if we redesign or reshape a system, the traditional methods are no longer suitable and this usually happens in strategic alliances.

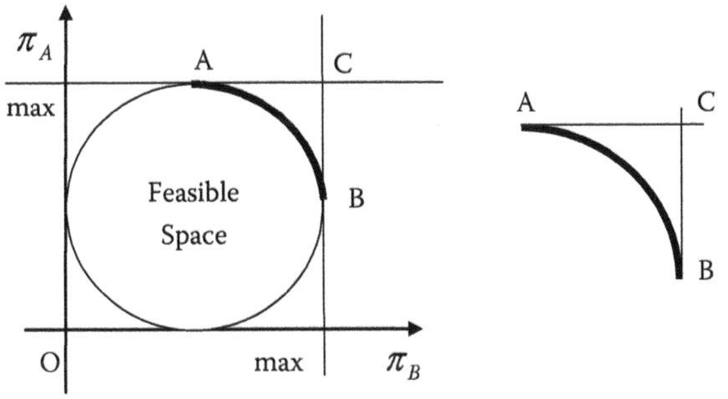

FIGURE 9.1 Feasible options using linear programming.

9.3 DE NOVO PERSPECTIVE OF STRATEGIC ALLIANCES

De novo programming may be extended. The extension is called the de novo perspective and it is useful for explaining the formation of strategic alliances. The de novo perspective combines transaction cost theory and the resource-based view to provide a holistic perspective for achieving an aspiration or desired level.

Based on transaction cost theory, if the minimum cost lies between the transaction cost and the production cost, a firm should seek strategic alliances. Here we add alliance cost (shared operation, negotiating, and risk costs) to explain the formation of strategic alliances. The rule of transaction cost theory can be modified. For example, if alliance cost $\leq \sum_{i=1}^{N}$ the individual firm's cost, the firm should seek alliances.

From the resource-based view, firms seek strategic capabilities by linking to partner resources to create synergies in a market. The rule of the resource-based view can be modified to reflect the situation. If alliance revenue $\geq \sum_{i=1}^{N}$ the individual firm's revenue, the firm should seek alliances.

We now combine transaction cost theory and the resource-based view to form the de novo perspective. If a firm only chooses hierarchy or alliances, for example, between two firms (S and T), the rule of strategic alliances can be utilized. If $V(S \cup T) - U(C_{ST}) > V(S) + V(T) - U(C_s) - U(C_T)$, the firm seeks alliances. If we extend this to a general form, the expression is:

$$\text{If} \quad V(S_1 \cup S_2 \ldots \cup S_N) - U(C_{alliance\ cost}) \geq \sum_{i=1}^{N} [V(S_i) - U(S_i)] \quad i = 1, 2, \ldots, N \quad (9.2)$$

where $V(\cdot)$ denotes the revenue function, $U(\cdot)$ denotes the cost function, C_S and C_T denote the total product cost in S and T, respectively, and C_{ST} denotes the alliance cost between S and T. The probability of firms, S and T seeking alliances can be expressed, respectively, as:

$$P(S) = \begin{cases} 1 & \lambda V(S \cup T) - \theta U(C_{ST}) > V(S) - U(C_S) \\ 0 & \lambda V(S \cup T) - \theta U(C_{ST}) < V(S) - U(C_S) \end{cases} \quad (9.3a)$$

and

$$P(T)=\begin{cases}1 & (1-\lambda)V(S\cup T)-(1-\theta)U(C_{ST})>V(T)-U(C_T)\\ 0 & (1-\lambda)V(S\cup T)-(1-\theta)U(C_{ST})<V(T)-U(C_T)\end{cases} \tag{9.3b}$$

where λ denotes the percentage of increasing alliance revenue in S and θ denotes the percentage of reducing alliance cost in S. Now we can shift our discussions to de novo programming. The problem of resource allocation can be expressed as:

$$\begin{aligned} \max \quad & V(S\cup T)-U(C_{ST}) \\ s.t. \quad & \boldsymbol{wx}\le B, \quad \boldsymbol{x}\ge 0 \end{aligned} \tag{9.4}$$

where $\boldsymbol{w}=\boldsymbol{pA}=(w_1,\ldots,w_n)\in R^n$ and $\boldsymbol{p}=(p_1,\ldots,p_m)\in R^m$ and $B\in R$ represent the unit price of resources and the total available budget, respectively. Then the knapsack solution is

$$\boldsymbol{x}^*=[0,\ldots,B/C_k,\ldots,0]^T \tag{9.5}$$

where

$$c_k/c_k^*=\max_j(c_j/c_j^*) \tag{9.6}$$

and the optimal solution to (9.4) is given by (9.5) and

$$b^*=Ax^* \tag{9.7}$$

The final alliance profit ($\Psi(S^*)$) in S is

$$\Psi(S^*)=\boldsymbol{i'b^*}-\mathrm{U}(C_{ST}) \tag{9.8}$$

where **i** is the identity column vector. Based on Equation (9.8), we can judge whether or a firm should seek alliances by Equation (9.2) and Equation (9.3). Furthermore, using de novo programming, we can easily allocate the optimal resources and create synergies between alliances.

9.4 NUMERICAL EXAMPLE

In this section, we use a numerical example modified from Zeleny (1995) to demonstrate the profit difference between hierarchy and alliances and propose criteria to determine whether a firm should enter strategic alliances.

For simplicity, we assume two firms designated S and T produce the same two products and have the same two production elements and total product costs denoted $U(C_S)$

and $U(C_T)$, respectively. Firm S can determine its optimal resource allocation by using mathematical programming:

$$\max \quad f_1 = 400x_1 + 300x_2$$

$$\max \quad f_2 = 6x_1 + 8x_2$$

$$\begin{aligned} \text{s.t.} \quad & 4x_1 \le 10 \\ & 2x_1 + 6x_2 \le 12 \\ & 12x_1 + 4x_2 \le 30 \\ & 3x_2 \le 5.25 \\ & 4x_1 + 4x_2 \le 13 \\ & x_1, x_2 \ge 0 \end{aligned}$$

where $p_1 = 30, p_2 = 40, p_3 = 9.5, p_4 = 20$, and $p_5 = 10$ are market prices (dollars per unit) of the resources b_1 through b_5, respectively. Functions f_l and f_2 denote the revenues of product 1 and product 2, respectively, and $B = 1300$ denotes the firm's total budget.

Using traditional mathematical programming, we can easily solve the optimal distribution of a resource portfolio at $x_1 = 2.125$ and $x_2 = 1.125$. Firm S can achieve total revenue by the summation function f_1 and f_2 equal 1187.5 + 21.75 = 1209.25. Then the profit of firm S can be expressed as 1209.25 – $U(C_S)$. Using the same procedure, the profit of firm T is 1209.25 – $U(C_T)$.

On the other hand, if the two firms enter an alliance, the problem of resource allocation can be solved by de novo programming as follows.

$$\max \quad f_1 = 400x_1 + 300x_2$$

$$\max \quad f_2 = 6x_1 + 8x_2$$

$$\begin{aligned} \text{s.t} \quad & 4x_1 \le 20 \\ & 2x_1 + 6x_2 \le 24 \\ & 12x_1 + 4x_2 \le 60 \\ & 3x_2 \le 10.5 \\ & 4x_1 + 4x_2 \le 26 \\ & x_1, x_2 \ge 0 \end{aligned}$$

Let $B = 2600$ denote the total alliance budget. First, we use traditional mathematical programming to solve the knapsack problem.

(i) For the max f_1, we solve

$$\max \quad f_1 = 400x_1 + 300x_2$$

$$s.t. \quad 354x_1 + 378x_2 \le 2600$$

$$x_1, x_2 \ge 0$$

and the answer can be formed as $x_1^1 = 2600/354 \approx 7.34$, $x_2^1 = 0$, $f_1 = 2937.85$, and $B^1 = 2600$.

(ii) For the max f_2, we solve

$$\max \quad f_2 = 6x_1 + 8x_2$$

$$s.t. \quad 354x_1 + 378x_2 \le 2600$$

$$x_1, x_2 \ge 0$$

and the solutions are $x_1^1 = 0$, $x_2^1 = 2600/378 \approx 6.88, f_2 = 55.03$, and $B^2 = 2600$.

After solving the above problems, we can find the ideal point $f^{**} = (2937.85, 55.03)$, the synthetic solution $x^{**} = (7.34, 6.88)$, and the synthetic budget $B^{**} = 5200$. The ratio r^{**} must be calculated to contract the synthetic solution to an optimal designed solution x^*. The results can be shown as follows:

$$r^{**} = B/B^{**} = 2600/5200 = 0.5$$

$$x^* = r^{**} \times x^{**} = (0.5 \times 7.34, \quad 0.5 \times 6.88) \approx (3.67, \quad 3.44)$$

Then, the alliance revenue can sum functions f_1 and f_2 as (400 × 3.67 + 300 × 3.44) + (6 × 3.67 + 8 × 3.44) = 2549.54. The alliance profit can be expressed as $\Psi(S^*) = 2549.54 - U(C_{ST})$. Although the alliance revenue is more than the total of the individual firm's profit, it does not necessarily go to strategic alliances because the necessary condition of strategic alliances is:

$$2549.54 - U(C_{ST}) > 1209.25 - U(C_S) + 1209.25 - U(C_T) \tag{9.9}$$

When the formulation

$$U(C_{ST}) \le 131.04 + (U(C_S) + U(C_T)) \tag{9.10}$$

is satisfied, the firm has stimulus to seek strategic alliances. However, Equation (9.10) does not ensure that the individual firm will enter strategic alliances, and the criterion depends on sufficient conditions. For firm S, the sufficient condition for strategic alliances is

$$\lambda \cdot 2549.54 - \theta U(C_{ST}) > 1209.25 - U(C_S) \tag{9.11}$$

and when Equation (9.11) is satisfied, firm S enters alliances.

Note that Shi (1995) provided six kinds of optimum path ratios to find the optimal solution in de novo programming, whereas we demonstrate only one of the six. Other kinds of optimum path ratios can also be calculated for alternatives in strategic alliances.

Extending our concept to a real-world case involves far more complex calculations than our example and may involve a large-scale alliance situation or multiple criteria and multiple constraint (MC^2) problems. In a large-scale alliances situation, we can adopt the large-scale MC^2 algorithms proposed by Hao and Shi (1996) or other heuristic methods such as multi-objective evolution algorithms (MOEAs) to overcome the problem. On the other hand, de novo can be extended to incorporate problems involving multiple criteria and multiple constraints:

$$\begin{aligned} \max \quad & \delta' C x \\ s.t. \quad & \boldsymbol{Ax} \le \boldsymbol{D}\gamma, \\ & \boldsymbol{x} \ge \boldsymbol{0}. \end{aligned} \tag{9.12}$$

where δ and γ denote the unknown relative and constraint level weight vector, respectively, and satisfy $\|\delta\| = 1$ and $\|\gamma\| = 1$.

While several MC^2 programming methods have been proposed to solve the above problem (Shi, 1999), de novo programming can be widely used to analyze various strategic alliances. Furthermore, if strategic alliances must consider debt situations, the contingency plan should be used (Shi, 2001).

9.5 DISCUSSION

As we know, reducing costs and enhancing revenues are the usual motives for seeking strategic alliances. However, no literature has provided a concrete equation that will help firms to determine whether to pursue alliances. In addition, resource allocation between alliance members is also a difficult issue and traditional mathematical programming seems to be unable to provide a sound solution.

In Section 9.4, we demonstrate a numerical example to solve questions regarding the criteria for forming strategic alliances and ensuring optimal resource allocations in alliances. Neither transaction cost theory nor the resource-based view provides criteria for firms to enter alliances. As we demonstrate, the criteria can be classified as necessary and sufficient conditions. If increasing profit is a necessary condition, the alliance must satisfy it. Businesses have various motives for forming strategic alliances but only when their sufficient conditions are satisfied should alliances be formed.

Based on the numerical example, Equation (9.10) can answer whether a firm should consider strategic alliances. It indicates that only when the alliance cost $U(C_{ST})$ is lower than $131.04 + U(C_S) + U(C_T)$ should a firm consider seeking strategic alliances. Equation (9.11) provides a concrete answer to the question of whether a firm should enter an alliance. That is, if $\lambda \cdot 2549.54 - \theta U(C_{ST}) > 1209.25 - U(C_S)$ is satisfied,

TABLE 9.1
Comparisons of Various Perspectives

Dimensions	Transaction Cost Theory	Resource-Based View	De Novo Perspective
Level of analysis	Firm	Firm	Firm
Unit of analysis	Transaction	Resource	Transaction and resource
Premise	Minimum transactions determine optimal governance structures	Heterogeneity of resources among firms	Both
Resource allocation method	NA	NA	De Novo programming
Motive for strategic alliances	Minimum cost	Maximum value creation	Both

NA = not applicable.

the firm will have economic benefit when it enters a strategic alliance. Table 9.1 compares transaction cost theory, the resource-based view, and the de novo perspective.

For the second problem, traditional mathematical programming appears unable to create synergies in alliances. In contrast, through de novo programming an optimal resource allocation is planned and the results show the effects of synergy.

To summarize, the de novo perspective provides a complete explanation for strategic alliances and the alliance criteria are offered by mathematic equations included in this chapter. Businesses can easily make the calculations for determining actions to take regarding strategic alliances, and the optimal distribution for alliance resources also can be determined.

9.6 CONCLUSIONS

Transaction cost theory uses minimum cost data to analyze the results of formation of strategic alliances. The resource-based view focuses on seeking valuable resources to achieve a global optimal result. This chapter proposes the use of the de novo perspective to evaluate the formation of strategic alliances and provide synergistic solutions for resource allocation in order to achieve the aspiration result.

Clearly, a strategic alliance is a multi-criteria optimal system design (MCOSD) problem rather than a multi-criteria optimal system analysis (MCOSA) problem. Productive resources should not be engaged individually because they do not contribute individually according to their marginal productivities. In alliance situations, the de novo approach is more suitable than traditional mathematical programming.

The most critical problem with the de novo approach is that the required budget will exceed the subject budget calculated via de novo programming in some situations. This may be a serious problem for individual firms, but the financial leverage available in alliances can overcome this difficulty. In addition, the profit from economies of scale can also be seen with De Novo programming.

In short, the De Novo perspective provides another view of strategic alliances and determines the optimal resource allocation. Unlike traditional mathematical programming, the De Novo approach does not have the limitation of element independence. This characteristic allows this operations research technique to extend to analysis of synergies, economies of scale, and other spill over effects.

10 Choosing Best Alliance Partners and Allocating Optimal Alliance Resources Using Fuzzy Multi-Objective Dummy Programming Model

A strategic alliance may be defined as a cooperative arrangement between two or more independent firms that exchange or share resources to gain a competitive advantage. Since the 1980s, strategic alliances have been widely discussed (Porter and Fuller, 1986; Harrigan and Newman, 1990; Auster 1994) and hundreds of papers have been published about this issue. The essential motives of strategic alliances are "synergy effects" as represented in the following equation:

$$V(\bigcup_k S_k) > \sum_k V(S_k); \qquad k = 1, \ldots, K \tag{10.1}$$

where $V(\cdot)$ denotes the satisfaction (or value) function and S_k denotes the kth alliance firm. When Equation (10.1) is satisfied, firms can share better satisfaction levels than their original states through strategic alliances.

Since some firms rush into strategic alliances without appropriate preparation or planning to choose appropriate partners and resource allocations, their alliances often fail (Dacin et al., 1997). Questions surrounding alliances and related criteria and complex and diversified, i.e., alliance firms have different goals, cultures, and resources. The best alliance partners and resource allocations may be incompatible. This chapter proposes a model for determining the best partner choices and optimal resource allocation for strategic alliances.

Although the criteria for choosing correct partners are widely known, for example, complementary strengths, commitment, coordination, and compatible goals (Brouthers et al. 1995; Arino and Abramov, 1997; Yoshino and Rangan, 1995; Gerlinger, 1991), these papers seem to ignore resource allocation. Clearly, only by using these alliance resources effectively can synergy effects ensue. The issue of resource allocation in operations research (Bretthauer and Shetty, 1995, 1997; Lai and Li, 1999;

Robinson et al., 1992) has been discussed extensively. Resource allocation is intended to produce maximum profits for an enterprise and satisfaction and/or utility for customers using limited resources. It is reasonable to incorporate both concepts to overcome the problems surrounding choices of alliance partners and resource allocation.

In order to overcome these problems and derive a useful model, several issues should be considered. First, since the problem of choosing alliance partners is part of a combinatorial problem, the scaling problem should be considered. Second, the objectives of a firm and the satisfaction expected from an alliance should be measured and calculated precisely. Third, based on market mechanisms, the unit prices of resources should be incorporated into the selected model. It is clear that optimal resource allocation varies with fluctuations in unit prices of resources. Finally, because real-world problems usually impose restrictions, resource allocation solutions should be carefully planned and supported.

This chapter proposes the fuzzy multi-objective dummy programming model to satisfy these issues and provide the best alliance cluster and optimal resource allocation combinations. In addition, two types of strategic alliances—joint ventures and mergers and acquisitions (M&A)—are demonstrated to allow a firm to choose the best alliance partners and allocate optimal alliance resources. We present a numerical example using the proposed method. On the basis of the results, we can conclude that the proposed method can provide an optimal alliance cluster and satisfaction for alliance partners.

10.1 REVIEW OF STRATEGIC ALLIANCES

A strategic alliance may be defined as a cooperative arrangement between two or more independent firms that exchange or share resources for competitive advantage. From the resource-based view (Barney, 1991; Grant, 1991; Wernerfelt, 1984; Barney et al., 2001; Barney, 2001), valuable resources that firms require but do not own act as the motive for entering strategic alliances. Many classifications for valuable resources have been proposed (Miller and Shamsie, 1996), and these resources are generally classified as tangible (financial and technological) and intangible (knowledge-based and managerial) resources.

In order to acquire competitive advantage and the ability to respond quickly in a dynamic environment, firms should consider how to construct and extend limited resources to develop the capability for sustainable competitive advantages (Teece et al., 1997). Through strategic alliances, firms can gain their partners' complementary resources to enhance or reshape their internal activities to create synergies and competitive advantage within a market (Nohria and Garcia-Pont, 1991). Based on varying degrees of vertical integration or independence (Lorange and Roos, 1992), Figure 10.1 depicts common forms of strategic alliances.

Two questions arise. First, how can we choose the correct partners? Second, how can we allocate the valuable resources? Clearly, choosing the correct partner is the first step of entering a successful strategic alliance. Correct selection requires careful screening and may be a time-consuming process. Furthermore, a successful alliance requires that firms can gain nothing unless they use their newly acquired resources effectively. In other words, the optimization of resource allocation is the

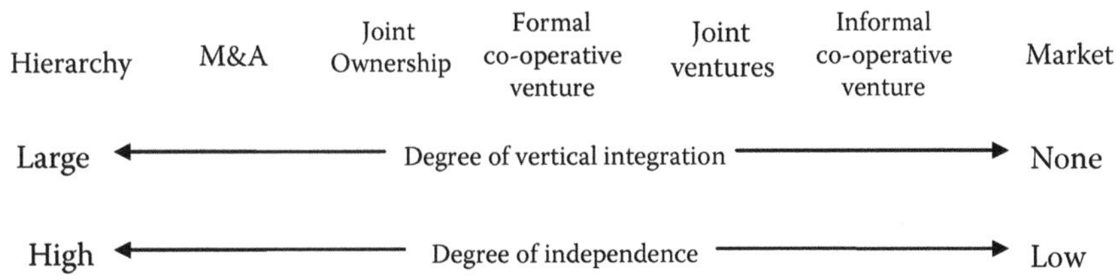

FIGURE 10.1 Spectrum of strategic alliances.

key to whether firms can create synergies and obtain competitive advantage. In this chapter, we demonstrate two types of strategic alliance: (1) joint ventures and (2) mergers and acquisitions (M&A) that can provide the best partners and determine the optimal resource allocation for the alliance.

A joint venture can be defined as the sharing of assets, risks, profits, or investment projects by more than one firm or an alliance cluster (Kough, 1988, 1991; Harrigan, 1985). Joint ventures are mechanisms for reducing the transaction costs incurred when acquiring other firms (Hennart and Reddy, 1997). Over the time that joint ventures have been used in practice, the number in the United States grew by 423% between 1986 and 1995 (Hitt et al., 1997). If a strategic alliance is a joint venture, a firm can share the surplus resources with alliance partners to increase total alliance satisfaction. Figure 10.2 depicts the joint venture processes.

Note that based on the concepts above, it is obvious that the equilibrium of a joint venture requires all alliance partners to achieve the same satisfaction.

Mergers and acquisitions represent an extreme strategic alliance in which two or more firms form a single enterprise that uses all partners' resources to optimize the goals of the organization (Whitelock and Rees, 1993). This means that the enterprise may eliminate products in an effort to optimize overall alliance satisfaction. A firm will favor the M&A choice over a joint venture when the assets it needs are not commingled with unneeded assets by the firm that holds them. The assets may be

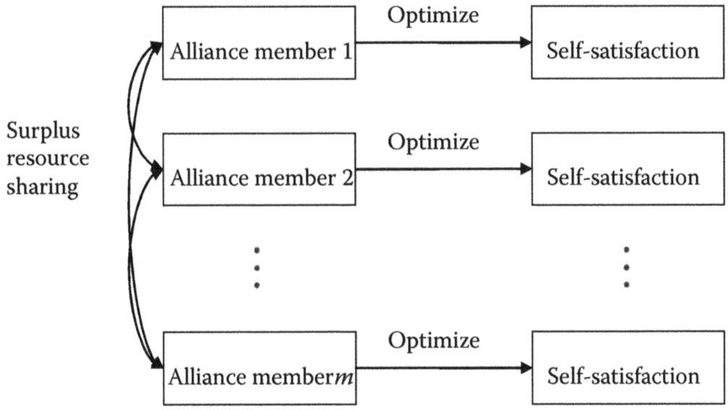

FIGURE 10.2 Joint venture concept.

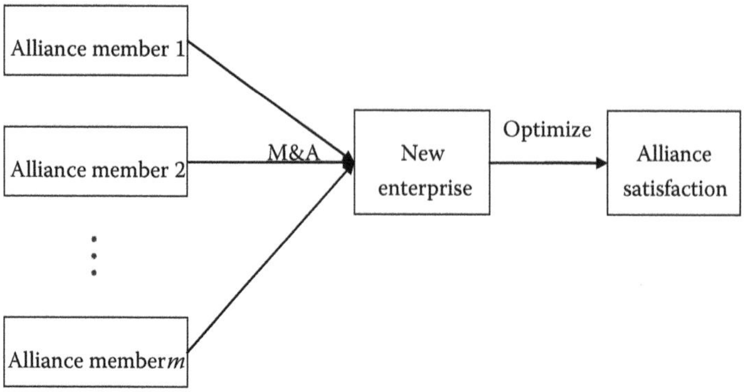

FIGURE 10.3 Concept of mergers and acquisitions.

acquired by buying the firm or a part of it (Hennart and Reddy, 1997). Figure 10.3 presents the processes of M&A for optimizing alliance satisfaction.

The main difference between joint ventures and M&A is that joint ventures emphasize the sharing of surplus resources to optimize alliance satisfaction (i.e., alliance members still develop their own products). M&A are concerned only with optimizing overall alliance satisfaction. To determine the way to allocate optimal alliance resources in a market mechanism, de novo programming is proposed.

10.2 FUZZY MULTIPLE OBJECTIVE DUMMY PROGRAMMING

In this section, we first describe the concepts of fuzzy sets so that the readers can better understand the method. However, we do not present all the issues concerning fuzzy sets and restrict the discussion to relevant issues of this chapter. The concepts of fuzzy sets were proposed to extend the classical crisp set to consider the certainty in the interval [0,1]. Since real-world problems usually are partly true and partly false, fuzzy sets are widely employed to deal with uncertainties, especially subjective uncertainties.

In order to determine the degree of uncertainty, the degree of membership is developed. Given a fuzzy set $\tilde{A}$ of a universe Y, the membership function of set A can be defined as:

$$\mu_{\tilde{A}}(y): Y \rightarrow [0,1]$$

where

$$\mu_{\tilde{A}}(y): 1 \text{ if } y \text{ is totally in } \tilde{A} \tag{10.2}$$

$$\mu_{\tilde{A}}(y): 0 \text{ if } y \text{ is not in } \tilde{A} \tag{10.3}$$

$$0 < \mu_{\tilde{A}}(y) < 1 \text{ if } y \text{ is partly in } \tilde{A} \tag{10.4}$$

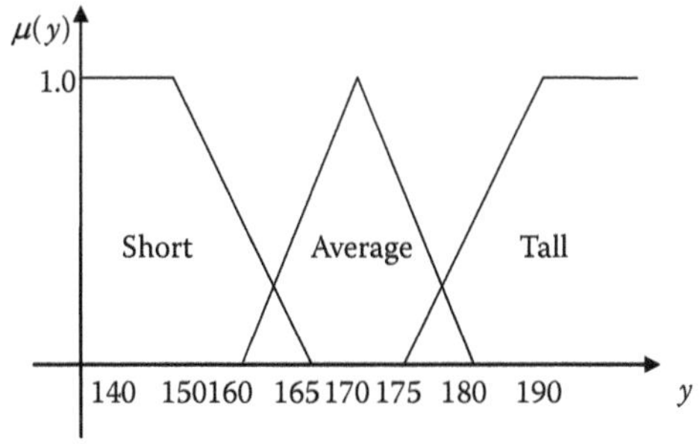

FIGURE 10.4 Fuzzy set concept.

Usually, the fuzzy set $\tilde{A}$ can be represented using a triangular or trapezoidal fuzzy number. Consider the following example in Figure 10.4. Three fuzzy sets (short, average, and tall) represent degrees of height. The fuzzy sets of short and tall can be represented by trapezoidal fuzzy numbers (140, 140, 150, 165) and (175, 190, 190, 190), respectively. The fuzzy set of average can be represented by triangular numbers 160, 170, 180.

To measure the satisfaction of strategic alliances, the concept of fuzzy sets is used. The conventional fuzzy programming problem (Zimmermann, 1978) can be represented as follows:

$$\begin{aligned} \max \quad & \boldsymbol{Cx} \tilde{\geq} z \\ s.t. \quad & \boldsymbol{Ax} \tilde{\leq} \boldsymbol{b}. \end{aligned} \tag{10.5}$$

where $\tilde{\geq}$ and $\tilde{\leq}$ are the fuzzification of $\geq$ and $\leq$, respectively. Then the satisfaction for each objective aspiration level can be represented using the following linear membership function:

$$\mu(z_q) = \begin{cases} 0 & , \quad \boldsymbol{c}_q\boldsymbol{x} \leq z_q^l \\ (z_q^u - \boldsymbol{c}_q\boldsymbol{x}) / d_q, & z_q^l \leq \boldsymbol{c}_q\boldsymbol{x} \leq z_q^u \\ 1 & , \quad \boldsymbol{c}_q\boldsymbol{x} \geq z_q^u \end{cases} \tag{10.6}$$

where z_q^u and z_q^l are the aspiration level and the minimum level, respectively and d_q denotes the, subjective perception of the minimum tolerant constants that usually assumes $d_q = z_q^u - z_q^l$, and the corresponding relation can also be depicted as shown in Figure 10.5.

Note that the minimum (maximum) level $z_q^l(z_q^u)$ can be obtained by solving each single objective mathematical programming model. To illustrate the two-objective mathematical programming problem, the first minimum (maximum) level $z_1^l(z_1^u)$ can be obtained by solving the following model:

$$\begin{aligned} \min\ (\max) \quad & z_1 = \boldsymbol{c}_1\boldsymbol{x} \\ s.t. \quad & \boldsymbol{Ax} \leq \boldsymbol{b}. \end{aligned} \tag{10.7}$$

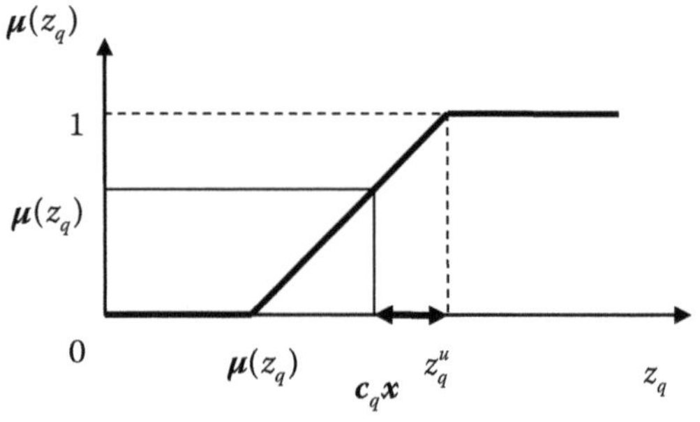

FIGURE 10.5 Membership function for z_q.

$\mu(z) = \min\{\mu(z_q) \mid q = 1,\ldots,Q\}$ denotes the overall satisfaction level while $u = 1 - \mu(z)$ denotes the overall regret level. We can model the joint venture and M&A strategic alliance types as follows. Without loss of generalization in the maximum problem, assume I firms form an alliance cluster and J firms are candidates to be chosen to enter the alliance. Then, according to the concept of joint ventures, we can propose a model as follows:

10.2.1 Joint Venture Model

$$
\begin{aligned}
\min \quad & u + e \\
s.t. \quad & \boldsymbol{A}_i \boldsymbol{x}_i \le \boldsymbol{b}_i, \qquad\qquad (10.8)\\
& S_j \cdot (\boldsymbol{A}_j \boldsymbol{x}_j) \le \boldsymbol{b}_j \quad \text{where} \quad S_j \in \{0,1\}, \\
& \boldsymbol{C}_i \boldsymbol{x}_i + n_i - p_i = z_i^u, \\
& S_j(\boldsymbol{C}_j \boldsymbol{x}_j) + n_j - p_j = S_j(z_j^u), \\
& \left.\begin{array}{l} u \ge n_i / (z_i^u - z_i^l), \\ u \ge S_j \cdot n_j / (z_j^u - z_j^l), \end{array}\right\} \text{indicate the equilibrium of joint ventures,} \\
& \left.\begin{array}{l} V\left(\boldsymbol{x}_i + \sum_{j=1}^{m} S_j \cdot \boldsymbol{x}_j\right) + e = B_i + \sum_{j=1}^{m} S_j \cdot B_j, \\ p\left(\boldsymbol{b}_i + \sum_{j=1}^{m} S_j \cdot \boldsymbol{b}_j\right) = V\left(\boldsymbol{x}_i + \sum_{j=1}^{m} S_j \cdot \boldsymbol{x}_j\right), \end{array}\right\} \text{consider market mechanisms,} \\
& n_i \ge 0; p_i \ge 0; n_i \cdot p_i = 0; n_j \ge 0; p_j \ge 0; n_j \cdot p_j = 0,
\end{aligned}
$$

$x \ge 0$; $b \ge 0$, where $\boldsymbol{x} = [\boldsymbol{x}_i, \boldsymbol{x}_j]; \boldsymbol{b} = [\boldsymbol{b}_i, \boldsymbol{b}_j]$, $i = 1,\ldots,I; j = 1,\ldots,J$,
$x,b \in$ *Integer* (where products and resources are undividable conditions),

and where n_i, n_j and p_i, p_j denote slack and surplus variables in alliance cluster and candidate partners, respectively, S_j denotes the dummy variable in the jth firm where 1 indicates entry into the strategic alliance, e denotes the unused budget that can be ignored in resource dividable systems but cannot be ignored in resource undividable systems. Note that in the minimum problem we can substitute $u \geq n_i / (z_i^u - z_i^l)$ and $u \geq S_j \cdot n_j / (z_j^u - z_j^l)$ with $u \geq p_i / (z_i^u - z_i^l)$ and $u \geq S_j \cdot p_j / (z_j^u - z_j^l)$, i.e., we can ignore p_i and p_j when dealing with the maximum problem.

On the other hand, the M&A model can also be derived based on Figure 10.3 to obtain the optimal alliance satisfaction as shown in the next section.

10.2.2 M&A Model

$$
\begin{aligned}
\min \quad & u+e \\
s.t. \quad & \boldsymbol{A}_i\boldsymbol{x}_i \leq \boldsymbol{b}_i, \qquad\qquad (10.9)\\
& S_j \cdot (\boldsymbol{A}_j\boldsymbol{x}_j) \leq \boldsymbol{b}_j \quad \text{where} \quad S_j \in \{0,1\}, \\
& \boldsymbol{C}_i\boldsymbol{x}_i + \sum_{j=1}^{m} S_j(\boldsymbol{C}_j\boldsymbol{x}_j) + n_i - p_i = z_i^u + \sum_{j=1}^{m} S_j(z_j^u), \\
& u \geq n_i \Big/ \left[(z_i^u - z_i^l) + \sum_{j=1}^{m} S_j \cdot (z_j^u - z_j^l) \right], \text{maximizes M\&A satisfaction}, \\
& \left.\begin{aligned}
& V\left(\boldsymbol{x}_i + \sum_{j=1}^{m} S_j \cdot \boldsymbol{x}_j\right) + e = B_i + \sum_{j=1}^{m} S_j \cdot B_j, \\
& p\left(\boldsymbol{b}_i + \sum_{j=1}^{m} S_j \cdot \boldsymbol{b}_j\right) = V\left(\boldsymbol{x}_i + \sum_{j=1}^{m} S_j \cdot \boldsymbol{x}_j\right),
\end{aligned}\right\} \text{consider market mechanisms}, \\
& n_i \geq 0; p_i \geq 0; n_i \cdot p_i = 0, \\
& x \geq 0;\ b \geq 0, \text{ where } \boldsymbol{x} = [\boldsymbol{x}_i, \boldsymbol{x}_j]; \boldsymbol{b} = [\boldsymbol{b}_i, \boldsymbol{b}_j],\ i = 1,\ldots,I;\ j = 1,\ldots,J, \\
& x,b \in \textit{Integer} \text{ (where products and resources are undividable conditions).}
\end{aligned}
$$

On the basis of the two models above, we can conclude the advantages of the proposed method. First, we can easily choose the correct alliance partners by setting a dummy variable S using conventional mathematical programming methods or other heuristic algorithms such as genetic algorithms or simulated annealing. Second, using the concept of fuzzy sets, we can easily measure alliance satisfaction. Next, by incorporating the concept of de novo programming, the unit prices of the resources can be considered in the proposed models.

If both the technological coefficients and resource portfolio are undividable, the programming can be easily rewritten as an integer programming problem. Since several algorithms such as the branch and bound algorithms (Bretthauer and Shetty, 1995), linear knapsack method (Mathur et al., 1986; Hochbaum, 1995), and dynamic programming algorithm (Glover, 1975), can be used to solve this integer fuzzy multi-objective dummy programming problem, it is more suitable for dealing with real-world alliance problems. In order to demonstrate the advantages of the proposed method, a numerical example is employed to display the satisfaction results for both joint ventures and M&A cases.

10.3 NUMERICAL EXAMPLE

Strategic alliances are widely adopted by firms to increase their competitive advantages. By exchanging or sharing alliance resources, all alliance members can obtain better satisfaction levels than their original levels. However, since every firm has its own products, objective functions, constraints, and capital, it is hard for firms to find the best alliance partners. In this section, the joint venture and M&A strategies are considered so that we can determine sound solutions.

Assume an enterprise considers entering strategic alliances with five candidate firms. For simplicity, these six firms all produce two products (x and y) and have the same objectives of revenue R, quality Q, and satisfaction S, and face the same production constraints of material M, channel C, promotion P, and expertise E. Table 10.1 provides additional information about all six firms including technology coefficients and capital.

TABLE 10.1
Objective and Production Data for Six Firms

		Objectives			Constraints				
	Products	***R***	***Q***	***S***	***M***	***C***	***P***	***E***	**Capital**
Enterprise	x	78	4.25	3.52	7	2	7	2	4,800
	y	125	2.74	4.86	6	8	5	3	
Firm 1	x_1	165	5.78	2.28	3	6	7	2	3,600
	y_1	100	3.03	2.84	5	4	5	3	
Firm 2	x_2	85	4.97	4.21	2	3	7	2	6,400
	y_2	140	2.77	7.42	5	2	5	3	
Firm 3	x_3	70	5.99	7.54	8	6	7	2	5,200
	y_3	120	3.93	3.44	6	6	5	3	
Firm 4	x_4	75	5.57	4.98	8	2	7	2	4,200
	y_4	125	2.36	7.48	8	6	5	3	
Firm 5	x_5	80	4.38	3.24	4	8	7	2	5,400
	y_5	130	3.69	2.87	6	6	5	3	
Unit price					20	15	30	10	

The aspiration level of the enterprise can be determined by solving the following equations:

$$\begin{aligned}
\max\quad & 78x+125y \\
\max\quad & 4.25x+2.74y \\
\max\quad & 3.52x+4.86y \\
s.t.\quad & 7x+6y\le b_{e1}, \\
& 2x+8y\le b_{e2}, \\
& 7x+5y\le b_{e3}, \\
& 2x+3y\le b_{e4}, \\
& x,y\ge 0, \\
& b_{e1},b_{e2},b_{e3},b_{e4}\ge 0, \\
& B_e=4800.
\end{aligned}$$

In order to increase the enterprise's objective values, the joint venture strategy is considered for choosing the best alliance partners. In addition, the corresponding resource allocation should be determined. Using Equation (10.8), we can obtain the best alliance cluster and optimal resource allocation for the case of joint ventures as shown in Table 10.2.

By removing certain alliance partners, we can obtain the optimal resource allocation for the enterprise and firm 3 using the fuzzy multi-objective dummy programming model as shown in Tables 10.3 and 10.4.

TABLE 10.2
Alliance Partners and Resource Allocation in Joint Venture

Resource Allocation	Alliance	Enterprise	Firm 3
X	11.93	6.15	5.78
Y	11.74	6.21	5.53
$\mathbf{b}_1$	159.74	80.33	79.41
$\mathbf{b}_2$	129.85	61.99	67.86
$\mathbf{b}_3$	142.22	74.12	68.10
$\mathbf{b}_4$	59.09	30.94	28.15
Revenue	2324.57	1256.20	1068.37
Quality index	99.51	43.17	56.34
Satisfaction index	114.42	51.84	62.58
$\mu(z)$	0.570	0.570	0.570
B	10,000	5,069.45	4,930.55

TABLE 10.3
Optimal Resource Allocation in Enterprise

Resource allocation	X	Y	b_{e1}	b_{e2}	b_{e3}	b_{e4}
Value	5.73	5.97	75.94	59.23	69.97	29.34
Resource allocation	Revenue	Quality index	Satisfaction index	$\mu(Z_e)$	B_e	
Value	1,193.39	40.71	49.19	0.43	4,800	

TABLE 10.4
Optimal Resource Allocation for Firm 3

Resource allocation	x	Y	b_{31}	b_{32}	b_{33}	b_{34}
Value	6.24	5.65	83.84	71.36	71.95	29.43
Resource allocation	Revenue	Quality index	Satisfaction index	$\mu(Z_3)$	B_3	
Value	1,114.99	59.60	66.51	0.65	5,200	

By comparing Tables 10.2 through 10.4, we see that firm 3 shares redundant resources with the enterprise, thus increasing alliance satisfaction. It is clear that the alliance satisfaction is larger than the average satisfaction, i.e., it satisfies the following equation:

$$\mu(z) > \frac{1}{2}[\mu(z_e) + \mu(z_3)] \text{ (i.e. } 0.57 > \frac{1}{2}(0.43 + 0.65) = 0.54)$$

The result indicates that due to the emergence of synergy effects, the firms have motives to enter joint ventures. Next, we use the integer M&A model (Equation 10.9) to determine the best alliance partners and optimal resource allocation for potential M&A as shown in Table 10.5.

TABLE 10.5
Alliance Partners and Resource Allocation in M&A

Resource Allocation	Alliances	Enterprise	Firm 2	Firm 3
X	26	12	8	6
Y	20	0	20	0
b_1	248	84	116	48
b_2	124	24	64	36
b_3	282	84	156	42
b_4	112	24	76	12
Revenue	4836	936	3480	420
Quality index	182.1	51	95.16	35.94
Satisfaction index	269.56	42.24	182.08	45.24
$\mu(z)$	0.60	0.60	0.60	0.60
B	16,400	4,800	8,720	2,880

On the basis of Table 10.5, the best alliance cluster in M&A is different from the result for joint ventures. However, the synergy effects can also be found using the same method to analyze the development of an M&A strategy. The next section discusses costs and other implementation issues.

10.4 DISCUSSION

Strategic alliances are widely used in business to obtain synergy effects. These synergy effects may come from economies of scale, economies of scope, learning effects, or other tangible and intangible assets. However, many firms fail in strategic alliances because of the lack of sound planning or screening to choose correct partners and devise resource allocations. This chapter discusses a new method to overcome these problems.

Two types of strategic alliances, joint ventures and M&A, are demonstrated here to allow a comparison of the proposed methods. From the numerical example, we can see that in a joint venture the surplus resources of firm 3 (5,200 – 4,930.55 = 269.45) are shared with the enterprise to increase the alliance satisfaction from 0.54 to 0.57. This is why the enterprise has motives for considering a joint venture strategy. The same situation can be found for the M&A strategy.

From our implementation, the M&A strategy seems to provide a better satisfaction level than the joint venture. However, this may not be true in practice. Since we have not considered alliance costs for coordination, control, and risk avoidance, an optimal alliance strategy cannot be determined. The alliance costs of joint venture and M&A plans are depicted in Figure 10.6.

Since the alliance costs for joint ventures and for M&A are very different, we cannot ignore their effects on alliance costs. It is clear that considering the different cost functions may result in a different optimum alliance strategy.

In addition, the most important issue in facilitating strategic alliances may be the determination of fair sharing criteria. For the joint ventures case, the satisfaction of firm 3 decreased from 0.65 to 0.57. It is impossible for firm 3 to enter alliances if its satisfaction level in an alliance is lower than its original level. Therefore, the rational way to assign synergy effects in our joint ventures case can be restricted such that and where and denote the true satisfaction level for the enterprise and firm 3, respectively, after a joint venture. The same method can be used to set the appropriate sharing mechanism for M&A arrangements. More

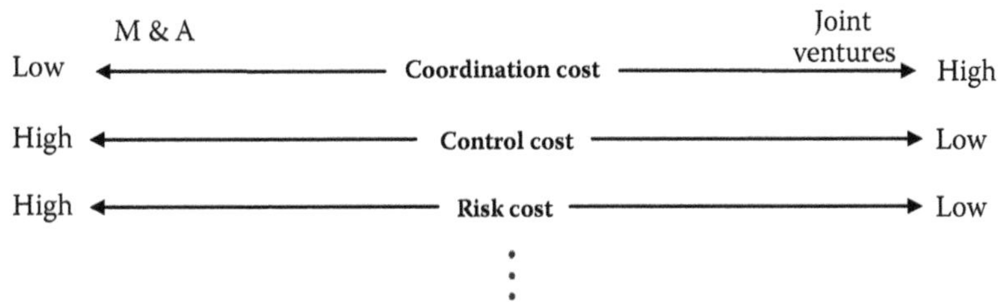

FIGURE 10.6 Alliance costs in joint ventures and mergers and acquisitions.

discussions about setting fair-sharing criteria can be found in one of our earlier papers (Huang et al., 2005).

10.5 CONCLUSIONS

The goal of a strategic alliance is to create and share maximum synergy effects among alliance partners. In order to achieve this goal, the correct alliance partners and the appropriate resource allocations are critical. In this chapter, we propose the fuzzy multi-objective dummy programming model to overcome these problems. On the basis of the numerical results, we can conclude that both the joint venture and the M&A model can provide the best alliance cluster, the maximum synergy effects, and optimal alliance satisfaction.

11 Multiple Objective Planning for Supply Chain Production and Distribution Model: Bicycle Manufacturer

Because of globalization, the supply chain has become increasingly important for many enterprises because it involves flow process controls and levels of management integration. Components of the supply chain include raw materials, production, distribution, retail sales, and after-sales service to consumers. The related primary objectives are to (1) lower the total costs of production along the supply chain and (2) increase production efficiency. Through the efforts of both government and industry, Taiwan was admitted into the World Trade Organization (WTO) on January 1, 2001. Within the context of the WTO, manufacturers face global competition and thus must accelerate the flows of materials and information. As a result, the development of an effective supply chain is essential for an industry to become successful.

Bicycle manufacturing, generally considered a conventional industry, demands massive labor and capital investments. Within the past 50 years, the industry supply chain has expanded to include up-, middle-, and down-stream manufacturers, making Taiwan one of the significant bicycle manufacturing centers of the world. In 1980, Taiwan was the world's largest exporter of bicycles, surpassing even Japan. Even though bicycles are not a high technology product, they follow trends of fashion, creativity, and diversity so speed of supply is a critical factor for remaining competitive in the face of changing demands of consumers.

Because Taiwan's bicycle industry is affected by international competition, manufacturers must develop global management abilities to confront changing situations. This chapter considers multiple objective production and distribution, focusing on Taiwan's conventional bicycle manufacturing industry. It discusses enterprise profit and customer service levels for multi-objective programming to develop a production-and-distribution model for bicycle manufacturers in Taiwan.

Five methods have been adopted for comparison: multi-objective compromise programming, fuzzy multi-objective programming, weighted multi-objective compromise programming, weighted fuzzy multi-objective programming, and two-phase fuzzy multi-objective programming for analyzing vector-valued optimization (Salukvadze, 1971a and b, 1979) through various methods. Finally, weighted multi-objective programming is used to conduct sensitivity analysis to obtain outcomes that manufacturers can use for reference in developing their supply chains.

The results show that increases of per-unit production costs decrease total profits in real empirical studies. If the unit inventory cost increases in an effort to improve customer service levels, total profit may increase, but not significantly. Furthermore, shortage cost has an interactive effect on an enterprise and increased inventory cost will lower the shortage cost for achieving satisfaction levels.

11.1 LITERATURE ON SUPPLY CHAIN AND MULTI-OBJECTIVE PROGRAMMING FOR PRODUCTION AND DISTRIBUTION

Since this study explores the issues of production and distribution of bicycles that relate to management operations and the industrial supply chain, we first review the literature discussing supply chain management. We then present studies on multi-objective programming, compromise programming, fuzzy multi-objective programming, weighted multi-objective programming, and two-phase fuzzy multi-objective programming because they concern the production and distribution of bicycles.

11.1.1 Relevant Supply Chain Literature

Chandra and Fisher (1994) dealt with the production and delivery processes of a single factory with a single product over multiple periods. Comparing the separation and integration models of production and transportation problems, they found that the integration model of production and delivery could help lower total costs.

Nagata et al. (1995) examined the production and transportation problems of multiple products and multiple factories over multiple periods, using multi-objective and fuzzy multi-objective models for programming. Their programming problem for multiple periods included utilizing uncertain information when considering a management plan so as to construct a reasonable multi-objective production and transportation model.

Tzeng et al. (1996) addressed practical issues, using fuzzy bi-criteria multi-index linear programming to deal with uncertain supply and demand environments related to coal procurement and delivery schedules for the Taiwan Power Company (TPC). The company had to deal with multiple destinations, varied goals, and several types of shipping vessels.

Petrovic et al. (1998, 1999) used fuzzy models and simulation for a supply chain and developed a decision-making system for an uncertain environment. They determined the inventory level and the order amounts using a supply chain model that considered time and cost constraints to simulate operation control.

Van der Vorst et al. (1998) considered that supply chain management should consider decreasing or limiting uncertainty so that the integral benefits of the chain could be improved. Sources of uncertainty include vertical order prediction, information input, administration management, decision-making procedures, and innate uncertainties. Their study used the food chain as an example to improve allocation and operation management structures.

Although many uncertainty factors were eliminated in that illustration, results indicate that a reasonable benefit from supply chain management occurred. Dhaenens-Flipo (2000) proposed clustering problems of production and transportation of multiple products from multiple factories for integrated planning. To deal with sophisticated problems involving spatial decomposition, he converted them into sub-problems and employed the resolution for the vehicle routing problem (VRP). This helped simplify the problem and lower total costs.

In summary, when the current study of the bicycle manufacturer considers the maximization of both enterprise profit and customer service levels, the resolution can also be obtained from developing a multi-objective programming model in view of the global supply chain system. The proposed model is described in the following section.

11.1.2 Multi-Objective and Fuzzy Multi-Objective Programming

Zeleny (1982) pointed out that programming has nothing to do with decision making when there is only a single objective, as decision making is already endowed in the estimation of the objective function value coefficient. Thus, when the objective function coefficient is determined, the decision maker can only accept or discard the outcome resolved by the model, and no other information can ever be obtained from the model. In contrast, the main aim of multi-objective programming is to find a feasible non-inferior solution set or compromise solution so that the decision maker can effectively focus on the trade-offs when several objectives are in conflict. This is generally indicated as follows:

$$\begin{aligned} \max \quad & \boldsymbol{f} = [f_1(\boldsymbol{x}), f_2(\boldsymbol{x}), \ldots, f_k(\boldsymbol{x})] \\ s.t. \quad & \boldsymbol{A}\boldsymbol{x} \le \boldsymbol{b} \\ & \boldsymbol{x} \ge \boldsymbol{0} \\ where \quad & \boldsymbol{b} = [b_1, b_2, \ldots, b_m]^T,\ \boldsymbol{x} = [x_1, x_2, \ldots, x_n]^T,\ \boldsymbol{A} = A_{m\times n}. \end{aligned} \tag{11.1}$$

11.1.2.1 Multi-Objective Compromise Programming

The ideal solution in multi-objective space was first introduced by Salukvadze (1971a and b) and has come to be known as Salukvadze's solution (Yu and Leitmann, 1974). The ideal solution is defined as indicating the optimal value f_i^* in the feasible solution domain $x \in X$ of every single objective, $f_i(\boldsymbol{x}), i = 1,2,\ldots,k$. Then, based on the concept of compromise programming, and according to the aforementioned definition of distance scale d_p, we have to locate a point that has the shortest distance to the ideal solution from the non-inferior solution set (i.e., the minimal gap that ensures the achieving level and criteria are equally weighted), which can be written as:

$$\min d_p$$
$$\textbf{\textit{s.t.}}\ x \in X \tag{11.2}$$

where $d_p = (\sum_{i=1}^{k}(\frac{f_i^* - f_i(x)}{f_i^* - f_i^-})^p)^{\frac{1}{p}}$, $1 \le p<\infty$ *and* p ranges from 1 to ∞. For $p = 1$ and $p = \infty$; the above expression (Yu, 1973, 1974, 1985; Salukvadze, 1971a, 1971b) becomes $d_{p=1} = \sum_{i=1}^{k}(f_i^* - f_i(\boldsymbol{x}))/(f_i^* - f_i^-)$ and $d_{p=\infty} = \max_i[(f_i^* - f_i(\boldsymbol{x}))/(f_i^* - f_i^-) \mid i = 1,2,\ldots,k]$.

In the above equation, the distance scale d_p varies with different values of p and has diverse meanings (Yu, 1973, 1985). Hwang and Yoon (1981) claimed that it is necessary to consider the shortest (closest) positive ideal solution (PIS) and the farthest negative ideal solution (NIS), so that the greatest profit can be obtained and the greatest risk avoided during decision making. To find a compromise solution, we must know the target (positive ideal) point and the parameter for the regret function $r(f(x);p)$. Then a generalized family of normalized distance measure dependent on power p can be expressed:

$$\begin{aligned} \min\quad & d \\ \text{s.t.}\quad & \frac{f_i^* - f_i(\boldsymbol{x})}{f_i^* - f_i^-} \le d; \quad i = 1,2,\ldots,k \\ & \boldsymbol{Ax} \le \boldsymbol{b} \\ & \boldsymbol{x} \ge \boldsymbol{0} \end{aligned} \tag{11.3}$$

11.1.2.2 Weighted Multi-Objective Programming

According to the multi-objective linear programming problem put forward by Martinson (1993), two distinct models of fuzzy multi-objective and multi-objective compromise programming can be used. The easiest way to deal with this issue is to settle the weight values of each objective by the preferences of the decision makers (i.e., using the concept of the minimal gap for achieving level and assigning different weights to the criteria).

If the weighted value of each objective is w_{G_i}, the resolution for the compromise programming of the multi-objective programming can be found as follows:

$$\begin{aligned} \min\quad & d \\ \textbf{\textit{s.t.}}\quad & \left[\frac{f_i^*(\boldsymbol{x}) - f_i(\boldsymbol{x})}{f_i^*(\boldsymbol{x}) - f_i^-(\boldsymbol{x})}\right] \le d, \quad i = 1,2,\ldots,k \\ & \boldsymbol{Ax} \le \boldsymbol{b} \\ & \boldsymbol{x} \ge \boldsymbol{0} \end{aligned} \tag{11.4}$$

11.1.2.3 Fuzzy Multi-Objective Programming

Bellman and Zadeh (1970) applied the notion of fuzzy sets for decision-making theory by considering conflicts between the constraint equation and the objective equation of general programming. They proposed the max–min operation method to determine the optimal decision from the two solutions. Tanaka et al. (1974) advanced fuzzy mathematics programming (FMP), which resulted in the widespread application of the technique on several practical levels.

Zimmermann introduced fuzzy set theory into the conventional linear programming problem in 1976 and combined the fuzzy linear programming model with multi-objective programming into fuzzy multi-objective linear programming (Zimmermann, 1978). The fuzzy linear programming employed in the study used the max–min-operation that turns multiple objectives into a single one (see Appendix A at the end of this chapter). First, the upper and lower bound limits of each objective and constraint equation are determined, and multiple objectives are turned into a single objective for solving the maximal achievement level λ.

$$\begin{aligned} \max_{x} \quad & \lambda \\ \textbf{s.t.} \quad & \lambda \leq \frac{f_i(\boldsymbol{x}) - f_i^-}{f_i^* - f_i^-}, \quad i = 1,2,\ldots,k \\ & \boldsymbol{Ax} \leq \boldsymbol{b} \\ & \boldsymbol{x} \geq \boldsymbol{0} \end{aligned} \tag{11.5}$$

11.1.2.4 Weighted Fuzzy Multi-Objective Programming

The membership function of the weighted fuzzy linear programming can be indicated as follows:

$$\mu_D(\boldsymbol{x}^*) = \max_{i} \min_{x} \left\{ w_{G_i} \mu_{G_i}(\boldsymbol{x}^*), \quad i = 1,2,\ldots,k \right\} \tag{11.6}$$

Converting multiple objectives into a single objective for resolution transformed this problem into the following exact LP problem for resolution (Martinson, 1993; Appendix B at the end of this chapter):

$$\begin{aligned} \min \quad & \lambda \\ \textbf{s.t.} \quad & \left[\frac{1}{w_{G_i}} \left(\frac{f_i(x) - f_i^-}{f_i^* - f_i^-} \right) \right] \geq \lambda, \quad i = 1,2,\ldots,k \\ & \boldsymbol{Ax} \geq \boldsymbol{b} \\ & \boldsymbol{x} \geq \boldsymbol{0} \end{aligned} \tag{11.7}$$

11.1.2.5 Two-Phase Fuzzy Multi-Objective Programming

Lee and Li (1993) proposed two-phase fuzzy multi-objective programming to deal with the defects of non-compensatory solutions in the linear fuzzy objective programming of Zimmermann (1978). They used Zimmermann's linear programming method to determine a non-compensatory solution (Phase I), and then employ the greatest average fuzzy membership function (Phase II) of an individual objective among objectives once the non-compensatory solution is found. It has been shown that the multi-objective solution obtained from the two-phase fuzzy multi-objective programming is a compensatory solution for raising the achieved level. The steps of resolution (basing the second phase on the first-phase solutions of fuzzy multi-objective programming) are explained below.

Phase I — Locate the non-compensatory solution $(\lambda^{(I)}, \boldsymbol{x}^{(I)})$ of Zimmermann's fuzzy linear programming.

Phase II —

$$\max_{x} \quad \bar{\lambda} = \frac{1}{k}\sum_{i=1}^{k} \lambda_i$$

$$\text{s.t.} \quad \lambda^{(I)} \le \lambda_i \le \left[\frac{f_i(\boldsymbol{x}) - f_i^-(\boldsymbol{x})}{f_i^*(\boldsymbol{x}) - f_i^-(\boldsymbol{x})}\right], \qquad i = 1,2,\ldots,k \tag{11.8}$$

$$\boldsymbol{Ax} \ge \boldsymbol{b}$$

$$\boldsymbol{x} \ge \boldsymbol{0}$$

This study uses the five kinds of multi-objective programming methods mentioned earlier to resolve and compare the supply-chain model constructed below. It also compares and analyzes these results and conducts sensitivity analysis using one of the models for a real case.

11.2 ESTABLISHING MODEL FOR BICYCLE SUPPLY CHAIN

The research on supply chains in this study mainly investigates the production and distribution problems confronted by a bicycle manufacturer competing in a global market. The production and distribution of bicycles are explored in an effort to establish a model for achieving two win–win objectives: (1) maximize profits and competitiveness and (2) maximize utility (such as service quality) by satisfying customers and maintaining appropriate prices.

11.2.1 Basic Assumptions, Definitions, and Establishment of Model

Generally, the goal of enterprise management is to obtain maximum profit. The manufacturing and retail service industries, however, have both gradually turned toward consumer service to develop stability and maintain competitiveness. This model incorporates the idea of improved consumer service into its design and allows

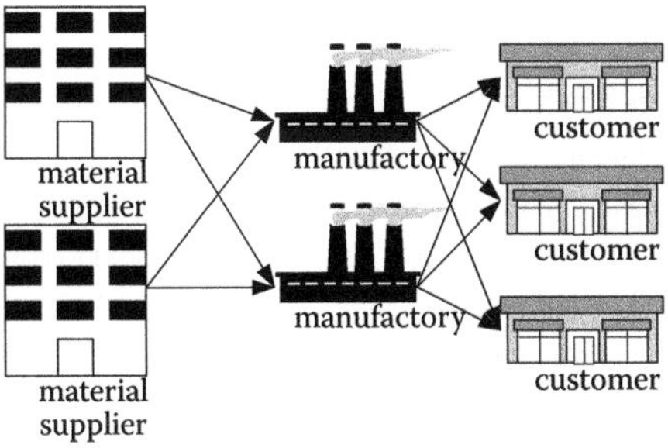

FIGURE 11.1 Supply chain concepts.

for flexible adjustments in order to consolidate marketing sources and respond to demand rather than following rigid production schedules. Ultimately, favorable consumer service for a stable clientele can be achieved with such a design. This real study based on conditions in the bicycle industry uses the following assumptions (see Figure 11.1):

1. To determine and control the key factors in a simplified manner, the material flow processes of bicycle manufacture are set as: (1) input of raw materials, (2) inventory of raw materials, (3) production, (4) product inventory, and (5) distribution. This study focuses only on the model structures of production, product inventory, and distribution.
2. For a bicycle manufacturer, this real study considers the problems of production and distribution for multiple products over multiple periods as demands vary from many factors such as the international economy and weather.
3. Since product demand will vary based on factors such as weather, seasons, and holidays, one year within the research period is designated the planning period and divided into four parts corresponding to annual seasons.
4. The strategy of the industry already is to build to order (BTO) when an order is received. Therefore, we assume that product demand is already known.
5. Only land–sea transportation is considered.

Table 11.1 defines the symbols used in the model. Table 11.2 lists the relevant decision-making variables.

11.2.2 Model Construction

11.2.2.1 Objective 1: Maximize Total Profit and Competitiveness

Achieving maximum profit and competitiveness is one of the business objectives. Equation (11.13) demonstrates the maximization formula. Profits and competitiveness involve factors such as revenues and costs. Cost factors include production (for regular and overtime working hours), transportation, inventory, shortages, maintaining market advantage, management, research and development, innovation,

TABLE 11.1
Definitions of Model Symbols

p_{kt}	Selling price of product k at period t
d_{jkt}	Estimated demand of product k at period t in locality j
α_{ijkt}	Unit delivering cost of product k at period t from factory i to point of demand j
β_{ikt}	Unit production cost of product k at period t in factory i during ordinary working hours
γ_{ikt}	Unit production cost of product k at period t in factory i during overtime working hours
ε_{ikt}	Unit inventory cost of product k at period t in factory i
δ_{jkt}	Unit shortage cost of product k during period t at locality of demand j
a_{ikrt}	Consumption of needed material r for production per product k at period t in factory i
f_{ikrt}	Upper bound value of useable material r at period t in factory i
b_{ikt}	Production time needed for production per product k at period t in factory i
η_{it}	Upper bound value of ordinary working hours at period t in factory i
c_{it}	Upper bound value of overtime working hours at period t in factory i
h_{ikt}	Minimal safety inventory for product k at period t in factory i
θ_{ikt}	Upper bound value of safety inventory at period t in factory i

knowledge competency, and other tangible and intangible items. In the real case portrayed in this chapter, the model is constructed for maximizing total profits from manufacturing over a short time as described below.

Maximum total profits = [total revenues – total costs (production cost during ordinary working hours + production cost during overtime working hours + transportation cost for product + inventory cost of product + shortage cost of product)].

$$\text{Max} \quad z_0 = \sum_{t=1}^{T}\sum_{k=1}^{K}\left\{ p_{kt}\sum_{i=1}^{m}\sum_{j=1}^{n} y_{ijkt} - \left(\sum_{i=1}^{m}\beta_{ikt}x_{ikt} + \sum_{i=1}^{m}\gamma_{ikt}\omega_{ikt} + \sum_{i=1}^{m}\sum_{j=1}^{n}\alpha_{ijkt}y_{ijkt} + \sum_{i=1}^{m}\varepsilon_{ikt}e_{ikt} + \sum_{j=1}^{n}\delta_{jkt}s_{jkt} \right) \right\} \tag{11.9}$$

TABLE 11.2
Decision-Making Variables

x_{ikt}	Amount of product k made in period t at factory i during ordinary working hours
ω_{ikt}	Amount of product k made in period t at factory i during overtime working hours
y_{ijkt}	Amount of product k delivered in period t at factory i to locality of demand j
e_{ikt}	Inventory of product k at period t in factory i
S_{jkt}	Shortage of product k during period t at locality of demand j

11.2.2.2 Objective 2: Maximize Level of Customer Service Quality per Period

In a real case, bicycle products must be of high quality and properly serve customers of the company. Therefore, under the assumption that both production time and transportation time per unit of product are fixed in this real case, a portion of production time is designated as flexible so that it can be adjusted to meet demand changes and satisfy the needs of global customers. By treating this flexible time (extra production time per period/total amount time available for production per period) as a crucial factor affecting the level of service quality, the assumption of a relationship between time and service level can be indicated by the following formula.

Level of service at period $t = f$ (spare production time at period t/total time available for production at period t)

$$\max \ S_t = f(W_t), \quad t = 1,2,...,T \tag{11.10}$$

where $W_t = \dfrac{\text{spare production time at period } t}{\text{total amount of time available for production at period } t}$ and W_t can be estimated as follows.

$$W_t = 1 - \sum_{i=1}^{m}\left(\sum_{k=1}^{K} b_{ikt}x_{ikt} + \sum_{k=1}^{K} b_{ikt}\omega_{ikt}\right) / (\eta_{it} + c_{it}). \tag{11.11}$$

If a manufacturer is confronted with problems such as unstable supply and demand or mismanagement, the result could be delayed product delivery. If the value of W_t is high, many factory facilities are idle instead of meeting the urgent needs of customers. The manufacturer faces low turnover even though it produces quality products. Collecting real data is required to fit the relationship of service level and production feasibility. The result is shown in Figure 11.2.

As a result, a trade-off relationship is created between the objective Equations (11.14) and (11.15).

11.2.3 Model Constraints

11.2.3.1 Raw Material Constraint

The amount of raw material needed for production should be less than or equal to the amount of raw material available:

$$\sum_{k=1}^{K} a_{ikrt}(x_{ikt} + \omega_{ikt}) \le f_{irt} \quad i = 1,\ldots,m; \quad r = 1,\ldots,R; \quad t = 1,\ldots,T \tag{11.12}$$

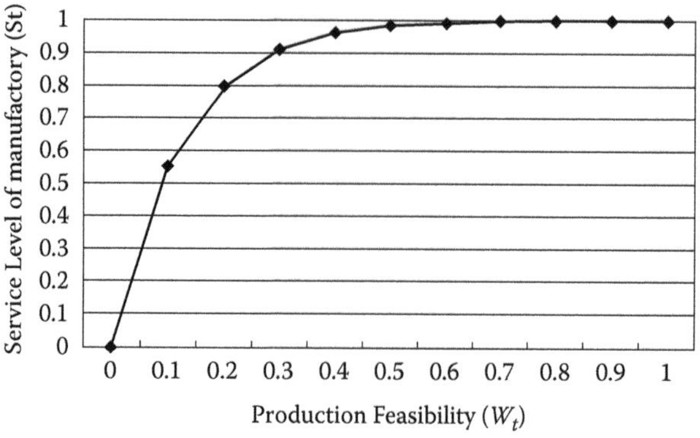

FIGURE 11.2 Relationship between service level and production feasibility in t period.

11.2.3.2 Productivity Constraint

Since a manufacturer must use plants and equipment, its production is limited to a certain maximum capability. This study uses base production time as its unit: normal working hours multiplied by time needed for production with per-unit normal working hours less than or equal to maximal productivity during normal working hours. Overtime hours are multiplied by the time needed for production with per-unit overtime less than or equal to maximal productivity during overtime working hours.

$$\sum_{k=1}^{K} b_{ikt} x_{ikt} \le \eta_{it} \qquad i = 1,\dots,m;\ t = 1,\dots,T \tag{11.13}$$

$$\sum_{k=1}^{K} b_{ikt} \omega_{ikt} \le c_{it} \qquad i = 1,\dots,m;\ t = 1,\dots,T \tag{11.14}$$

11.2.3.3 Inventory Constraint

Due to the constraint of inventory space, the total amount of product inventory in every period must be smaller than maximum inventory capacity. Furthermore, the product inventory for a period should be equal to the total inventory for the last period and amount for the current period less the inventory distributed.

Amount of inventory product ≤ maximal capacity of inventory

Amount of inventory this period = inventory capacity of last period

+ production amount of current period

− amount of product distributed

$$\sum_{k=1}^{K} e_{ikt} \leq \theta_{it} \quad i = 1,...,m;\ t = 1,...,T \tag{11.15}$$

$$e_{ikt} = e_{ik(t-1)} + (x_{ikt} + \omega_{ikt}) - \sum_{j=1}^{n} y_{ijkt} \quad i = 1,\ldots,m;\ k = 1,\ldots,K;\ t = 1,\ldots,T \tag{11.16}$$

11.2.3.4 Relationship of Product Distribution and Demand

The actual amount of product distributed must be less than or equivalent to demand, as predicted in each locality. The difference between demand and the amount distributed represents the shortage.

Amount of product distributed < = Estimated amount of demand
Shortage product = Amount of Product needed – Amount of Product distributed

$$\sum_{j=1}^{n} y_{ijkt} \leq d_{jkt} \quad j = 1,\ldots,n;\quad k = 1,\ldots,K;\quad t = 1,\ldots,T \tag{11.17}$$

$$s_{jkt} = d_{jkt} - \sum_{i}^{m} y_{ijkt} \quad j = 1,\ldots,n;\ k = 1,\ldots, K;\ t = 1,\ldots,T \tag{11.18}$$

11.2.3.5 Non-Negative Constraint

$$x_{ikt} \geq 0 \quad i = 1,\ldots,m;\ k = 1,\ldots,K;\ t = 1,\ldots,T \tag{11.19}$$

$$\omega_{ikt} \geq 0 \quad i = 1,\ldots,m;\ k = 1,\ldots,K;\ t = 1,\ldots,T \tag{11.20}$$

$$y_{ijkt} \geq 0 \quad i = 1,\ldots,m;\ j = 1,\ldots,n;\ k = 1,\ldots,K;\ t = 1,\ldots,T \tag{11.21}$$

$$e_{ikt} \geq 0 \quad i = 1,\ldots,m;\ k = 1,\ldots,K;\ t = 1,\ldots,T \tag{11.22}$$

11.3 REAL EMPIRICAL CASE OF A BICYCLE MANUFACTURER

In this section, the real empirical case of a bicycle manufacturer in Taiwan is used to demonstrate that this model can effectively provide good ideas and approaches to supply chain problems. The latest data for the management scenarios were obtained from manufacturers and incorporated into the model. Some information was modified from actual data to avoid revealing business secrets.

11.3.1 Problem Description and Definitions

The operation problem of this practical example is designed to cover three points of supply and seven points of demand. The duration of production programming is divided into four periods per year. As learned from case interviews, the demand for bicycle products varies seasonally. Thus, marketing strategy must compensate for the uneven supply and demand.

Unlike manufacturers in other industries, bicycle makers produce according to the season. In our study, we assigned the four seasons as (1) July to September, (2) October to December, (3) January to March of the next year, and (4) April to June. Our study concerns only bicycles without heavy engines. The seven types of products considered for programming are mountain bikes, aluminum alloy bikes, lightweight bikes, children's bikes, sport bikes, racing bikes, and carbon fiber bikes.

The working hours of the manufacturing plant are designated as the 5 usual work days per week, with x hours worked 24 hours per day. Based on the known productivity of three manufacturers in different countries, it takes an average of 30 to 40 hours to assemble a bicycle, and the productivity constraint applies during the ordinary working hours in every season. The overtime hours are incurred during holidays.

11.3.2 Results and Analysis

The operation in this subsection includes a (1) multi-objective compromise programming solution (MOCP); (2) weighted multi-objective programming solution (WMOP); (3) fuzzy multi-objective programming solution (FMOP); (4) weighted fuzzy multi-objective programming solution (WFMOP); and (5) two-phase fuzzy multi-objective programming solution (TPFMOP). The results appear in Tables 11.3 and 11.4.

As can be seen from Table 11.3, the value of the total profits by the weighted multi-objective model is far larger than the profit calculated by the multi-objective compromise model. It is the same for the service levels for all four periods. The magnitude of value change of the service level in the fourth period as programmed in the fifth model is rather limited, indicating more stable service quality. It is thus understood that the multi-objective weight obtained depends largely on the importance with which decision makers endow objectives. Furthermore, the size of the weighted value will have to be adjusted according the nature of the specific problem.

TABLE 11.3
Solutions from Multi-Objective Programming

Outcome \ Model	MOCP	WMOP[a]	FMOP	WFMOP[a]	TPFMOP
Total profit (100 NT$)	310387	456411	311336	493296	310925
1st period	0.93	0.92	0.93	0.89	0.93
2nd period	0.95	0.93	0.95	0.89	0.95
3rd period	0.79	0.68	0.79	0.58	0.79
4th period	0.88	0.87	0.88	0.76	0.88

[a] Weights = 4:1:1:1:1 according to decision makers.

Comparing the results of fuzzy multi-objective, weighted fuzzy multi-objective, and two-phase fuzzy multi-objective programming, the maximum of the total profit value is from weighted multi-objective compromise programming. Comparing the result to a single objective, the maximization of profit will be about the same. The objective values of the service level in each programming period in the weighted fuzzy multi-objective model are similar to the programmed results of the profit maximization of a single objective. It can thus be said that the model has almost lost its multiple-objective significance.

Since the programmed results of fuzzy multi-objective and two-phase fuzzy multi-objective are close, the total achieved value in the two-phase multi-objective programming is larger by only 0.001. If individual objectives are analyzed—the objective of the two-phase fuzzy multi-objective programming—its resultant value would be somewhat smaller than that of fuzzy multi-objective programming, while the achievement of the service level in each period is enhanced by 0.001, with an insignificant level of improvement.

Comparing the model results in Table 11.4, we can see that material represents the highest manufacturing cost (70%) for bicycles in terms of total production cost. The next highest item is manufacturing cost, which amounts to 11% of production cost. Thus, the production cost amounts to 81% of total cost. This indicates that bicycle production is a labor-intensive industry involving high material costs. Despite the fact that the bicycle production, distribution, and selling activities have already been internationalized, analysis of its cost factors indicates that bicycles are not high-tech products that require immediate delivery. Therefore, they are usually distributed by inexpensive sea freight so the ratio of transportation cost to total cost is almost negligible.

Customers' costs, however, can reach 10% of the total and constitute the third highest cost. In summary, the most important factors for bicycle production are in-plant manufacturing and material costs, followed by export costs from the country of manufacture and import taxes at the site of demand. Of these three cost items, material cost is beyond the control of the manufacturer, so manufacturing cost as are import and export taxes where the manufacturing plant is located.

After comparison of the multi-objective compromise model to two-phase fuzzy multi-objective programming, we note large differences in objectives of multi-objective compromise model and weighted multi-objective model programming although the objective value of profit in multi-objective compromise is low. Even though the service level in each period of the weighted multi-objective model is not as high as that for the multi-objective compromise model, the programmed results of service level from the first to fourth periods of the weighted multi-objective model (except for the third period when the level was about 0.7), are all well over 0.8.

In addition, the programmed results of the fuzzy multi-objective model and two-phase fuzzy multi-objective model are much the same. The outcome of the weighted fuzzy multi-objective programmed model, however, does not differ significantly from the programmed resolution of a single objective. Consequently, if the decision makers who analyzed the results of each model wish to achieve programmed results of multi-objective production and distribution with maximum profit and highest quality service, the weighted multi-objective model would be more compatible with the results.

TABLE 11.4
Cost-Benefit Analysis of Multi-Objective Programming Model

	Model	MOCP		WMOCP[a]		FMOP		WFMOP[a]		TPFMOP	
	Outcome[b]	Cost-Profit	%	Cost-Profit	%	Cost-Profit	%	Cost-Profit	%	Cost-Profit	%
	Total profit	2571618	–	2662507	–	2570337	–	2720545	–	2570034	–
	Total cost	2261231	–	2237054	–	2259001	–	2227248	–	2259109	–
Production cost	Material cost	1568831	0.69	1655888	0.74	1610677	0.71	1684356	0.76	1610528	0.71
(ordinary)	Manufacturing cost	221504	0.10	227610	0.10	231246	0.10	229880	0.10	231149	0.10
Production cost											
(overtime)	Material cost	42474	0.02	0	0	0	0.00	0	0.00	0	0.00
	Manufacturing cost	12565	0.01	0	0	0	0.00	0	0.00	0	0.00
Transportation cost	Tax cost	212911	0.09	234169	0.10	212999	0.09	245300	0.11	213110	0.09
	Ocean shipping cost	53630	0.02	61130	0.03	53535	0.02	67288	0.03	53471	0.02
	Inventory cost	388	0.00	220	0	336	0.00	424	0.00	339	0.00
	Shortage cost	148927	0.07	58038	0.03	150208	0.07	0	0.00	150511	0.07

[a] Weights = 4:1:1:1:1 according to decision makers.
[b] Unit = Hundreds of New Taiwan dollars (NT$).

TABLE 11.5
Results of Changes to Unit Manufacturing Cost

Objective	Total Profit (100 NT$)	Service Level 1st Period	2nd Period	3rd Period	4th Period
Present situation	425453	0.90	0.92	0.69	0.82
Increase 20%	387295	0.90	0.91	0.70	0.83
Variation	–0.090	0.00	–0.01	0.01	0.01
Increase 40%	347342	0.90	0.91	0.69	0.82
Variation	–0.184	0.00	–0.01	0.00	0.00
Increase 60%	307922	0.89	0.90	0.69	0.82
Variation	–0.276	–0.01	–0.02	0.00	0.00
Increase 80%	269729	0.89	0.90	0.68	0.82
Variation	–0.366	–0.01	–0.02	–0.01	0.00
Increase 100%	233360	0.89	0.90	0.68	0.81
Variation	–0.452	–0.01	–0.02	–0.01	–0.01

11.3.3 Sensitivity Analyses and Discussions

Based on the conclusions above, the weighted multi-objective programming model is useful for sensitivity analysis for the following situation that has three separate parts: (1) changes of unit production cost (manufacturing cost increases of 20, 40, 60, 80, and 100%); (2) changes of unit inventory cost (increases of 20, 40, and 60%); and (3) changes of unit shortage cost (decreased by 20%, then increased by 20 and 40%).

Changes of production cost — As can be seen from Table 11.5, an increase of production cost can impact the value of each objective. With every increment of 20% added to the manufacturing cost, total profit decreases approximately 9%. Production cost, however, has much less impact on the service level for each period.

Changes of inventory cost — As indicated from Table 11.6, after inventory cost is increased by 20, 40, and 60%, it will conversely increase both total profits and service levels in the third and fourth periods.

TABLE 11.6
Results of Changes to Unit Inventory Cost

Objective	Total Profit (100 NT$)	Service Level 1st Period	2nd Period	3rd Period	4th Period
Present situation	425453	0.90	0.92	0.69	0.82
Increase 20%	426902	0.90	0.91	0.70	0.83
Variation	0.003	0.00	–0.01	0.01	0.01
Increase 40%	426919	0.90	0.91	0.70	0.83
Variation	0.003	0.00	–0.01	0.01	0.01
Increase 60%	427384	0.90	0.91	0.70	0.83
Variation	0.005	0.00	–0.01	0.01	0.01

TABLE 11.7
Results of Changes to Unit Shortage Cost

Objective	Total Profit (100 NT$)	Service Level 1st Period	2nd Period	3rd Period	4th Period
Present situation	425453	0.90	0.92	0.69	0.82
Increase 20%	430538	0.90	0.91	0.71	0.84
Variation	0.012	0.00	−0.01	0.02	0.02
Increase 40%	424601	0.89	0.91	0.69	0.82
Variation	−0.002	−0.01	−0.01	0.00	0.00
Increase 60%	424562	0.89	0.90	0.69	0.82
Variation	-0.002	-0.01	-0.02	0.00	0.00

Changes of shortage cost — As seen in Table 11.7, the changes to shortage cost will have insignificant impacts on the five objectives studied because we considered the feasible time and space issues for adjusting the demand change.

To summarize the analyses above, if the unit production cost increases, total profit will be reduced. In contrast, an increase of unit inventory cost will cause total profit to rise, although unit shortage cost has no impact on total profit. On the other hand, the increase of these three unit cost items will lead to a decrease of shortage cost and increase of partial inventory cost. This indicates that shortage cost will be reduced because inventory will be increased to meet demand. This result shows that the inventory cost and shortage cost seem to exert interactive behaviors whereas the enhancement of inventory cost reduces shortage cost.

11.4 CONCLUSIONS AND RECOMMENDATIONS

Currently, most Taiwan bicycle manufacturers are original equipment manufacturing (OEM) plants known for high quality. Enhancement of service level has been one of their goals for keeping their customers. In addition, both the manufacturing and service industries are gradually becoming customer-service oriented because stable customer service helps enterprise survival. As a result, this study uses the notion that production elasticity based on demand is more favorable than full production. This allocates demand space for adjustment to customer elasticity and increases the quality of customer service. Although such practice would lower enterprise profit because of the trade-off relationship between these two objectives, this study uses multi-objective programming to model this system.

This study uses five kinds of multi-objective programming methods for resolutions: multi-objective compromise programming, fuzzy multi-objective programming, weighted multi-objective programming, weighted multi-objective fuzzy programming, and two-phase fuzzy multi-objective programming. A comparison of these five kinds of programming reveals that the result from weighted multi-objective programming would be well received by decision makers.

The results of the study indicate that production cost is the highest (81%) of all the total cost items for bicycle production and that bicycle production is a labor-intensive and high material cost industry. Although bicycle manufacturing involves multinational production, distribution, and sales, its cost of time is relatively unimportant in comparison with the costs for high-tech industries.

Distribution is achieved by sea freight and transportation in terms of total cost is almost insignificant. Export and import taxes, however, represent the third-highest cost. Therefore, the factors exerting the greatest effects on production and distribution for the bicycle industry are material costs of the product and manufacturing costs of the plant, followed by export taxes imposed by the country where the plant is located and import taxes at the distribution site.

Total profit will decrease when unit production cost, unit inventory cost, and unit shortage cost are increased. An increase of inventory cost, however, will increase total profit, while shortage costs impact total profits insignificantly. If all three of these unit cost items are increased, the shortage cost will decrease and the partial inventory cost will increase, indicating that shortage quantities will decrease because of cost; thus, inventory quantity should be increased to meet demand. Results show that inventory cost and shortage cost interact and the result of an increase of inventory cost is reduced shortage cost.

This study offers several recommendations for subsequent research:

1. Although this study considered the service level of the manufacturer, assumptions were used for the function relationship between service level and production elasticity. Nevertheless, many other factors affect the service level of manufacturer. Subsequent studies could investigate them further.
2. Planning a model for study, it is necessary to consider the weight relationships among each of the objectives. The preferences of the decision maker are extremely important.
3. For practical management, transaction price and manufacturing cost are not linear. Transaction behavior allows discounts and differential pricing; manufacturing cost requires a factor such as economic scale to ensure the model is realistic. In this study, we investigated the problems of import and export, whose influences are widespread and constitute important issues in enterprise management.
4. The fuzzy multi-objective programming employed in this study is simply a fuzzy objective equation; fuzzy constraint equations can be added in subsequent studies.
5. This model considered only the issues of production and distribution because the upper stream part supply is already well known and was not included. Thus, considering relevant variables created by changes of upper stream manufacturing suppliers would make the model even more comprehensive.

APPENDIX A: MAX–MIN OPERATION

A.1 Determination of Ceiling and Bottom Limits of Objectives and Constraint Equations

If we assume that the decision maker has the most satisfactory and ideal ceiling limit value $f_i(\mathbf{x})$ and the bottom limit value $f_i^*(\mathbf{x})$ toward the ith objective $f_i^-(\mathbf{x})$, he or she can decide the values of the ceiling and bottom limit according to preference. Or the decision maker can take such objective as the function of the feasible solution space and determine the values through calculation of the pay-off table.

A.2 Establishment of Membership Function

The membership function toward the ith fuzzy objective is as follows:

$$\mu_{G_i}(\mathbf{x}) = \begin{Bmatrix} 0 & f_i(\mathbf{x}) \le f_i^-(\mathbf{x}) \\ \dfrac{f_i(\mathbf{x}) - f_i^-(\mathbf{x})}{f_i^*(\mathbf{x}) - f_i^-(\mathbf{x})} & f_i^-(\mathbf{x}) \le f_i(\mathbf{x}) \le f_i^*(\mathbf{x}) \\ 1 & f_i(\mathbf{x}) \ge f_i^*(\mathbf{x}) \end{Bmatrix}$$

as shown in Figure A.1.

A.3 Setting Membership Function for Decision Making Set $\mu_D(x)$

$$\mu_D(\boldsymbol{x}^*) = \min_{i,j}\{\mu_{G_i}(\boldsymbol{x}),\ i = 1,\ldots,k\}$$

From the max–min operation equation, the feasible fuzzy set can be found at the intersection of the objective and constraint equations. Since the decision maker needs precise recommendations, the maximum value of the membership in this decision making set is required. As a result, the maximum is utilized, and the corresponding membership function is thus obtained.

$$\begin{aligned} \mu_D(\mathbf{x}^*) &= \max\min\{\mu_{G_i}(\mathbf{x}^*),\ i = 1,\ldots,k\} \\ &\ge \max\min\{\mu_{G_i}(\mathbf{x}),\ i = 1,\ldots,k\} \end{aligned}$$

A.4 Turning Multiple Objectives into Single Objective for Resolution

Finally, this problem can be transformed into a precise LP problem for resolution:

$$\max \quad \lambda$$

$$\text{s.t.}$$

$$\left[\frac{f_i(x) - f_i^-(x)}{f_i^*(x) - f_i^-(x)}\right] \geq \lambda, \quad i = 1, \ldots, k$$

$$\boldsymbol{Ax} \leq \boldsymbol{b}$$

$$\boldsymbol{x} \geq \boldsymbol{0}$$

Thus, general linear programming can be used for resolution.

APPENDIX B: CONCEPT OF WEIGHTED FUZZY MULTI-OBJECTIVE PROGRAMMING

This method differs from weighted multi-objective programming for compromise solutions of practical problems. In general, the larger the gap for achieving a level, the greater the weight (relative importance). On the other hand, where the achievement level λ is higher, the weight (relative importance) is lower. We worked on a real empirical study to improve the environmental quality for metropolitan Taipei. We found that if the indicator of environmental quality was low (high), the residents of the area would be dissatisfied (satisfied) and the weight of the indicator would be low (high). See Equation (11.7). When we compared three cities—Taipei, Tokyo, and Seoul— the results were all the same.

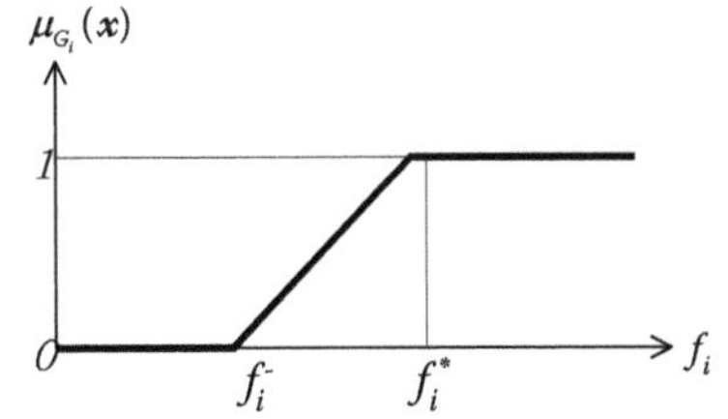

FIGURE A.1 Membership function of objective.

12 Fuzzy Interdependent Multi-Objective Programming

Since Bellman and Zadeh (1970) originally proposed the concepts of decision making in a fuzzy environment, much research has been proposed to guide study in the field of fuzzy multi-objective programming (FMOP). The first step of FMOP is to view objectives and constraints as fuzzy sets and characterize them by their individual membership functions. Then, a crisp (non-fuzzy) solution is generated by transforming FMOP into multi-objective programming (MOP) and determining the optimal solution to achieve the highest degree of satisfaction in the decision set. For further discussions, readers can refer to Zimmermann (1978); Verners (1987); Martinson (1993); and Lee and Li (1993). As with MOP, the problem of FMOP can be defined by calculating the following model:

$$\begin{aligned} &\max_{\tilde{x}\in X}/\min \quad \{f_1(\tilde{x}), f_2(\tilde{x}),\ldots, f_n(\tilde{x})\} \\ &s.t. \quad X = \{\tilde{x} \in X \mid g_k(\tilde{x}) \le 0, \quad k = 1,\ldots,m\}. \end{aligned} \tag{12.1}$$

Much effort has been directed to this problem, both in theory and in practice (Sakawa, 1993; Sakawa et al., 1995; Shibano et al., 1996; Shih et al., 1996; Ida and Gen, 1997; Shih and Lee, 1999). What seems lacking, however, is considering the problem of interdependence between objectives. As we know, the supportive and the conflicting objectives usually occur in realistic decision-making problems. From the view of optimization, because the optimal solution may be different while objectives are interdependent, the problem of interdependence between objectives in FMOP problems should not be overlooked.

Carlsson and Fullér (1994, 1996) proposed two methods to reshape the membership function to deal with the problem of FMOP with interdependence. Östermark (1997) extended their method to consider temporal interdependence between objectives. However, several shortcomings of their methods should be overcome before employing fuzzy interdependent multi-objective programming (FIMOP) in practice.

First, one method proposed by Carlsson and Fullér (1995) does not precisely measure the supportive or the conflicting grade between the objectives and can deal only with one-dimensional decision space. In contrast, a later method (Carlsson and Fullér (2002) can be employed only in linear fuzzy independent multi-objective programming (LFIMOP). Since real-life problems are usually complex, a general method should be devised for dealing with all kinds of FIMOP problems.

In this chapter, we propose another FIMOP to overcome these problems. First, a new index is developed to measure the interdependent grade between fuzzy objectives precisely. This method is suitable not only for the many-dimensional decision spaces but also for nonlinear FIMOP problems. A numerical example is used to demonstrate the method and compared with conventional FMOP. On the basis of the numerical results, we concluded that the proposed method can extend FMOP for considering the issue of interdependences between fuzzy objectives.

The problem of interdependence with objectives in multi-objective programming is discussed in Section 12.2. Fuzzy interdependent multi-objective programming is proposed in Section 12.3. In Section 12.4, we present a numerical example to demonstrate the proposed method and compare the results with the conventional FMOP model. Discussions are presented in Section 12.5 and conclusions are in the final section.

12.1 INTERDEPENDENCE WITH OBJECTIVES

The main problem of the conventional MOP model is the impractical assumption of independence of objectives. To demonstrate the impact, we can transform the MOP model into a single-objective programming (SOP) model by using the following compromise programming (Yu, 1985):

$$\begin{aligned} \min \quad & r(y;p) = \| \boldsymbol{y} - \boldsymbol{y}^* \|_p \\ s.t. \quad & \boldsymbol{X} = \{\boldsymbol{x} \in \boldsymbol{X} \mid g_k(\boldsymbol{x}) \le 0, \quad k = 1,\ldots,m\}, \end{aligned} \tag{12.2}$$

where $r(y;p)$ is a measurement of regret from y to $\boldsymbol{y}^*$ according to the l_p-norm, y denotes the objective vector, and $\boldsymbol{y}^*$ denotes the ideal point vector.

Assume a two-objective problem and let $p = \infty$. Consider a case depicted in Figure 12.1. The optimal solution should be $\boldsymbol{y}^{\infty}$ if and only if $f_1(\boldsymbol{x})$ is independent of $f_2(\boldsymbol{x})$. However, if $f_1(\boldsymbol{x})$ supports $\hat{f}_2(\boldsymbol{x})$, the optimal solution should transfer from $\boldsymbol{y}^{\infty}$ to $\hat{y}_{\infty}$. Therefore, if we want to extend MOP to interdependent multi-objective programming (IMOP), we should consider the interdependence between objectives,

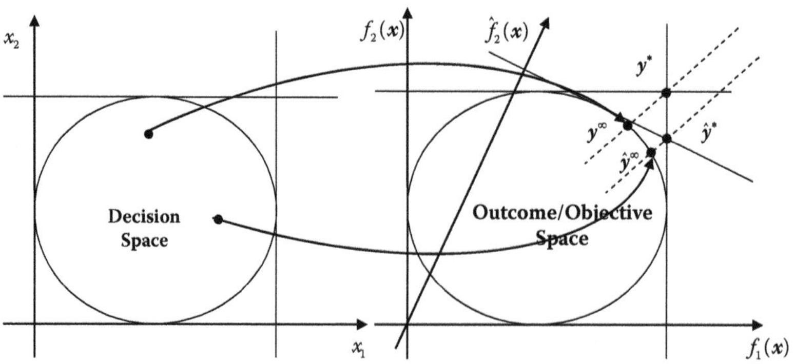

FIGURE 12.1 Optimal solution between independent and interdependent objectives.

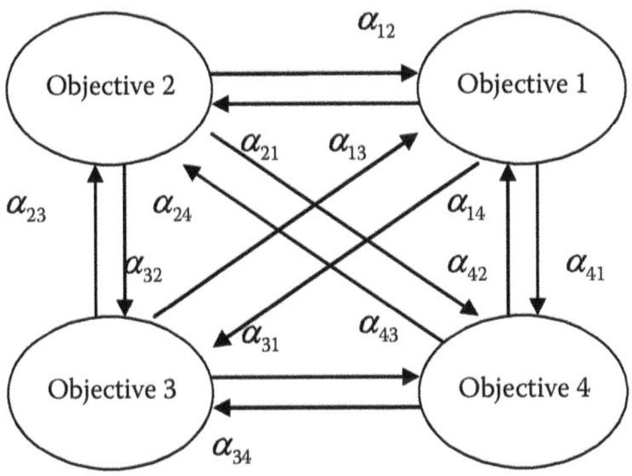

FIGURE 12.2 Interdependence with multiple objectives.

as shown in Figure 12.2. On the basis of the figure, we can see that the objectives in MOP should be modified as:

$$\hat{f}_i(\boldsymbol{x}) = f_i(\boldsymbol{x}) + \sum_{j=1, j\neq i}^{n} \alpha_{ij}[f_1(\boldsymbol{x}), f_2(\boldsymbol{x}), \ldots, f_n(\boldsymbol{x})], \quad \forall \quad 1 \leq i, j \leq n \tag{12.3}$$

where α_{ij} denotes the grade of interdependence from the jth objective to the ith objective. Based on Equation (12.3), MOP is a special case while $\alpha_{ij} = 0$ in IMOP. Therefore, the MOP problem should be reformulated as:

$$\begin{aligned} &\max/\min_{x \in X} \quad \{\hat{f}_1(\boldsymbol{x}), \hat{f}_2(\boldsymbol{x}), \ldots, \hat{f}_n(\boldsymbol{x})\} \\ &s.t. \quad \boldsymbol{X} = \{\boldsymbol{x} \in \boldsymbol{X} \mid g_k(\boldsymbol{x}) \leq 0, \qquad k = 1, \ldots, m\}. \end{aligned} \tag{12.4}$$

To solve this IMOP problem, the grade of interdependence should be derived first (Carlsson and Fullér, 1994, 1995, and 1996; Östermark, 1997). The grade of interdependence of the ith objective is defined using the following equation:

$$\Delta(f_i) = \sum_{j=1, j\neq i}^{n} \text{sign}(\alpha_{ji}), \tag{12.5}$$

where $\Delta(f_i)$ is the number of objectives supported by the ith objective minus the number of objectives hindered by the ith objective.

Using Equation (12.5), we can reformulate the conventional MOP model into the IMOP model. However, according to Equation (12.5), two main shortcomings should be overcome so that we can solve the IMOP problem more precisely. First, it can be seen that Equation (12.5) does not precisely reflect the supportive or conflicting degree of the objectives due to the sign operation. Furthermore, that method can deal only with one-dimensional decision space.

To overcome this problem, Carlsson and Fullér (1996) proposed another index to measure the interdependent grade between objectives in the many-dimensional decision space. Assume two linear objectives, $f_i(\boldsymbol{x}) = <\boldsymbol{c}_i, \boldsymbol{x}>$ and $f_j(\boldsymbol{x}) = <\boldsymbol{c}_j, \boldsymbol{x}>$ where $\|\boldsymbol{c}_i\|$, $\|\boldsymbol{c}_j\| = 1$, and the grade of interdependence between f_i and f_j can be defined by:

$$\alpha_{ij} = \alpha_{ji} = \frac{1 - <c_i, c_j>}{2}, \tag{12.6}$$

where $< >$ denotes the dot product and $\|\cdot\|$ denotes the length of the specific vector.

According to Equation (12.6), however, it can be seen that this index can deal only with the linear case. In addition, α_{ij} is not always equal to α_{ji} in real-world problems. To overcome these problems and extend the method to FIMOP, we propose another approach to measure the grade of interdependence between (fuzzy) objectives. In addition, two models are proposed to deal with the FIMOP problems in the next section.

12.2 FUZZY INTERDEPENDENT MULTI-OBJECTIVE PROGRAMMING

To propose the FIMOP model, we should first consider the crisp case so that the fuzzy case can be extended naturally. As mentioned above, the critical point of dealing with IMOP is to obtain the grade of interdependence between objectives so that we can reformulate the MOP model. Therefore, to overcome the problem of interdependence between objectives in MOP, we first define the notations. Let $f_i(\boldsymbol{x})$ and $f_j(\boldsymbol{x})$ be two objectives. We say that:

1. $f_i(\boldsymbol{x})$ supports $f_j(\boldsymbol{x})$ on $\boldsymbol{X}$ (denoted by $f_i(\boldsymbol{x}) \uparrow f_j(\boldsymbol{x})$) if $f_i(\boldsymbol{x}') \geq f_j(\boldsymbol{x})$ entails $f_j(\boldsymbol{x}') \geq f_i(\boldsymbol{x})$, for all $\boldsymbol{x}', \boldsymbol{x} \in \boldsymbol{X}$; $\partial f_j(\boldsymbol{x}) / \partial f_i(\boldsymbol{x}) > 0, \forall \boldsymbol{x} \in \boldsymbol{R}^n$;
2. $f_i(\boldsymbol{x})$ is in conflict with $f_j(\boldsymbol{x})$ on $\boldsymbol{X}$ (denoted by $f_i(\boldsymbol{x}) \downarrow f_j(\boldsymbol{x})$) if $f_i(\boldsymbol{x}') \geq f_j(\boldsymbol{x})$ entails $f_j(\boldsymbol{x}') \leq f_i(\boldsymbol{x})$, for all $\boldsymbol{x}', \boldsymbol{x} \in \boldsymbol{X}$; $\partial f_j(\boldsymbol{x}) / \partial f_i(\boldsymbol{x}) < 0, \forall \boldsymbol{x} \in \boldsymbol{R}^n$;
3. $f_i(\boldsymbol{x})$ is independent with $f_j(\boldsymbol{x})$ on $\boldsymbol{X}$ (denoted by $f_i(\boldsymbol{x}) \perp f_j(\boldsymbol{x})$), otherwise.

If $\boldsymbol{X} \in \boldsymbol{R}^n$, we say that $f_i(\boldsymbol{x})$ supports (or is in conflict with) $f_j(\boldsymbol{x})$ globally. If the objective functions are differentiable on $\boldsymbol{X}$, we may define:

1. $f_i(\boldsymbol{x}) \uparrow f_j(\boldsymbol{x})$ on $\boldsymbol{X} \Leftrightarrow \partial_{\boldsymbol{e}} f_i(\boldsymbol{x}) / \partial_{\boldsymbol{e}} f_j(\boldsymbol{x}) \geq 0, \forall \boldsymbol{e} \in \boldsymbol{R}^n, \boldsymbol{x} \in \boldsymbol{X}$,
2. $f_i(\boldsymbol{x}) \downarrow f_j(\boldsymbol{x})$ on $\boldsymbol{X} \Leftrightarrow \partial_{\boldsymbol{e}} f_i(\boldsymbol{x}) / \partial_{\boldsymbol{e}} f_j(\boldsymbol{x}) \leq 0, \forall \boldsymbol{e} \in \boldsymbol{R}^n, \boldsymbol{x} \in \boldsymbol{X}$,

where $\partial_{\boldsymbol{e}} f_i(\boldsymbol{x})$ denotes the derivative of $f_i(\boldsymbol{x})$ with respect to the direction $\boldsymbol{e} \in \boldsymbol{R}^n$ at $\boldsymbol{x} \in \boldsymbol{R}^n$. If for a given direction $\boldsymbol{e} \in \boldsymbol{R}^n$, $\partial_{\boldsymbol{e}} f_i(\boldsymbol{x}) \partial_{\boldsymbol{e}} f_j(\boldsymbol{x}) \geq 0$ holds for all $\boldsymbol{x} \in \boldsymbol{X}$, then we say that $f_i(\boldsymbol{x})$ supports $f_j(\boldsymbol{x})$ with respect to the direction e on $\boldsymbol{X}$.

Clearly, it is hard to calculate the interdependence between objectives because it may vary at different decision points. To calculate α_{ji}, a compromise is to ask the

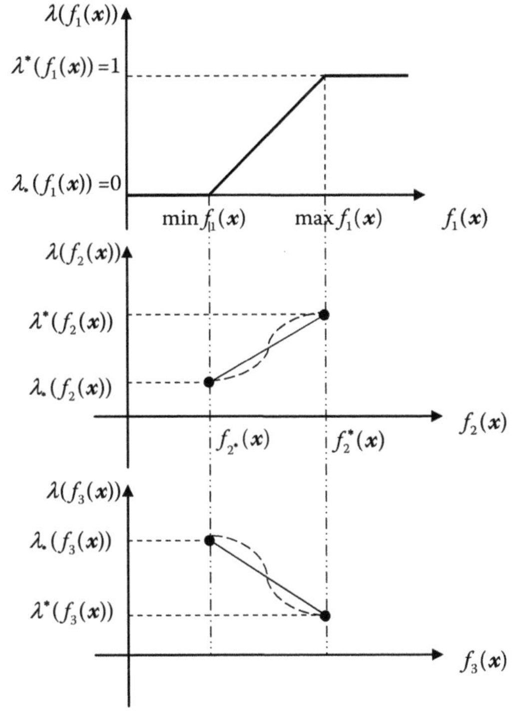

FIGURE 12.3 Concept of α_{ji}.

question: If the satisfaction level of $f_i(\boldsymbol{x})$ moves from worst to best, how does the average satisfaction level of $f_j(\boldsymbol{x})$ vary?

To explain the above concept, consider three objectives: ($f_1(\boldsymbol{x})$, $f_2(\boldsymbol{x})$, and $f_3(\boldsymbol{x})$). Figure 12.3 illustrates the concept. If we want to measure α_{21} and α_{31}, we first calculate the maximal and minimal values of $f_1(\boldsymbol{x})$ and the corresponding values of $f_2(\boldsymbol{x})$ and $f_3(\boldsymbol{x})$. Note that $f_{2^*}(\boldsymbol{x})$ ($f_{3^*}(\boldsymbol{x})$) and $f_2^*(\boldsymbol{x})$ ($f_3^*(\boldsymbol{x})$) in Figure 12.3 denote the corresponding value of $f_2(\boldsymbol{x})$ ($f_3(\boldsymbol{x})$) while maximizing and minimizing $f_1(\boldsymbol{x})$, respectively.

In addition, $\lambda^*(f_2(\boldsymbol{x}))$ ($\lambda_*(f_3(\boldsymbol{x}))$) and $\lambda_*(f_2(\boldsymbol{x}))$ ($\lambda^*(f_3(\boldsymbol{x}))$) are the satisfaction levels of $f_2(\boldsymbol{x})$ ($f_3(\boldsymbol{x})$) corresponding to $f_{2^*}(\boldsymbol{x})$ ($f_{3^*}(\boldsymbol{x})$), and $f_2^*(\boldsymbol{x})$ ($f_3^*(\boldsymbol{x})$), respectively. In theory, the satisfaction level from $f_{2^*}(\boldsymbol{x})$ ($f_{3^*}(\boldsymbol{x})$) to $f_2^*(\boldsymbol{x})$ ($f_3^*(\boldsymbol{x})$) may be nonlinear, such as the *S*-curve in our example, and result in a calculation problem.

Therefore, as long as the satisfaction level from $f_{2^*}(\boldsymbol{x})$ ($f_{3^*}(\boldsymbol{x})$) to $f_2^*(\boldsymbol{x})$ ($f_3^*(\boldsymbol{x})$) is close to an average value, we can calculate $\alpha_{21} \cong [\lambda^*(f_2(\boldsymbol{x})) - \lambda_*(f_2(\boldsymbol{x}))] / [\lambda^*(f_1(\boldsymbol{x})) - \lambda_*(f_1(\boldsymbol{x}))]$ and $\alpha_{31} \cong [\lambda^*(f_3(\boldsymbol{x})) - \lambda_*(f_3(\boldsymbol{x}))] / [\lambda^*(f_1(\boldsymbol{x})) - \lambda_*(f_1(\boldsymbol{x}))]$ for simplicity. Figure 12.3 shows that $\alpha_{21} > 0$ indicates $f_1(\boldsymbol{x})$ supports $f_2(\boldsymbol{x})$, and $\alpha_{31} < 0$ indicates $f_1(\boldsymbol{x})$ is in conflict with $f_3(\boldsymbol{x})$.

Next, we provide two examples to show the corresponding conditions and explain why the proposed method is justified.

Example 12.1 Consider the following two-objective mathematical programming:

$$\max\{f_1(\boldsymbol{x}), f_2(\boldsymbol{x})\},$$

$$\text{where} \quad f_1(\boldsymbol{x}) = 2x_1^2 - 3x_2^2 \quad \text{and} \quad f_2(\boldsymbol{x}) = -3x_1 + x_2^2.$$

$$s.t. \quad x_1 + x_2 \le 20,$$

$$x_1, x_2 \ge 0.$$

Then, we use Monte Carlo simulation and generate 5,000 samples to derive the feasible solution of the problem, as shown in Figure 12.3. We can see from the figure that Objective 1 and Objective 2 are in conflict with each other and the block line of the feasible solution indicates the efficient set that represents the grade of interdependence between the objectives. Since the interdependent grade is a non-linear curve, we can estimate the approximate interdependent grade between the objectives by a straight line using the proposed method (dashed line in Figure 12.4).

Example 12.2 Consider the following two-objective mathematical programming:

$$\max\{f_1(\boldsymbol{x}), f_2(\boldsymbol{x})\},$$

$$\text{where } f_1(\boldsymbol{x}) = x_1^2 + 3x_2^2 \quad \text{and} \quad f_2(x) = 2x_1 + 3x_1x_2.$$

$$s.t. \quad 0 \le x_1, x_2 \le 2.$$

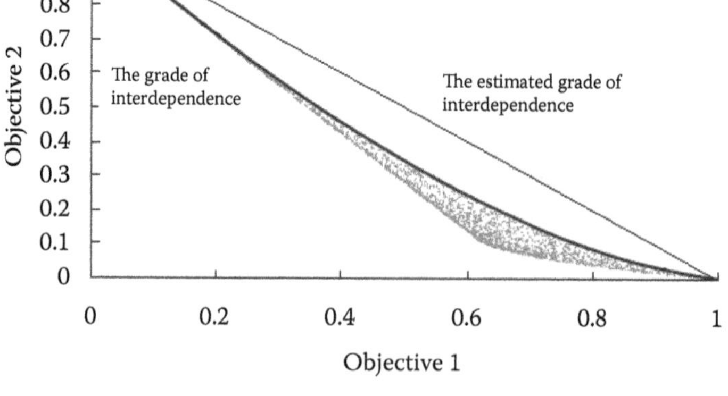

FIGURE 12.4 Interdependent grade between objectives.

Using a procedure like Example 12.1, we devised Figure 12.5 to represent the interdependent relationship of Objectives 1 and 2. Based on the figure, the objectives are supportive of each other.

According to Figures 12.4 and 12.5, if the interdependent grade function is monotonic and slightly non-linear, the proposed method (comparing the slope of the interdependent grade) is acceptable. In addition, since Carlsson and Fullẽr's method cannot deal with the above problems, the proposed method provides a satisfactory solution.

According to these concepts, the procedures to calculate the grade of interdependence between objectives can be developed. First, we employ SOP to obtain the maximal and minimal values of each objective and the corresponding values of other objectives. For example, to derive the maximal and minimal values of the jth objective and the corresponding values of other objectives, we solve:

$$\begin{aligned} &\max(\min) \quad f_j(\boldsymbol{x}) \\ &s.t. \quad \boldsymbol{X} = \{\boldsymbol{x} \in \boldsymbol{X} \mid g_k(\boldsymbol{x}) \leq 0, \quad k = 1,\ldots,m\}, \end{aligned} \tag{12.7}$$

then calculate

$$f_i^*(\boldsymbol{x}) \ (f_{i*}(\boldsymbol{x})), \quad \forall \quad i \neq j; \quad 1 \leq i, j \leq n,$$

where $f_i^*(\boldsymbol{x})$ denotes the corresponding value while maximizing $f_j(\boldsymbol{x})$ and $f_{i*}(\boldsymbol{x})$ is the corresponding value while minimizing $f_j(\boldsymbol{x})$. Note that if multiple optimal solutions exist, two methods can be considered. First, the optimal solution can always be obtained by incorporating the preference of the decision maker, such as the "more

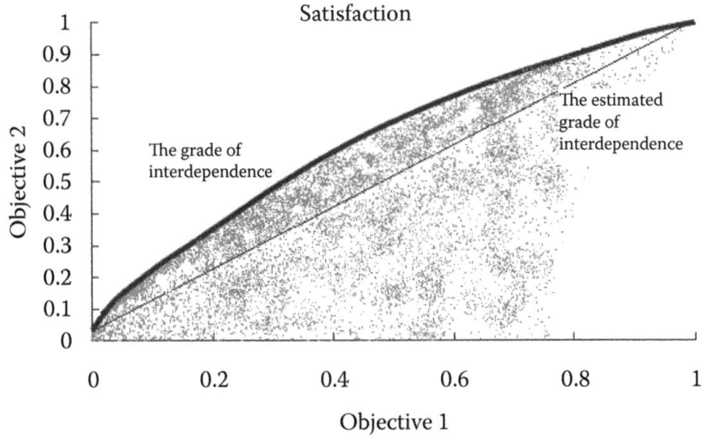

FIGURE 12.5 Interdependent grade between objectives.

average, the more preferred." Therefore, $f_i^*(\boldsymbol{x})$ and $f_{i*}(\boldsymbol{x})$ can be calculated. Second, if r multiple optimal solutions exist for both maximizing and minimizing $f_j(\boldsymbol{x})$, we can determine $f_i^*(\boldsymbol{x})$ and $f_{i*}(\boldsymbol{x})$ using:

$$f_i^*(\boldsymbol{x}) = \max\{f_{i1}^*(\boldsymbol{x}), f_{i2}^*(\boldsymbol{x}), \ldots, f_{ir}^*(\boldsymbol{x})\}; \tag{12.8}$$

$$f_{i*}(\boldsymbol{x}) = \min\{f_{i1*}(\boldsymbol{x}), f_{i2*}(\boldsymbol{x}), \ldots, f_{ir*}(\boldsymbol{x})\}, \tag{12.9}$$

where $f_{ik}^*(\boldsymbol{x})$ denotes the kth corresponding value while maximizing $f_j(\boldsymbol{x})$ has r multiple optimal solutions, and $f_{ik*}(\boldsymbol{x})$ denotes the kth corresponding value while minimizing $f_j(\boldsymbol{x})$ has r multiple optimal solutions. We can then calculate the approximate grade of interdependence between the objectives using the following equation:

$$\begin{aligned}\alpha_{ij} &\equiv \partial f_i(\boldsymbol{x}) / \partial f_j(\boldsymbol{x}) \\ &\cong [\lambda^*(f_i(\boldsymbol{x})) - \lambda_*(f_i(\boldsymbol{x}))] / [\lambda^*(f_j(\boldsymbol{x})) - \lambda_*(f_j(\boldsymbol{x}))]\end{aligned} \tag{12.10}$$

where $f_i^*(\boldsymbol{x})$ denotes the corresponding value while maximizing $f_j(\boldsymbol{x})$, $f_{i*}(\boldsymbol{x})$ is the corresponding value while minimizing $f_j(\boldsymbol{x})$, and $\lambda(\cdot)$ denotes the satisfaction level and can be calculated by:

$$\lambda(f_j(\boldsymbol{x})) = \begin{cases} 1 & \text{if } f_j(\boldsymbol{x}) \geq M_j, \\ \dfrac{f_j(\boldsymbol{x}) - m_j}{M_j - m_j} & \text{if } m_j < f_j(\boldsymbol{x}) < M_j, \\ 0 & \text{if } f_j(\boldsymbol{x}) \leq m_j, \end{cases} \tag{12.11}$$

where M_j and m_j are the maximum and the minimum values of the corresponding objective, respectively.

Next, on the basis of Equations (12.7) through (12.11), we can extend the concepts above to consider the following fuzzy situation. First, with the specific α-cut, the maximum and minimum values of the jth fuzzy objective and corresponding values of the ith fuzzy objectives can be derived:

$$\begin{aligned}&\max(\min) \quad f_j(\tilde{\boldsymbol{x}}, \alpha) \\ &s.t. \quad \boldsymbol{X} = \{\boldsymbol{x} \in \boldsymbol{X} \mid g_k(\tilde{\boldsymbol{x}}, \alpha) \leq 0, \qquad k = 1, \ldots, m\},\end{aligned} \tag{12.12}$$

then calculate

$$f_i^*(\tilde{\boldsymbol{x}}, \alpha) \ (f_{i*}(\tilde{\boldsymbol{x}}, \alpha)), \ \forall \ i \neq j; \ 1 \leq i, j \leq n,$$

where $f_i^*(\tilde{\boldsymbol{x}}, \alpha)$ denotes the corresponding value while maximizing $f_j(\tilde{\boldsymbol{x}}, \alpha)$ and $f_{i*}(\tilde{\boldsymbol{x}}, \alpha)$ is the corresponding value while minimizing $f_j(\tilde{\boldsymbol{x}}, \alpha)$. Then the (approximate) grade of interdependence between the fuzzy objectives can be derived as:

$$\alpha_{ij} = [H(f_i^*(\tilde{x},\alpha)) - H(f_{i*}(\tilde{x},\alpha))] / [H(f_j^*(\tilde{x},\alpha)) - H(f_{j*}(\tilde{x},\alpha))] \qquad (12.13)$$

where $f_i^*(\tilde{x},\alpha)$ denotes the corresponding value while maximizing $f_j(\tilde{x},\alpha)$, $f_{i*}(\tilde{x},\alpha)$ is the corresponding value while minimizing $f_j(\tilde{x},\alpha)$, and $H(\cdot)$ denotes the satisfaction level under the fuzzy situation that can be calculated by:

$$H(f_i(\tilde{x},\alpha)) = \begin{cases} 1 & \text{if} \quad f(\tilde{x},\alpha) \geq M_i, \\ \dfrac{M_i - f_i(\tilde{x},\alpha)}{M_i - m_i} & \text{if} \quad m_i < f(\tilde{x},\alpha) < M_i, \\ 0 & \text{if} \quad f(\tilde{x},\alpha) \leq m_i. \end{cases} \qquad (12.14)$$

Then the FIMOP problem can be considered to deal with the following FMOP problem:

$$\begin{aligned} &\max_{\tilde{x}\in X}/\min \quad \{\hat{f}_1(\tilde{x},\alpha), \hat{f}_2(\tilde{x},\alpha), \ldots, \hat{f}_n(\tilde{x},\alpha)\} \\ &s.t. \quad X = \{x \in X \mid g_k(\tilde{x},\alpha) \leq 0, \quad k = 1,\ldots,m\} \end{aligned} \qquad (12.15)$$

where

$$\hat{f}_i(\tilde{x},\alpha) = f_i(\tilde{x},\alpha) + \sum_{j=1, j\neq i}^{n} \alpha_{ij}[f_1(\tilde{x},\alpha), f_2(\tilde{x},\alpha), \ldots, f_n(\tilde{x},\alpha)], \quad \forall \quad 1 \leq i, j \leq n \qquad (12.16)$$

Now, two approaches are proposed to deal with the FIMOP problem as follows. The first approach is to convert the grade of interdependence into the degree of satisfaction. Since the degree of satisfaction of the specific fuzzy objective can be measured by Equation (12.12), the degree of satisfaction of the ith modified fuzzy objective (incorporating the grade of interdependence caused by other fuzzy objectives) can be defined by $H(f_i(\tilde{x},\alpha)) + \Sigma_{j=1, j\neq i}^{n} \alpha_{ij}[H(f_j(\tilde{x},\alpha))]$. Then the interdependent relationships between the ith fuzzy objective and other fuzzy objectives can be defined by:

$$\begin{cases} H(f_i(\tilde{x},\alpha)) + \sum_{j=1,j\neq i}^{n} \alpha_{ij}[H(f_j(\tilde{x},\alpha))] > H(f_i(\tilde{x},\alpha)), \text{if } f_i(\tilde{x},\alpha) \text{ is supported by other fuzzy objectives;} \\ H(f_i(\tilde{x},\alpha)) + \sum_{j=1,j\neq i}^{n} \alpha_{ij}[H(f_j(\tilde{x},\alpha))] < H(f_i(\tilde{x},\alpha)), \text{if } f_i(\tilde{x},\alpha) \text{ is hindered by other fuzzy objectives;} \\ H(f_i(\tilde{x},\alpha)) + \sum_{j=1,j\neq i}^{n} \alpha_{ij}[H(f_j(\tilde{x},\alpha))] = H(f_i(\tilde{x},\alpha)), \text{if } f_i(\tilde{x},\alpha) \text{ is independent with other fuzzy objectives.} \end{cases}$$

(12.17)

The corresponding concepts above can be depicted as shown in Figure 12.6.

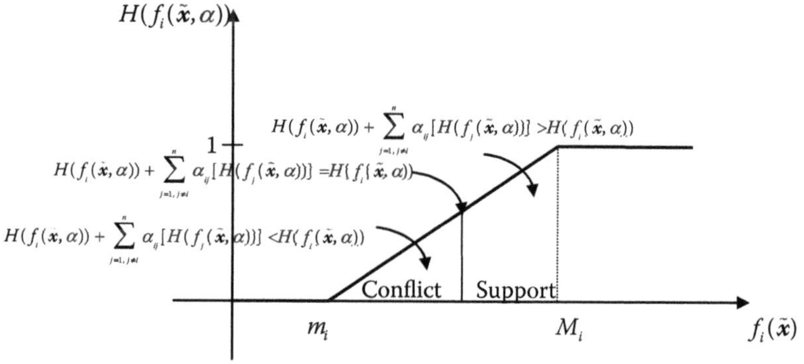

FIGURE 12.6 Degree of satisfaction by incorporating grade of interdependence.

The second approach adopts the ideas of Carlsson and Fullér (1994, 1995, and 1996) to reshape the degree of satisfaction by incorporating the grade of interdependence. The grade of interdependence of the ith fuzzy objective can be defined as $\tilde{\Delta}(f_i) = \Sigma_{j=1,j\neq i}^{n} \alpha_{ji}$. Then, the degree of satisfaction can be reformulated as:

$$\text{For } \tilde{\Delta}(f_i) \geq 0, \quad G(\tilde{x}_i, \tilde{\Delta}(f_i), \alpha) = \begin{cases} 1 & \text{if} \quad f_i(\tilde{x}, \alpha) \geq M_i \\ \left(\dfrac{f_i(\tilde{x}, \alpha) - m_i}{M_i - m_i}\right)^{1/(\tilde{\Delta}(f_i)+1)} & \text{if} \quad m_i < f_i(\tilde{x}, \alpha) < M_i \\ 0 & \text{if} \quad f_i(\tilde{x}, \alpha) \leq m_i \end{cases} \tag{12.18}$$

$$\text{For } \tilde{\Delta}(f_i) < 0, \quad G(\tilde{x}_i, \tilde{\Delta}(f_i), \alpha) = \begin{cases} 1 & \text{if} \quad f_i(\tilde{x}, \alpha) \geq M_i \\ \left(\dfrac{f_i(\tilde{x}, \alpha) - m_i}{M_i - m_i}\right)^{|\tilde{\Delta}(f_i)|+1} & \text{if} \quad m_i < f_i(\tilde{x}, \alpha) < M_i \\ 0 & \text{if} \quad f_i(\tilde{x}, \alpha) \leq m_i \end{cases} \tag{12.19}$$

The concepts above for reshaping the degree of satisfaction can also be depicted as shown in Figure 12.7. Now, with the specific α-cut, we can incorporate these two approaches to extend the conventional FMOP model to the following two FIMOP models as:

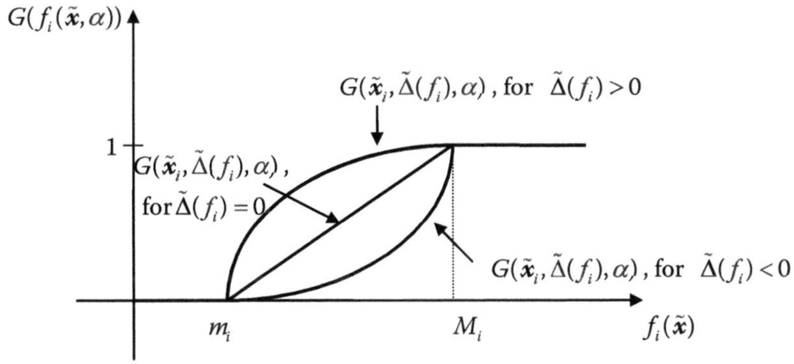

FIGURE 12.7 Reshaping degree of satisfaction.

Model 1

$$\begin{aligned} &\max \quad \nu \\ &s.t. \quad \boldsymbol{X}=\{\boldsymbol{x}\in \boldsymbol{X} \mid g_k(\tilde{\boldsymbol{x}},\alpha)\le 0, \quad k=1,\ldots,m\}; \\ &\qquad H(f_i(\tilde{\boldsymbol{x}},\alpha))+\sum_{j=1,j\ne i}^{n}\alpha_{ij}[H(f_j(\tilde{\boldsymbol{x}},\alpha))]\ge \nu, \quad \forall \quad 1\le i,j\le n. \end{aligned} \tag{12.20}$$

Model 2

$$\begin{aligned} &\max \quad \nu \\ &s.t. \quad \boldsymbol{X}=\{\boldsymbol{x}\in \boldsymbol{X} \mid g_k(\tilde{\boldsymbol{x}},\alpha)\le 0, \quad k=1,\ldots,m\}; \\ &\qquad G(\tilde{\boldsymbol{x}}_i,\tilde{\Delta}(f_i),\alpha)\ge \nu, \quad \forall \quad 1\le i,\ j\le n. \end{aligned} \tag{12.21}$$

Note that the main difference between Models 1 and 2 is that Model 1 changes the degree of satisfaction and Model 2 changes the shape of the satisfaction function. Both models, however, can reflect the interdependence between fuzzy objectives in FMOP. Next, we give a numerical example to demonstrate the proposed method and then discuss the results in Section 12.4.

12.3 NUMERICAL EXAMPLE

In this section, we use a numerical example to demonstrate the proposed method to deal with the FIMOP problem. Note that the fuzzy sets used in our paper are normal, convex, continuous, and bounded. In addition, only triangular fuzzy numbers are used. The membership function of each fuzzy number can be defined by:

$$\mu_A(x) = \begin{cases} 0 \ , & x \le a, \\ \dfrac{x-a}{b-a}, & a \le x \le b, \\ \dfrac{c-x}{c-b}, & b \le x \le c, \\ 0 \ , & c \le x, \end{cases}$$

where $\tilde{A} = (a,b,c)$ denotes the triangular fuzzy number and a, b, and c are the lower, center, and upper values, respectively. In addition, to concentrate on the problem of interdependence between fuzzy objectives, only crisp constraints are utilized. Consider a fuzzy three-objective programming problem as follows:

$$\begin{aligned} f_1(\tilde{x}) &= (4,6,7)x_1x_2 + (3,6,8)x_2 - (6,7,8)x_3, \\ f_2(\tilde{x}) &= -(1,3,5)x_1^2 + (9,10,12)x_2^{1/2} + (7,8,10)x_3, \\ f_3(\tilde{x}) &= -(2,3,4)x_1 - (1,2,5)x_2 - (3,5,7)x_3^{1/2}. \\ s.t. \quad & x_1^2 + x_2^2 + x_3^2 \le 100, \\ & x_1 + x_2 + x_3 \ge 10, \\ & x_1, x_2, x_3 \ge 0. \end{aligned}$$

In order to consider the problem of interdependence between fuzzy objectives, we should first derive the maximum and minimum values of the ith fuzzy objective and the corresponding values of other fuzzy objectives that are subject to constraints, as shown in Equation (12.12). While many fuzzy mathematical programming methods can be used to deal with this fuzzy single-objective optimization problem, we adopt Carlsson and Korhonen's method (1986) because of its simplicity and generality. The procedures of their method are explained in the appendix at the end of this chapter.

The membership function of each fuzzy variable in this chapter relates to Equations (A.3) through (A.5) in the appendix. According to Equation (A.1), we can derive the grade of interdependence between fuzzy objectives and the special case with α-cut = 0 and μ = 0.5 and 1 as shown in Tables 12.1 and 12.2. Note that other situations can also be considered with different α-cut and μ values using the same procedures.

Next, by employing Model 1 (M1) and Model 2 (M2) as expressed in Equations (12.20) and (12.21), respectively, we can derive the optimal solutions of FIMOP with μ = 0.5 and 1 as shown in Table 12.3. A comparison of FMOP and FIMOP appears in Table 12.3.

TABLE 12.1
Grade of Interdependence between Fuzzy Objectives (α-cut = 0, μ = 0.5)

$\max f_1(x) = 314.56$	$f_2(x) = -133.61$	$f_3(x) = -42.40$	$\alpha_{21} = -0.49$	$\alpha_{31} = -0.80$
$\min f_1(x) = -50.00$	$f_2(x) = 85.00$	$f_3(x) = -15.81$		
$\max f_2(x) = 99.31$	$f_1(x) = -29.09$	$f_3(x) = -25.02$	$\alpha_{12} = -0.08$	$\alpha_{32} = 0.15$
$\min f_2(x) = -350.00$	$f_1(x) = 0.00$	$f_3(x) = -30.00$		
$\max f_3(x) = -15.81$	$f_1(x) = -50.00$	$f_2(x) = 85.00$	$\alpha_{13} = -0.87$	$\alpha_{23} = 0.42$
$\min f_3(x) = -49.14$	$f_1(x) = 268.86$	$f_2(x) = -103.65$		

TABLE 12.2
Grade of Interdependence between Fuzzy Objectives (α-cut = 0, μ = 1)

$\max f_1(x) = 221.49$	$f_2(x) = -70.60$	$f_3(x) = -21.02$	$\alpha_{21} = -0.50$	$\alpha_{31} = -0.68$
$\min f_1(x) = -40.00$	$f_2(x) = 70.00$	$f_3(x) = -9.49$		
$\max f_2(x) = 82.43$	$f_1(x) = -27.74$	$f_3(x) = -12.54$	$\alpha_{12} = -0.11$	$\alpha_{32} = 0.44$
$\min f_2(x) = -200$	$f_1(x) = 0.00$	$f_3(x) = -20.00$		
$\max f_3(x) = -9.49$	$f_1(x) = -40.00$	$f_2(x) = 70.00$	$\alpha_{13} = -0.69$	$\alpha_{23} = 0.60$
$\min f_3(x) = -26.56$	$f_1(x) = 140.32$	$f_2(x) = -99.02$		

TABLE 12.3
Comparison of FMOP and FIMOP

Method	x_1	x_2	x_3	$f_1(\tilde{x})$	$f_2(\tilde{x})$	$f_3(\tilde{x})$
FMOP (μ = 0.5)	5.630	4.370	0.000	159.35	–88.99	–30
M1: FIMOP (μ = 0.5)	3.658	6.926	0.000	177.44	–19.20	–31.75
M2: FIMOP (μ = 0.5)	4.851	5.851	0.000	188.29	–56.96	–32.16
FMOP (μ = 1)	3.383	8.265	0.000	136.637	2.985	–15.031
M1: FIMOP (μ = 1)	3.825	9.240	0.000	169.092	–1.904	–16.890
M2: FIMOP (μ = 1)	3.739	8.978	0.000	161.209	–0.993	–16.465

On the basis of Table 12.3, we can conclude that the optimal decision variables and objectives of FIMOP are obviously different from those of the FMOP model. Furthermore, the significant differences between FIMOP and FMOP result from the interdependence between the objectives. Next, we present further discussions based on this numerical example to clarify the rationality of FIMOP.

12.4 DISCUSSION

FMOP approaches have been widely applied in various areas to deal with multi-objective decision-making problems. However, the problem of interdependence between objectives has not been considered in FMOP for effectively dealing with real-life problems. If we ignore the problem of interdependence between objectives, the result may differ greatly from the optimal solution. Many algorithms have been proposed to consider optimization in MOP and FMOP, for example, the evolutionary algorithms of Coello et al. (2002) and Deb (2001). We should not ignore the impact of interdependence between objectives.

Although Carlsson and Fullér discussed the problem of MOP with interdependence with objectives and proposed two methods to deal with the problems, their method can deal only with simple FIMOP problems including the one-dimensional decision space and linear cases, and cannot measure the grade of interdependence between objectives precisely in more realistic problems.

In this chapter, we propose two models for dealing with the above problems so that the proposed method may be suitable for both many-dimensional decision spaces and nonlinear cases. Furthermore, according to the results of the numerical example, we can see that although the two models yield different results, the directions of the decision variables and objectives are the same. The results also indicate the consistency of the two models.

The differences between Model 1 and Model 2 can be described. In Model 1, the interdependent grade between objectives is reflected by modifying the degree of satisfaction. Conversely, Model 2 incorporates the information of the interdependent grade between objectives to alter the shape of satisfaction function. Both models, however, can reflect the interdependence between fuzzy objectives in FMOP. In addition, from the view of numerical results, Model 1 reflects the influence of the interdependent grade between objectives more sharply than Model 2. Therefore, a decision maker can select one of the models according to his or her assessment of the interdependent grade between objectives.

The proposed method has a number of advantages. First, instead of being used only in one-dimensional decision spaces, the proposed method can also be used in many-dimensional decision spaces. Second, the grade of interdependence between the fuzzy objectives can be measured precisely. Finally, the proposed method can deal with nonlinear cases. However, it should be noted that the proposed index measures only the average grade of interdependence between the fuzzy objectives since it should vary based on the positions of decision variables. Therefore, a more accurate index to measure the grade of interdependence between objectives should be explored in future research.

12.5 CONCLUSIONS

This chapter considers the problem of FMOP with interdependent objectives. A novel index is proposed to measure the precise fuzzy grade of interdependence between fuzzy objectives, and two models are developed to solve FIMOP problems. The numerical results show that the problem of interdependence between (fuzzy) objectives is significant and the proposed method can provide a satisfactory solution for optimizing FIMOP solutions.

APPENDIX

According to Carlsson and Korhonen's method (1986), a fuzzy single-objective programming model can be defined by:

$$\begin{aligned} \max \quad & f(\boldsymbol{x}) = \tilde{\boldsymbol{c}}'\boldsymbol{x}, \\ s.t. \quad & \boldsymbol{X} = \{\boldsymbol{x} \in \boldsymbol{X} \mid \tilde{\boldsymbol{A}}\boldsymbol{x} \leq \tilde{\boldsymbol{b}}\}. \end{aligned} \tag{A.1}$$

By setting the specific α-cut and membership function of each fuzzy variable, the problem above can be reformulated as the following crisp mathematical programming model:

$$\begin{aligned} \max \quad & \sum_{j=1}^{n} \mu_{\tilde{c}_j}^{-1}(\alpha) x_j, \\ s.t. \quad & \sum_{j=1}^{n} \mu_{\tilde{a}_{ij}}^{-1}(\alpha) x_j \leq \mu_{\tilde{b}_i}^{-1}(\alpha), \quad \forall i = 1,\ldots,m;\ j = 1,\ldots,n, \end{aligned} \tag{A.2}$$

where α denotes the α-cut operation and μ^{-1} denotes the inverse membership function. The membership function of each fuzzy number can be defined by:

$$\mu_{a_{ij}} = (a_{ij} - a_{ij}^{u}) / (a_{ij}^{l} - a_{ij}^{u}); \tag{A.3}$$

$$\mu_{b_i} = (b_i - b_i^{u}) / (b_i^{l} - b_i^{u}); \tag{A.4}$$

$$\mu_{c_j} = (c_j - c_j^{u}) / (c_j^{l} - c_j^{u}) \tag{A.5}$$

where $a_{ij}^{u}, b_i^{u}, c_j^{u}, a_{ij}^{l}, b_i^{l}$, and c_j^{l} denote the upper and the lower values corresponding to the fuzzy numbers $\tilde{a}_{ij}, \tilde{b}_i$, and, $\tilde{c}_j$, respectively. Note that the membership function should be monotonic in order to avoid the problem that the inverse function may be unavailable. Other membership functions can also be employed using the same concepts.

13 Novel Algorithm for Uncertain Portfolio Selection

The mean-variance approach was proposed by Markowitz (1952) to deal with the portfolio selection problem. A decision maker can determine the optimal investing ratio for each security based on the historical return rate. The formulation of the mean-variance method can be described as follows (Markowitz, 1952, 1959, and 1987):

$$
\begin{aligned}
\min \quad & \sum_{i=1}^{n}\sum_{j=1}^{n}\sigma_{ij}x_i x_j \\
s.t. \quad & \sum_{i=1}^{n}\mu_i x_i \geq E, \\
& \sum_{i=1}^{n} x_i = 1, \\
& x_i \geq 0 \qquad \forall i = 1,\ldots,n.
\end{aligned}
\tag{13.1}
$$

where μ_i denotes the expected return rate of the ith security, σ_{ij} denotes the covariance coefficient between the ith security and the jth security, and E denotes the acceptable least rate of the expected return.

It is clear that the accuracy of the mean-variance approach depends on the accurate values of the expected return rate and the covariance matrix. Several methods have been proposed to forecast the appropriate acceptable return rate and variance matrix such as the arithmetic mean method (Markowitz, 1952, 1959, and 1987) and the regression-based method (Elton and Gruber, 1995). However, these methods derive only the precise expected return rate and covariance matrix and do not consider the problem of uncertainty.

Since the decision maker wants to determine the optimal portfolio strategy to gain maximum profits, how can we ignore future uncertainty? We should note that the possible area of the return rate and the covariance matrix should be derived to allow the decision maker to determine the future optimal portfolio selection strategy. In addition, these methods are based on the large sample theory and cannot provide satisfactory solutions in small sample situations (Elton et al., 1978).

In this chapter, the possible area of the return rate and the covariance matrix are derived using asymmetrical possibilistic regression. Then, the Mellin transformation is employed to calculate the uncertain return rate and the variance with specific distribution. Finally, the optimal portfolio selection model can be reformulated based on these concepts. In addition, a numerical example is used to illustrate the proposed method and compared with the conventional mean-variance method. On the basis of the simulated results, we can conclude that the proposed method can provide a better portfolio selection strategy than the conventional mean-variance method by considering uncertainty.

13.1 POSSIBILISTIC REGRESSION

The possibilistic regression model was first proposed by Tanaka and Guo (2001) to reflect the fuzzy relationship between the dependent and the independent variables. The upper and the lower regression boundaries are used in possibilistic regression to reflect the possibilistic distribution of the output values. By solving the linear programming (LP) problem, the coefficients of the possibilistic regression can be obtained easily.

Next, we describe the use of the possibilistic regression model (Tanaka and Guo, 2001) to obtain the uncertain return rate and the variance. To obtain accurate results, we extend the symmetrical fuzzy numbers to the asymmetrical fuzzy numbers. The general form of a possibilistic regression can be expressed as:

$$y = A_0 + A_1x_1 + \cdots + A_nx_n = \boldsymbol{A'x} \tag{13.2}$$

where A_i is an asymmetrical possibilistic regression coefficient denoted as $(a_i - c_{iL}, a_i, a_i + c_{iR})$. To achieve minimum uncertainty, the fitness function of the possibilistic regression can be defined as:

$$\min_{a,c} \quad J = \sum_{j=1,\ldots,m} (\boldsymbol{c}_L' \,|\boldsymbol{x}_j| + \boldsymbol{c}_R' \,|\boldsymbol{x}_j|) \tag{13.3}$$

In addition, the dependent variable should be restricted to satisfy the following two equations:

$$y_j \geq \boldsymbol{a}'\boldsymbol{x}_j - \boldsymbol{c}_L' \,|\boldsymbol{x}_j|, \tag{13.4}$$

$$y_j \leq \boldsymbol{a}'\boldsymbol{x}_j + \boldsymbol{c}_R' \,|\boldsymbol{x}_j|. \tag{13.5}$$

On the basis of the concepts above, we can obtain the formulation of a possibilistic regression model:

$$\begin{aligned} \min_{a,c} \quad & J = \sum_{j=1,\ldots,m} (\boldsymbol{c}_L' \,|\boldsymbol{x}_j| + \boldsymbol{c}_R' \,|\boldsymbol{x}_j|) \\ s.t. \quad & y_j \geq \boldsymbol{a}'\boldsymbol{x}_j - \boldsymbol{c}_L' \,|\boldsymbol{x}_j|, \\ & y_j \leq \boldsymbol{a}'\boldsymbol{x}_j + \boldsymbol{c}_R' \,|\boldsymbol{x}_j|, \qquad j = 1, \ldots, m \\ & \boldsymbol{c}_L, \boldsymbol{c}_R \geq \boldsymbol{0}. \end{aligned} \tag{13.6}$$

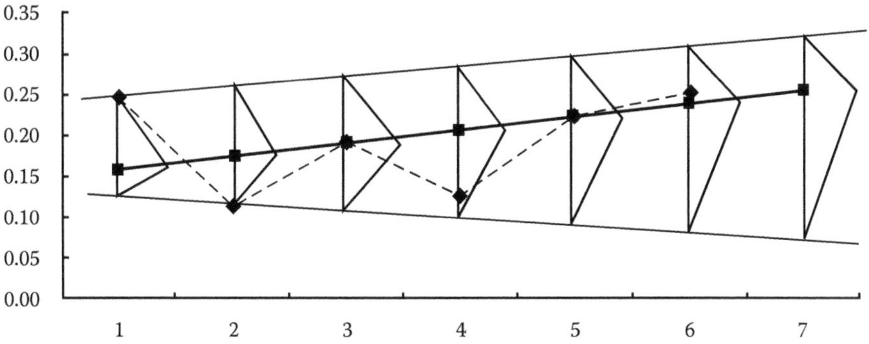

FIGURE 13.1 Possibilistic area of return rate and variance.

By solving the mathematical programming model above, we can obtain the uncertain return rate and the variance of the security with specific distribution in the future.

Next, we devise a graph (Figure 13.1) to describe the concept of the proposed method. Suppose we have six period return rates of stocks and we want to determine the optimal investing rate for each stock in period 7. Let the broken line denote the trend of the return rate of a stock. We can then perform the upper, lower, and center possibilistic regressions using Equation (13.6) to derive the possibilistic area of the return rate of period 7. Note that the triangular possibilistic distribution is used in this example. However, other possibilistic distributions can be employed using the same concepts.

We should highlight that the triangular area in period 7 denotes the distribution of the possible return rate and variance of the stock. That is, the decision maker should incorporate the information above to determine the optimal investing rate for each stock. However, since the possibilistic area may be a triangular, uniform, or other form of distribution, the problem is how to efficiently and effectively calculate the possible return rate and variance. We now describe the Mellin transformation for overcoming this problem.

13.2 MELLIN TRANSFORMATION

Given a random variable $x \in R^{+}$, the Mellin transformation $M(s)$ of a probability density function (pdf) designated ($f(x)$) can be defined as:

$$M\{f(x);s\} = M(s) = \int_0^{\infty} f(x)x^{s-1}\,dx \tag{13.7}$$

Let h be a measurable function on R into R and $Y = h(x)$ act as a transformed random variable. Some properties of the Mellin transformation can be described as shown in Table 13. 1. For example, if $Y = ax$, then the scaling property can be expressed as:

$$M\{f(ax);s\} = \int_0^{\infty} f(ax)x^{s-1}\,dx = a^{-s}\int_0^{\infty} f(ax)(ax)^{s-1} = a^{-s}M(s)$$

Next, let X represent a continuous non-negative random variable. The nth moment of X can be defined as:

$$E(X^n) = \int_0^\infty x^n f(x)\,dx \tag{13.8}$$

Then, by setting $n = 1$, the mean of X can be expressed as:

$$E(X) = \int_0^\infty x f(x)\,dx \tag{13.9}$$

and the variance of X can be calculated:

$$\sigma_x^2 = E(X^2) - [E(X)]^2 \tag{13.10}$$

Since the relationship between the nth moment and the Mellin transformation of X can be linked using the equation:

$$E(X^n) = \int_0^\infty x^{(n+1)-1} f(x)\,dx = M\{f(x); n+1\} \tag{13.11}$$

the mean and the variance of X can be calculated by:

$$E(X) = M\{f(x); 2\} \tag{13.12}$$

$$\sigma_x^2 = M\{f(x); 3\} - \{M\{f(x); 2\}\}^2 \tag{13.13}$$

From Equations (13.12) and (13.13), we can efficiently calculate the mean and variance of any distribution using the Mellin transformation. In practice, the uniform, triangular, and trapezoidal distributions are usually used and their corresponding

TABLE 13.1
Properties of Mellin Transformation

Property	$Y = h(x)$	$M(s)$
Scaling	ax	$a^{-s}M(s)$
Multiplication by x^a	$x^a f(x)$	$M(s+a)$
Rising to real power	$f(x^a)$	$a^{-1}M\left(\frac{s}{a}\right), a>0$
Inverse	$x^{-1}f(x^{-1})$	$M(1-s)$
Multiplication by $ln\ x$	$ln\ x f(x)$	$\frac{d}{ds}M(s)$
Derivative	$\frac{d^k}{ds^k}f(x)$	$\frac{\Gamma(s)}{\Gamma(s-k)}$

TABLE 13.2
Mellin Transformation of Three Probability Density Functions

Distribution	Parameter	M(s)
Uniform	*UNI(a,b)*	$\frac{b^s - a^s}{s(b-a)}$
Triangular	*TRI(l,m,u)*	$\frac{2}{(u-l)s(s+1)}\left[\frac{u(u^s - m^s)}{(u-m)} - \frac{l(m^s - l^s)}{(m-l)}\right]$
Trapezoidal	*TRA(a,b,c,d)*	$\frac{2}{(c+d-b-a)s(s+1)}\left[\frac{(d^{s+1} - c^{s+1})}{(d-c)} - \frac{(b^{s+1} - a^{s+1})}{(b-a)}\right]$

Mellin transformations can be summarized as shown in Table 13.2. More Mellin transformations for other probability density functions can be found in Yoon (1996).

On the basis of Table 13.2, we can efficiently derive the values of the mean and variance respect to the specific distribution by calculating $M(\mathbf{2})$ and $M(\mathbf{3})$. Next, we can reformulate the conventional mean-variance method as shown in the following mathematical programming model to consider the impact of uncertainty:

$$\min \sum_{i=1}^{n} x_i x_i \cdot [M_i(3) - M_i(2)^2] + \sum_{i=1}^{n}\sum_{j=1}^{n} x_i x_j \sigma_{ij} \tag{13.14}$$

$$s.t. \sum_{i=1}^{n} x_i M_i(2) \geq E,$$

$$\sum_{i=1}^{n} x_i = 1,$$

$$x_i \geq 0 \qquad \forall i = 1,\ldots,n.$$

The first part of the objective function denotes the next risk period of the security and the second part represents the unsystematic risk considered in the mean-variance model. Next, we use a numerical example to illustrate the proposed method and compare it with the conventional method.

13.3 NUMERICAL EXAMPLE

For simplicity, the possibilistic area of the return rate is represented as the triangular form in this example. Suppose the historical return rates of the five securities from periods *t*-6 to *t*-1 are represented as shown in Table 13.3. The corresponding time chart for the five securities is shown in Figure 13.2. Our concern here is to obtain the optimal portfolio selection strategy in the next period *t*.

First, we use the conventional mean-variance model to obtain the optimal portfolio selection strategy. To do this, the arithmetic mean and the covariance matrix can be calculated as shown in Tables 13.4 and 13.5.

TABLE 13.3
Historical Return Rates of Five Securities

Return Rate	t-6	t-5	t-4	t-3	t-2	t-1
Security 1	0.1686	0.1117	0.1149	0.1293	0.1397	0.1406
Security 2	0.1330	0.1466	0.1741	0.1131	0.1022	0.1552
Security 3	0.1698	0.1528	0.1302	0.1471	0.1139	0.1177
Security 4	0.1750	0.1026	0.1543	0.1475	0.1158	0.1148
Security 5	0.1291	0.1192	0.1491	0.1318	0.1377	0.1450

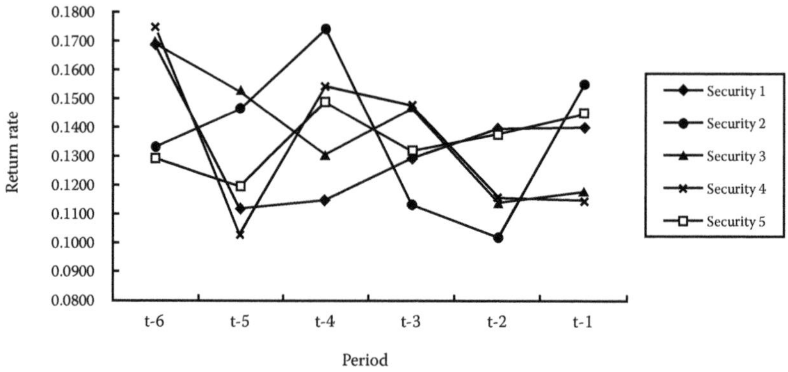

FIGURE 13.2 Time chart for five securities.

TABLE 13.4
Arithmetic Means of Expected Returns

Security	1	2	3	4	5	Average
Arithmetic mean	0.1341	0.1374	0.1386	0.1350	0.1353	0.136

TABLE 13.5
Covariance Matrix

	Security 1	Security 2	Security 3	Security 4	Security 5
Security 1	0.00036				
Security 2	–0.00017	0.00060			
Security 3	0.00010	0.00000	0.00039		
Security 4	0.00024	0.00004	0.00027	0.00066	
Security 5	0.00000	0.00009	–0.00014	0.00004	0.00010

By letting the acceptable least rate of the expected return equal its average return rate, we can obtain the optimal portfolio selection using Equation (13.1) as shown in Table 13.8.

Next, we use the proposed method to obtain the optimal portfolio selection. In order to obtain the possibilistic area of the five securities, possibilistic regression is employed. Then, using the Mellin transformation we can forecast the means and risks of the securities by considering the situation of uncertainty as shown in Table 13.6.

Furthermore, we incorporate the information of the forecasting mean to derive the second part of the objective function in Equation (13.14), i.e., the covariance matrix, as shown in Table 13.7.

Finally, with the same acceptable least rate of expected return we can obtain the optimal portfolio selection under the uncertain situation using Equation (13.14). The comparison of the conventional and proposed methods can be described as shown in Table 13.8.

Table 13.8 indicates that the main difference is the portfolio selection for Securities 1 and 4. In the next section, we will discuss the irrational reasoning using the conventional method in our numerical example.

13.4 DISCUSSION

The mean-variance method is widely used in the finance area to deal with the portfolio selection problem. However, the conventional method does not consider future uncertainty and usually fails n small sample situations. We now describe the shortcomings of the purpose and theory of conventional method.

The purpose of the mean-variance approach is to determine the t period optimal investing rate for a security based on the historical return rate. The key is to forecast the t period return rate as accurately as possible. However, it is clear that the arithmetic mean reflects only the average states of the past return rate instead of the future. Although many regression-based methods have been proposed to overcome the problem, they obey the assumptions of the large sample theory and cannot be used in theoretical small sample situations.

In addition, the regression-based methods cannot reflect degrees of uncertainty. Since we want to determine a future optimal portfolio selection, information about future uncertainties should not be ignored in models. In this chapter, we employ the possibilistic regression model to derive the possible mean and variance in the future. Then the Mellin transformation is used to determine the mean and the risk in the future by considering uncertain situations. Finally, we can use the information obtained to reformulate the mean-variance method to obtain optimal uncertain portfolio selection.

To highlight the shortcomings of the conventional method and compare it with the proposed method, we used a numerical example in Section 13.3. We can now prepare a time chart for Securities 1 and 4 to describe the irrational results using the conventional method as shown in Figure 13.3.

From the time chart, we can see for Security 1 an increase in period t-4 to t-1. It is rational to suppose Security 1 also has a positive return rate in period t. On the

TABLE 13.6
Possibilistic Area, Mean, and Variance

	Security 1	Security 2	Security 3	Security 4	Security 5
Possibilistic area	(0.1117, 0.1117, 0.1578)	(0.0868, 0.1407, 0.1741)	(0.0890, 0.0890, 0.1241)	(0.0646, 0.0646, 0.1294)	(0.1500, 0.1500, 0.1737)
Mean	0.1306	0.1339	0.1007	0.0862	0.1579
Variance	0.000118	0.000323	0.000068	0.000233	0.000031

TABLE 13.7
New Covariance Matrix

	Security 1	Security 2	Security 3	Security 4	Security 5
Security 1	0.00031				
Security 2	–0.00015	0.00052			
Security 3	0.00010	0.00001	0.00051		
Security 4	0.00023	0.00006	0.00046	0.00086	
Security 5	–0.00001	0.00007	–0.00022	–0.00010	0.00015

TABLE 13.8
Comparisons of Portfolio Selections

Portfolio Strategy	1	2	3	4	5	Return Rate	Portfolio Risk
Conventional	0.000	0.069	0.195	0.303	0.433	0.136	0.000056
Proposed	0.136	0.070	0.141	0.118	0.535	0.136	0.000073

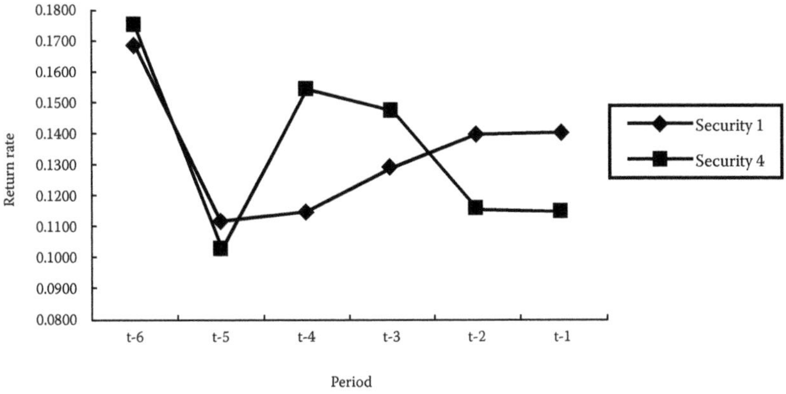

FIGURE 13.3 Time chart for Securities 1 and 4.

other hand, Security 4 shows a decrease since period t-4; optimal portfolio selection should eliminate the investing ratio for Security 4. On the basis of the numerical results, we can conclude that it is irrational to determine uncertain portfolio selection using the conventional method. However, the proposed method can accurately reflect the deductions above and can also provide more flexible portfolio alternatives. A decision maker can determine the optimal possibilistic distribution based on his or her domain knowledge or the empirical results to obtain an effective portfolio selection strategy.

13.5 CONCLUSIONS

Portfolio selection has been a continuing problem in the finance area since the 1950s. The conventional mean-variance method cannot provide satisfactory solutions involving uncertainties and small sample situations. In this chapter, the possibilistic regression model is employed to derive the possibilistic area of a future return rate.

The Mellin transformation is then used to obtain the mean and risk by considering uncertainty. Using this information, we propose a revised mean-variance model that incorporates a degree of uncertainty to deal with portfolio selection. A numerical example is used to demonstrate the proposed method. On the basis of the results, we can conclude that the proposed method can provide more flexible and accurate results than the conventional method under conditions of uncertainty in portfolio selection.

14 Multi-Objective Optimal Planning for Designing Relief Delivery Systems

Although minor earthquakes occur nearly every day, the effects of a strong earthquake are devastating. The Shaanxi earthquake, the deadliest in history, killed 830,000 in rural China in 1556. Recent fatal earthquakes took place in Taiwan in September 1999, India in January 2001, Southeastern Iran in December 2003, Sumatra in December 2004, and Pakistan in October 2005. Earthquakes have been some of humankind's major enemies in the battle against natural disasters.

The United Nations and the public and private sectors established many disaster-prevention and disaster-recovery agencies and programs. The difficulty with natural disasters like earthquakes is that even though thousands of networked seismograph stations are installed around the world and powerful computers continuously analyze collected data, we are still unable to predict when and where an earthquake will strike. Therefore, the most effective method to reduce the damage of a disaster is prevention through research, monitoring, dissemination of information, and education. Information coordination among related organizations is valuable and available but more is required.

After an earthquake occurs, effective disaster recovery efforts can reduce the death toll and damage and bring relief to surviving victims. These efforts include the establishment of a rescue command center; collection of information about the disaster area; identification of appropriate sites for shelters; determination of the best evacuation routes; arranging transportation for evacuation and delivery of relief materials; and installation of medical, fire prevention, and emergency construction facilities. This study focuses on fair and effective distribution of relief materials—making best efforts to ensure that required relief is distributed to all demand points.

Sato and Ichii (1996) investigated the efficiency of evacuations. Li et al. (1997) investigated crisis management procedures such as traffic control on highways. Tzeng and Chen (1999) conducted a study on scheduling programming for restoration construction and salvaging for road networks. Although these studies provide insights to various disaster recovery efforts, they make no mention of distribution of emergency relief.

This chapter will concentrate on the effectiveness and fairness of the overall distribution system to avoid the oversight of critical but difficult-to-reach areas of the real world. A fuzzy multiple objective model was used for this study and applied to a case study. Based on this case study, the corresponding measures needed for implementing the model can be put forward and allow additional scenario simulation. The model can be used to create local operational procedures and serve as part of a larger integrated relief distribution application. Finally, further study can be conducted to integrate this model into a comprehensive decision support system for disaster relief.

14.1 CHARACTERISTICS OF RELIEF DISTRIBUTION SYSTEMS

General physical distribution systems for businesses consider required materials, costs of materials, numbers and types of vehicles, modes of transportation, depots and locations, demands for materials, transportation networks, travel times, and other operational issues. One objective of a physical distribution system is finding a combination of those variables that minimizes travel time and size of required vehicle fleet, maximizes service capacity, and minimizes fixed and variable costs.

Similar to general physical distribution systems, relief distribution systems also involve three factors: demand, supply, and transportation. The collection points of commodities in non-devastated areas play the role of supply. The demand points are the devastated areas where relief is provided to victims who play the role of customers. Additionally, large-scale commodity distribution depots near demand points or devastated areas act as distribution centers. One difference between business and relief distribution systems is that relief systems provide temporary storage instead of permanent distribution warehouses maintained by businesses.

The two systems differ in one more characteristic. Instead of driving for business profits, the operators of disaster recovery operations are often government agencies or nonprofit organizations pursuing efficiency and fairness for disaster victims.

In the event of a disaster, decisions must be made in a very short time and are based on limited and often incomplete information. Since a relief distribution system may involve rapid changes of circumstances, an operator may have to take immediate emergency measures to minimize further damage and calm those affected through the issue of emergency orders, confiscation of civilian vehicles for emergency use, and closing of unsafe roads and structures. Table 14.1 presents a comparison of the features of relief distribution and regular distribution systems.

14.2 RELIEF DISTRIBUTION MODEL

A mathematical model of the disaster recovery distribution systems will be presented in this section.

14.2.1 Assumptions

A relief system should be based on the five assumptions listed below:

1. We consider only devastated areas that still are accessible through the current road network and disregard devastated areas that are completely isolated and would require helicopters or other extraordinary means of relief distribution.

TABLE 14.1
Comparison of General Distribution and Relief Distribution Systems

Comparison Items	General Distribution Systems	Relief Distribution Systems
System objectives	Maximize profit	Fairness and efficiency
Dimensional role	Factories	Commodity collection points
	Distribution centers	Commodity transfer depots
	Customers	Commodity demand points
Facility characteristics	Regular facilities	Temporary facilities
	Substantial/tangible existing	
Scheduling plan	Long term: location	
	Medium term: vehicle and fleet size	
	Short term: scheduling	Urgent decisions based on available information
Trade-offs between algorithm efficiency and optimization	Paying attention to optimization	Emphasis on algorithm efficiency
Delivery models	Round trip	Round trip
	Circulating	

2. Relief distribution considered in this system consists only of regular daily commodities and not materials that must be kept cold or require special transportation equipment.
3. We assume the availability and accessibility of information such as the quantities of materials needed, number of the people in each devastated area, plans for relief distribution, and road condition data and schedule for restoration if damaged.
4. Changing characteristics of disaster recovery such as the needs of the affected people and the availability of roads are considered constants within a discrete time slot. The time slot defined will be sufficient to allow distribution and allocation of all relief supplies in a given shipment but not so extended that delays and procrastination can occur.
5. The operator has the authority to mobilize enough military or civilian vehicles to assist relief distribution. Thus, no limit is imposed on the scale of the vehicle fleet.

14.2.2 Model Establishment

The design of the relief distribution systems is shown in Figure 14.1, and the relief transfer depots are treated as bridges between the upper-stream and lower-stream distribution systems. The model encompasses T planning periods, K items of relief commodities, I collection points, and J demand points. The purpose of the design is to resolve locations such as transfer depots for each of L candidates so that we can identify optimal distribution systems and investigate their efficiency.

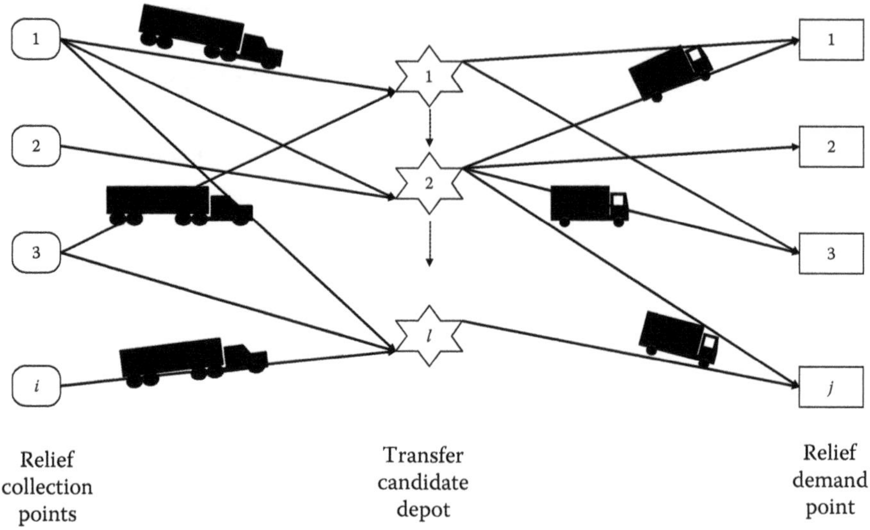

FIGURE 14.1 Relief distribution systems.

14.2.3 Symbol Explanation

Most of the parameters and variables employed in this model are time-related. The exceptions are the set-up costs *FC* of the candidate points for relief transfer depots, the weights of the relief items *W*, and the binary variable *z* that indicates whether a relief candidate location is selected to be a transfer depot.

14.2.3.1 Parameters and Variables

$c_i(t)$: Available truck capacity in period t at relief collection point i (capacity/car)
$c_l(t)$: Available truck capacity in period t at transfer point l (capacity/car)
$C_{il}(t)$: Unit transportation cost in period t from collection point i to transfer point l (dollars/car)
$C_{lj}(t)$: Unit transportation cost in period t from transfer point l to demand point j
$D_{k,j}(t)$: Amount of commodity k needed for demand point j in period t (unit of calculation for k material)
FC_l: Set-up cost (dollars) of relief transfer depot I
$R_{il}(t)$: Travel time in period t from collection point i to transfer point l (hours)
$R_{lj}(t)$: Travel time in period t from transfer depot l to demand point j (hours)
$S_{k,i}(t)$: Amount of item k collected in period t at collection point i
W_k: Package size (volume) of each package of commodity k

14.2.3.2 Variables Used

$AD_{k,j}(t)$: In period t, amount of item k actually required by each demand point j
$AS_{k,i}(t)$: In period t, amount of item k actually available at collection point i
$ms_k(t)$: In period t, least satisfaction score among demand points with regard to item k after relief distribution

$s_{k,j}(t)$: In period t, satisfaction score for commodity k at demand point j

$TC_{i,l}(t)$: In period t, total transportation cost (dollars) from collection point i to transfer candidate depot l

$TC_{lj}(t)$: In period t, total transportation cost (dollars) from transfer candidate point l to demand point j

$T_{il}(t)$: In period t, travel time from collection point i to transfer candidate depot l (If actual commodity is sent from collection point i to transfer candidate depot l, $T_{il}(t) = R_{il}(t)$; otherwise value is 0.)

$T_{lj}(t)$: In period t, travel time from transfer candidate depot l actually sent to demand point (If transfer depot l has material to be sent to demand point j, $T_{lj}(t) = R_{lj}(t)$; otherwise its value is 0.)

$x_{k,lj}(t)$: In period t, amount of item k transported from transfer candidate depot l to demand point j

$y_{k,il}(t)$: In period t, amount of relief k at collection point i sent to transfer candidate depot l

z_l: Whether candidate point l is chosen as transfer depot, with 0 indicating *not to be chosen* and 1 indicating *to be chosen*

14.2.4 Distribution Model

The model is constructed to achieve three objectives: least total cost f_1, minimum travel time f_2, and finally maximum satisfaction or fairness f_3.

$$\min\ f_1 = \sum_l FC_l \times z_l + \sum_t \sum_i \sum_l TC_{il}(t) + \sum_t \sum_i \sum_j TC_{lj}(t) \tag{14.1}$$

$$\min\ f_2 = \sum_t \sum_i \sum_l T_{il}(t) + \sum_t \sum_i \sum_j T_{lj}(t) \tag{14.2}$$

$$\max\ f_3 = \sum_t \sum_k ms_k(t) \tag{14.3}$$

s.t.

$$\sum_l \sum_j x_{k,lj}(t) \le \sum_j AD_{k,j}(t)\ and \sum_l \sum_j x_{k,lj}(t) \le \sum_i AS_{k,i}(t) \quad \forall\, t,k \tag{14.4}$$

$$\sum_i y_{k,il}(t) = \sum_j x_{k,lj}(t) \quad \forall\, t,k,l \tag{14.5}$$

$$\sum_l x_{k,lj}(t) \le AD_{k,j}(t) \quad \forall\, t,k,j \tag{14.6}$$

$$y_{k,il}(t) \le M \times z_l \quad \forall\, t,k,i,l \tag{14.7}$$

$$x_{k,lj}(t) \le M \times z_l \quad \forall\, t,k,l,j \tag{14.8}$$

$$\sum_{l} y_{k,il}(t) \le AS_{k,i}(t) \quad \forall\, t,k,i \tag{14.9}$$

$$y_{k,il}(t) \in \{0,1,2,\ldots\} \quad \forall\, t,k,i,l \tag{14.10}$$

$$x_{k,lj}(t) \in \{0,1,2,\ldots\} \quad \forall\, t,k,l,j \tag{14.11}$$

$$z_l \in \{0,1\} \quad \forall\, l \tag{14.12}$$

The three objectives in the proposed relief distribution model are indicated by Equations (14.1) to (14.3), respectively, and are explained in detail in the following paragraphs.

Objective 1: Minimizing total cost (economy objective f_1) — Costs include set-up and operating costs of transfer depots and transportation of relief commodities among the supply and demand points. Given that the sizes of relief shipments vary, their actual size must be known so that the calculation units of relief materials can be standardized. The transportation cost could be determined after calculating the frequency of shipments. The upper-stream transportation cost TC_{il} is computed as found in Equation (14.13), so the lower-stream transportation cost is TC_{lj}.

$$TC_{il}(t) = C_{il}(t) \times \left\lceil \frac{\sum_{k} W_k \times y_{k,il}(t)}{c_i(t)} \right\rceil \quad \forall\, t,i,l \tag{14.13}$$

where $\lceil x \rceil$ indicates upper-bound (ceiling) function, the smallest integer larger than or equal to x, such as $\lceil 4.8 \rceil = 5$.

Objective 2: Minimizing total travel time (effectiveness of distribution f_2) — Since travel times among collection, transfer, and demand points are already known, the calculation of travel time is required only if shipments move between collection points and transfer points. The total time for movements between those points represents actual travel time used. The calculation of the transportation time T_{il} upper-stream is found in Equation (14.14); the T_{lj} of the lower stream can be found in a similar fashion.

$$T_{il}(t) = \begin{cases} 0, & \text{if} \quad \sum_{k} W_k \times y_{k,il}(t) = 0 \\ R_{il}(t), & \text{if} \quad \sum_{k} W_k \times y_{k,il} > 0 \end{cases} \tag{14.14}$$

Objective 3: Maximizing satisfaction (f_3) — The primary purpose of this objective is to maximize satisfaction of fairness and minimize unfair distribution.

In this model, no limit is set for the satisfaction score because "if there is higher satisfaction in certain relief distribution, there must be some concession in other relief items," and we will treat each relief item independently. Thus, the weighting method is employed to sum the least satisfaction value of each relief item in every period of time to reduce the number of objective equations. The calculation of satisfaction and least satisfaction is as follows.

$$s_{k,j}(t) = \frac{\sum_{l} x_{k,lj}(t)}{AD_{k,j}(t)} \qquad \forall\, t,k,j \tag{14.15}$$

$$ms_k(t) = \min_{j} \{s_{k,j}(t)\} \qquad \forall t,k \tag{14.16}$$

Now we will explain the constraints. Equation (14.4) means scarce goods are not allowed to lie idle and an agency cannot ship what it does not have. The equation can be written as

$$\sum_{l}\sum_{j} x_{k,lj}(t) \le \min\left[\sum_{j} AD_{k,j}(t), \sum_{i} AS_{k,i}(t)\right] \qquad \forall\, t,k$$

Equation (14.5) deals with all goods shipped to and from transfer depots in the same period. Equation (14.6) prevents over-shipping of any one item. Equations (14.7) and (14.8) determine the selection of transfer depots among candidate locations. Equation (14.9) means only available goods are shipped to collection points; Equations (14.10) and (14.11) mean quantities are predetermined for each item.

In a relief system, the top priority is meeting the needs of victims. Although cost remains a consideration, it is unacceptable to have relief supplies remain idle in a system in an effort to save transportation costs or travel time. Therefore, during every period t, the total amount received at every relief demand point should be equal to the total amount shipped from the collection points for every item as indicated in Equation (14.5). In planning, we assume that the only items the victims need are relief supplies. If the provisions are not delivered in that period, however, the shortage can be made up in the next period as there would not be any so-called giving up time validity. Equations (14.17) and (14.18) explain in every period t the supply capability at each commodity supply point i and the calculation of the actual demand at each relief demand point j.

$$\begin{cases} AS_{k,i}(t) = S_{k,i}(t) & \forall k,i \quad \text{when} \quad t=1 \\ AS_{k,i}(t) = S_{k,i}(t) + [AS_{k,i}(t-1) - \sum_{l} y_{k,il}(t-1)] & \forall t,k,i \quad \text{when} \quad t \ge 2 \end{cases} \tag{14.17}$$

$$\begin{cases} AD_{k,j}(t) = D_{k,j}(t) & \forall k,j \quad when \quad t=1 \\ AD_{k,j}(t) = D_{k,j}(t) + [AD_{k,j}(t-1) - \sum_{l} x_{k,lj}(t-1)] & \forall t,k,j \quad when \quad t \geq 2 \end{cases} \tag{14.18}$$

14.2.5 Model Modification and Resolution

After the model was constructed based on real behavior, we needed an efficient method to reach a solution in the new era of evolutionary computation. Fuzzy programming combines the idea of fuzzy logic and provides a new method for uncertainty analysis in mathematical formulae. Although the importance of Objective 3 is known to be higher than those of the other two objectives, the weight relationship among these three objectives cannot be clearly defined. As a result, this study employed fuzzy multi-objective linear programming (Chen and Tzeng, 1999; Tzeng and Chen, 1998 and 1999) of a max–min operation to rewrite the mathematical equation for resolution. Thus, after the resolution for a single objective has been conducted to establish a multi-objective pay-off table, the membership function of the optimal (best) value (f_i^+) and the worst value (f_i^-) of each objective can be found.

Since Objective 3 is the maximization of the least satisfaction, its ideal value must be the even distribution of relief to each demand point regardless of cost. Therefore, should the value of the ceiling limit (upper boundary) of Objective 3 determined, every kind of relief item among all of the relief demand points in every period will suffice. Its value will be *KT*, where *K* is the total amount of relief items and *T* represented the number of planned periods.

Within relief distribution systems, the most important goals are satisfying the current needs of surviving victims as much as possible and reducing the damage following devastation. Clearly, time and money are not the ultimate aims; they are merely soft constraints for ensuring that resources are utilized effectively. Hence, f_3^+ is designated for *KT* so that Objective 3 will become the critical path for system resolution, making the system achieve its optimal result. As follows, the objective Equations (14.1) to (14.3) of the distribution model in the previous section can be rewritten as Equations (14.19) to (14.22) for achieving the maximal satisfaction level, while the original constraints remain intact as Equations (14.4) to (14.12):

$$Max\ \lambda \tag{14.19}$$

$$s.t.$$

$$\frac{f_1^- - f_1}{f_1^- - f_1^+} \geq \lambda \tag{14.20}$$

$$\frac{f_2^- - f_2}{f_2^- - f_2^+} \geq \lambda \tag{14.21}$$

$$\frac{f_3 - f_3^-}{f_3^+ - f_3^-} \geq \lambda \tag{14.22}$$

14.3 RELIEF DISTRIBUTION OPERATION: CASE ANALYSIS

Disaster recovery is like field combat: the final outcome of s strategy depends on whether it can be carried out effectively. Thus, this section will make use of scenario simulation to demonstrate how to utilize the constructed model for integral relief distribution, so that the operation procedures and related issues can be established for further studies.

14.3.1 Information Content and Data Collection

The values of parameters of the model must be established before planning. Data collection can be performed as a routine matter before a disaster; feedback gathered during the disaster, and need analysis based on feedback afterward. The work items are classified in Table 14.2.

14.3.2 Pre-Operation Stage

Data collection in the pre-operation stage involves data collected during routine non-disaster days concerning commodity measurements, travel and alternative routes, coordination capability, and truck capacities.

1. Calculate each relief commodity size volume equivalent. Among the daily commodities, choose the size of one to be used as the basic unit for measuring volume (the size of a sleeping bag is used as the criterion for measuring volume equivalents in this study). The volume size equivalents of common commodities can found in Table 14.3.
2. Establish an electronic map. Electronic maps made of Taiwan's entire highway network identify the shortest routes and alternative routes quickly and easily. This study employs the "Taiwan Island 1/25/2000 Transportation Network Numerical Value Map" prepared by the Taiwanese Transportation Institute of the Ministry of Transportation and Communications.
3. Investigate the coordination capability for emergency relief in all affected areas. Coordination capability for emergency relief is one criterion for predicting supply and demand for each area after a disaster. Asking questions about daily commodities is one place to start. What is the area's primary staple crop? Are any warehouses available for storage? Does the area have warehouses or factories that produce daily necessities?

TABLE 14.2
Information Classification

Stage	No.	Work Content	Dependencies
Pre-operation	1a	Calculate each relief commodity size volume equivalent (W_k)	No
	1b	Establish electronic map	No
	1c	Investigate coordination capability for emergency relief of whole area	No
	1d	Calculate capacity equivalent for each kind of truck	No
Disaster information transmission	2a	Estimate road destruction and the time needed for restoration	No
	2b	Survey degree of damage in each area	No
	2c	Forward location of each commodity demand point (j)	2b
	2d	Identify number of victims who need care at each demand point	2c
Planning and analysis	3a	Predict commodity demands for each demand point ($D_{k,j}(t)$)	1c, 2d
	3b	Select location for each commodity demand point (i)	2b
	3c	Determine commodity supply capability for each commodity collection point ($S_{k,i}(t)$)	1c, 3b
	3d	Determine best support vehicle category for delivering relief commodities to each collection point	3b
	3e	Determine support truck capacity required to deliver relief commodities to each commodity collection point ($c_i(t)$)	1d, 3d
	3f	Select locations for candidates for commodity transfer depots (l)	2c, 3b
	3g	Analyze shortest routes from commodity transfer candidate points to relief demand points	1b, 2a, 2c, 3f
	3h	Find shortest route from relief collection points to commodity transfer candidate points	1b, 2a, 3b, 3f
	3i	Analyze set-up costs of relief transfer depot candidates (FC_l)	3f
	3j	Determine support vehicle category for delivering commodities to commodity transfer depot candidates	3f
	3k	Determine support vehicle capacity for delivering commodities to commodity transfer depot candidates ($c_l(t)$)	1d, 3j
	3l	Calculate travel times from commodity transfer depot candidates to each relief demand point ($R_{lj}(t)$)	3g
	3m	Calculate unit transportation costs from commodity transfer depot candidates to each relief demand point ($C_{lj}(t)$)	3g
	3n	Calculate travel time from each relief collection point to commodity transfer depot candidates (R_{il}(t))	3h
	3o	Calculate unit transportation costs from relief demand points to commodity transfer depot candidates (C_{il}(t))	3h

TABLE 14.3
Commodity Size and Volume Equivalents

Item	Calculation Unit	Volume (cm³)	Volume Equivalent
Sleeping bag	Each (nylon sleeping bag)	45 × 25 × 11 = 12,375	1.00
Tent	Each (yurt for 6–8 people)	70 × 26 × 15 = 27,300	2.21
Mineral water	Box (1410 ml,12 bottles)	36 × 26 × 30 = 28,080	2.27
Rice	Pack (5 kg)	38 × 25 × 5.5 = 5,225	0.42
Instant noodles	Box (12 bowls)	43 × 29 × 17 = 21,199	1.71
Dry food	Box (30 packs nutrition biscuits)	38 × 27 × 18 = 18,468	1.49
Canned food	Box (12 cans)	29 × 21 × 5.8 = 3,532	0.29

4. Calculate truck capacity equivalents. The carrying capacity of a vehicle is affected by more than its tonnage. Whether a truck has a canvas cover is one of many issues to consider. The primary military vehicles for distribution are Hummers and 10.5-ton trucks (both with covers).

Measurements of the maximum distribution capacities for these vehicles are provided by the government. Vehicles provided by civilians, however, are usually small trucks without covers. Their maximum heights for carrying are dictated by Article 4 of Regulation 79 of the "Traffic Safety Regulations of the Thoroughfare" stating that the height of carried goods "should not exceed 4 meters, or 2.5 meters for small vehicles, as measured from ground level" (see Table 14.4).

14.3.3 Disaster Information Transmission

The second type of information needed varies with the development of the disaster, and includes the work contents and the data collection types. The following steps are suggested:

1. Estimate the extent of road destruction and the time required for restoration. In addition to investigative reports from the damaged areas, a bird's eye view can be obtained by helicopter immediately after the disaster to

TABLE 14.4
Truck Capacity Equivalents

Vehicle	Carry Space (cm³)	Carry Equivalent
Hummer	280 × 200 × 140 = 7,840,000	634
Military truck, 10.5 tons	600 × 250 × 175 = 26,250,000	2,121
Civilian truck, 1.5 tons	231 × 150 × 130 = 4,504,500	364

Note: 1 sleeping bag volume = 1 equivalent.

determine the state of the traffic network in advance of deploying ground vehicles. Furthermore, time needed for road restoration for important areas can then be estimated based on the manpower and equipment dispatched and the degree of damage.

2. Survey each area. This establishes a feedback system for assessing devastation. A survey of all areas should be conducted and information from villages and towns transmitted to the central command unit to determine needs for relief. The numbers of men, women, elderly, children, and the victims who need care should be reported also.

14.3.4 Route Planning and Network Analysis

The third level of information is obtained by planning and analysis using a database previously established from data forwarded by devastated areas. The primary work contents are as follows:

1. Predict commodity demands. Estimates of possible demand for all daily necessities over specific periods should be made based on emergency coordination capability, the extent of devastation, and the age and sex data for victims who need care at the relief demand points. For example, it is important to determine how much food and water an average adult needs to survive.
2. Plan commodity collection depots. After the extent of disaster is known, affected and non-affected areas can be delineated. Suitable locations such as village or township offices and county or city administration centers that area residents know well can serve as relief collection points in non-devastated areas. The private sector and members of the public can be encouraged to donate relevant daily necessities. In addition, once the location is chosen, the supply capacity of the area could be estimated based on emergency coordination data for each area and allocation of support vehicles can be determined.
3. Set up transfer depots. After locations of demand points are set and collection points are chosen, several large-scale transfer depots can be established based on the extent of road damage and time required for restoration. The two principles for establishing a transfer depot are (a) the location should be prominent and be accessible via alternative roads and (b) the location should have enough space to store, coordinate, and package relief commodities. After transfer depots are selected, establishment costs can be estimated and vehicle allocations can be planned.
4. Select the quickest route. Based on the distribution of commodity demand, collection, and transfer points after analysis of data about road destruction and estimated restoration time, GIS software (MapInfo, TransCad, ARC/INFO)

can be used on the electronic map to find the quickest distribution route. Because relief distribution is often an emergency, the government could enact stringent traffic control measures. The advantage of stringent control is that better time and speed estimates may be made. In normal traffic, volumes and speeds are variable and times are unpredictable. Also, vehicle specifications (size, mode, etc.) should be considered when estimating unit transportation costs between two points.

14.4 CASE ILLUSTRATION AND DATA ANALYSIS

14.4.1 Case Illustration

The case analysis focuses on Taichung, Nantou City, and Nantou County, Taiwan, that experienced a major earthquake on September 21, 1999. The demand and supply points are shown in Table 14.5. The commodities were gathered at Fongyuan for the northern areas (Taipei, Taoyuan, and Hsinchu) and at Douliou, Yunlin County, for the southern areas (Kaohsiung and Pingdong).

The study covers five collection points, eight demand points, and four transfer depots. Some areas performed multiple duties. For example, Fongyuan served

TABLE 14.5
Research Case Locations

Demand Point		Supply Point		Transfer Depot	
Taichung city	Baseball field	Fongyuan city	Fongyuan stadium	Fongyuan city	Fongyuan stadium
Taiping city	Fire department	Dongshih township	Dongshih elementary school	Caotun township	Farmer association warehouse
Dali city	City government	Chunghua city	Chunghua county government	Nantou city	Nantou county stadium
Nantou city	County stadium	Yuanlin township	Yuanlin township government	Mingjian township	Mingjian elementary school
Puli township	Fire department	Douliou city	Chung Shiou Temple		
Kuoshin township	Kuoshin Street				
Chungliou township	Township government				
Chushan township	Township government				

as both a supply point and a transfer depot, while Nantou was a transfer station and a demand point. Relief supplies for distribution included sleeping bags, tents, mineral water, and four kinds of instant noodles distributed over four planned periods.

The case study considered the highway network of county and provincial roads and each road section was assigned a designated travel time. The quickest route was determined by using TransCAD software. The study simplified restoration work on damaged roads by dividing the work into three categories. As a result, the quickest travel time between two places was subject to change. Data showing quickest travel times, travel distances, numbers of victims in need of care in each area, demand modes for every item of relief, and supply information are shown in Tables 14.6 and 14.7.

14.4.2 Data Analysis and Discussion

Using LINGO for analysis, Fongyuan and Nantou decided the sites for transfer depots based on low set-up costs and geographic locations. The results indicated unit transportation costs and travel times from most demand and supply points, such as from Taichung, Taiping, Dali, and Dongshih to Fongyuan, and from Yuanlin to Nantou were minimal. In addition, since Fongyuan also acted as a collection point and Nantou as a demand point, the choice of these two places enhanced the integral performance of the system.

Note the disparity of supply sources in Table 14.7. The Fongyuan transfer depot had to move some of its relief supplies to the Nantou transfer depot to meet the demands of Nantou and collect more supplies. Considering travel time and transportation cost, the Fongyuan transfer depot mainly collected and distributed relief to Taichung, Taiping, Dali, and Dongshih, while Nantou was responsible for Puli, Kuoshin, Chungliou, Chushan, and Yuanlin. Chunghwa was not selected for transport because it sits between Taichung and Nantou and would not have affected the system significantly.

The respective rankings of satisfaction of the three objective values were 0.93, 0.82, and 0.65, while the minimal satisfaction applied to Objective 3. Therefore, Objective 3 was the bottleneck of the system in line based on the premise that the model will not compromise the equity of relief distribution because of cost and time constraints.

Table 14.7 indicates that for Objective 3 the result shows even distribution when supply is over demand. In other words, the relief given to all of demand points in every period revealed a certain fixed ratio to actual demand. Thus, the satisfaction of all of the demand points in each period for a kind of commodity will be integral. Although the satisfaction of Objective 3 was only 0.6527, it was modified ($f_3^+ = TK = 16$) and the fuzzy multi-objective programming was

TABLE 14.6
Commodity Distribution Results for Supply Points and Transfer Depots

Supply Point	Item		T1 Fongyuan	T1 Nantou	T2 Fongyuan	T2 Nantou	T3 Fongyuan	T3 Nantou	T4 Fongyuan	T4 Nantou
Fongyuan	Amount of Delivery (millions)	Sleeping Bags	14.66	0	37.54	0	46.92	0	0	0
		Tents	3.24	0	14.52	0	4.23	16.34	0.83	0
		Mineral Water	12.00	0	15.36	0	0	15.36	13.82	0
		Instant Noodles	26.99	0	34.55	0	33.88	0.67	3.50	0
	Distributed Equivalent		95.21	0	163.58	0	114.22	72.13	39.19	0
	Times of Delivery		5.00	0	8.00	0	6.00	4.00	7.00	0
Dongshih	Amount of Delivery (millions)	Sleeping Bags	0	0.51	1.23	0	1.44	0	0	0
		Tents	0.25	0.04	0.52	0	0	0	0	0
		Mineral Water	0.30	0	0.36	0	0	0.36	0	0
		Instant Noodles	1.54	0	1.85	0	0	1.85	0	0
	Distributed Equivalent			0.61	6.36	0	1.44	3.98	0	0
	Times of Delivery			1.00	2.00	0	1.00	1.00	0	0
Chunghwa	Amount of Delivery (millions)	Sleeping Bags	0	5.15	0	14.41	0	18.01	2.43	0
		Tents	0	0.79	0.16	2.01	0	0	0	0
		Mineral Water	0.74	7.79	6.16	5.78	1.36	10.58	10.26	0.48
		Instant Noodles	14.54	8.59	21.69	10.70	0	0.53	0	1.52
	Distributed Equivalent			39.25	51.44	50.37	3.08	42.92	25.74	3.69
	Times of Delivery			2.00	3.00	3.00	1.00	7.00	5.00	1.00

(Continued)

TABLE 14.6
Commodity Distribution Results for Supply Points and Transfer Depots (Continued)

Supply Point	Item		T1 Fongyuan	T1 Nantou	T2 Fongyuan	T2 Nantou	T3 Fongyuan	T3 Nantou	T4 Fongyuan	T4 Nantou
Yuanlin	Amount of Delivery (millions)	Sleeping Bags	0	2.06	0	4.65	0	5.81	0	0
		Tents	0	0.43	0	0.73	0	0.95	0	0.10
		Mineral Water	0	0.45	0	0.51	0	0.51	0	0.46
		Instant Noodles	0	11.57	0	13.07	0	13.07	0	0
	Distributed Equivalent			23.82	0	29.77	0	31.42	0	1.27
	Times of Delivery			4.00	0	5.00	0	5.00	0	1.00
Douliou	Amount of Delivery (millions)	Sleeping Bags	0	10.29	0.75	32.18	1.44	0	0	1.60
		Tents	0	2.74	0.06	10.88	0	0	0	0.25
		Mineral Water	0	6.74	0	10.78	10.78	0	0	9.70
		Instant Noodles	0	29.30	0.31	46.57	5.21	0	0	0
	Distributed Equivalent			81.75	1.40	160.35	34.82	0	0	24.17
	Times of Delivery			4.00	1.00	8.00	2.00	0	0	4.00
Extra supplies (millions)		Sleeping Bags	–108.08[a]		–57.53		39.82		71.45	
		Tents	–33.91		–16.80		36.06		40.65	
		Mineral Water	–22.25		–33.57		–39.85		–39.99	
		Instant Noodles	–58.27		–30.07		73.53		94.26	

[a] Minus signs indicate item shortages.

TABLE 14.7
Distribution Results from Transfer Depots to Demand Points

Planning Stage	Transfer Depot	Taichung: Delivery Units (millions) Sleeping Bags	Tent	Mineral Water	Instant Noodles	Distributed Equivalent	Deliveries	Satisfaction (%) Sleeping Bags	Tent	Mineral Water	Instant Noodles	Taiping: Delivery Units (millions) Sleeping Bags	Tent	Mineral Water	Instant Noodles	Distributed Equivalent	Deliveries	Satisfaction (%) Sleeping Bags	Tent	Mineral Water	Instant Noodles
T1	Fongyuan	4.4	1.0	3.8	13	37	2	23	18	56	61	2.6	0.6	2.3	7.5	22	2	23	18	56	61
	Nantou	0	0	0	0	0	0					0	0	0	0	0	0				
T2	Fongyuan	12	3.9	5.3	3.5	39	2	61	63	53	81	7.3	2.3	3.2	10	37	2	61	63	53	81
	Nantou	0	0	0	14	24	6					0	0	0	0	0	0				
T3	Fongyuan	9.9	2.9	5.3	7.5	41	2	100	100	49	100	6.0	0	3.2	4.5	21	1	100	100	49	100
	Nantou	0	0	0	0	0	0					0	1.7	0	0	3.8	2				
T4	Fongyuan	0.5	0.2	0	0.7	2.1	1	100	100	47	100	0.3	0.1	2.8	0.4	7.7	2	100	100	47	100
	Nantou	0	0	4.7	0	11	3					0	0	0	0	0	0				

Planning Stage	Transfer Depot	Dali: Delivery Units (millions) Sleeping Bags	Tent	Mineral Water	Instant Noodles	Distributed Equivalent	Deliveries	Satisfaction (%) Sleeping Bags	Tent	Mineral Water	Instant Noodles	Nantou: Delivery Units (millions) Sleeping Bags	Tent	Mineral Water	Instant Noodles	Distributed Equivalent	Deliveries	Satisfaction (%) Sleeping Bags	Tent	Mineral Water	Instant Noodles
T1	Fongyuan	1.8	0.9	3.5	12	32	2	23	18	56	61	0	0	0	0	0	0	23	18	56	61
	Nantou	2.3	0	0	0	2.3	1					8.5	2.0	7.3	24	–	–				
T2	Fongyuan	0	2.7	4.9	16	45	3	61	63	53	81	0	0	0	0	0	0	61	63	53	81
	Nantou	11	1.0	0	0	14	4					24	7.5	10	34	–	–				
T3	Fongyuan	9.3	0	1.3	3.0	17	1	100	100	49	100	12	0	0	14	36	2	100	100	49	100
	Nantou	0	2.7	3.6	4.0	21	6					7.7	5.6	10	0	–	–				
T4	Fongyuan	0.5	0	4.4	0	11	2	100	100	47	100	1.1	0.3	9.1	1.3	25	4	100	100	47	100
	Nantou	0	0.2	0	0.6	1.4	1					0	0	0	0	–	–				

(Continued)

TABLE 14.7
Distribution Results from Transfer Depots to Demand Points (Continued)

		Chungliao										Chushan									
		Delivery Units (millions)						Satisfaction (%)				Delivery Units (millions)						Satisfaction (%)			
Planning Stage	Transfer Depots	Sleeping Bags	Tent	Mineral Water	Instant Noodles	Distributed Equivalent	Deliveries	Sleeping Bags	Tent	Mineral Water	Instant Noodles	Sleeping Bags	Tent	Mineral Water	Instant Noodles	Distributed Equivalent	Deliveries	Sleeping bags	Tent	Mineral Water	Instant Noodles
T1	Fongyuan	1.8	0	0	0	1.8	1	23	18	56	61	2.0	0.5	1.7	5.7	17	1	23	18	56	61
	Nantou	0	0.4	1.5	5.0	13	4					0	0	0	0	0	0				
T2	Fongyuan	0	0	0	0	0	0	61	63	53	81	0	0	0	0	0	0	61	63	53	81
	Nantou	4.9	1.6	2.1	7.0	25	7					5.6	1.8	2.4	8.0	29	7				
T3	Fongyuan	4.0	0	0	3.0	9.2	1	100	100	49	100	4.6	1.3	0	3.4	13	1	100	100	49	100
	Nantou	0	1.2	2.1	0	7.4	3					0	0	2.4	0	5.5	2				
T4	Fongyuan	0	0	0	0	0	0	100	100	47	100	0	0	0	0	0	0	100	100	47	100
	Nantou	0.2	0.1	1.9	0.3	5.2	2					0.3	0.1	2.2	0.3	5.9	2				

		Puli										Koushin									
		Delivery Units (millions)						Satisfaction (%)				Delivery Units (millions)						Satisfaction (%)			
Planning Stage	Transfer Depots	Sleeping Bags	Tent	Mineral Water	Instant Noodles	Distributed Equivalent	Deliveries	Sleeping Bags	Tent	Mineral Water	Instant Noodles	Sleeping Bags	Tent	Mineral Water	Instant Noodles	Distributed Equivalent	Deliveries	Sleeping Bags	Tent	Mineral Water	Instant Noodles
T1	Fongyuan	0	0	0	0	0	0	23	18	56	61	2.0	4.6	1.7	5.7	17	1	23	18	56	61
	Nantou	7.2	1.6	6.1	20	59	15					0	0	0	0	0	0				
T2	Fongyuan	20	6.4	8.5	29	102	5	61	63	53	81	0	0	0	0	0	0	61	63	53	81
	Nantou	0	0	0	0	0	0					5.6	1.8	2.4	7.9	28	7				
T3	Fongyuan	0	0	0	0	0	0	100	100	49	100	4.5	0	2.4	3.4	16	1	100	100	49	100
	Nantou	16	4.7	8.6	12	67	19					0	1.3	0	0	2.9	1				
T4	Fongyuan	0	0.3	7.7	1.1	20	4	100	100	47	100	0	0	0	0	0	0	100	100	47	100
	Nantou	0.9	0	0	0	0.9	1					0.3	0.1	2.1	0.3	5.8	2				

changed into a single objective for resolution. Consequently, Objective 3 reached its optimal state.

As a whole, the final resolution allowed Objective 3 to reach its optimum level and Objectives 1 and 2 reached certain levels of satisfaction despite the constraints. Therefore, the results derived are reasonable and could be used by decision makers. If the ideal value of the objective to maximize minimal satisfaction is adjusted to (total planning period) × (total items for distribution), even when supply is below demand, the result would still follow the same ratio of distribution to each of the demand points. After this, a suitable distribution route would be located.

The resulting analysis of the objective is in line with the results we have seen in reality. The decision maker can, based on need, select an optimal value of Objective 3 and split it between the evenly distributed result and (total period of planning) × (total items of distribution). In doing so, the importance of Objective 3 is greater than those of the other two, yet neither of the two is dominant. The actual satisfaction level of Objective 3 will not reach 100%, while the achievement values of the other two objectives can be improved.

14.5 CONCLUSIONS AND RECOMMENDATIONS

Relief distribution is one of the most important aspects of disaster recovery. The features inherent in relief distribution systems found in this study are based on the assumption that the government has the authority to expropriate enough military or civilian vehicles to help with relief distribution and control traffic during the period of relief distribution. We used fuzzy multi-objective programming to create an emergency relief distribution model for the reference of decision makers. As a part of a disaster recovery system, sufficient correct data as specified above must be collected and available before a model starts to operate.

The data content needed to be processed for each of the parameters is listed in Table 14.2, and Figure 14.2 elaborates on the priority and relationship of each procedure related to implementation. Data collection was divided into the pre-operation stage, post-disaster stage, analysis of the information compiled, and final results of relief distribution. To test the feasibility and effectiveness of the study method, a case study was used to illustrate the concepts described.

Compatible measures needed for the execution of this model were put forward for use in emergency relief distribution during and after a natural disaster. Further in-depth study is needed to provide steps and recommendations for each of the procedures and develop a more representative method of estimation. This study provided some insights for the decision support system that must include a database containing pre-operation plans and geographic information.

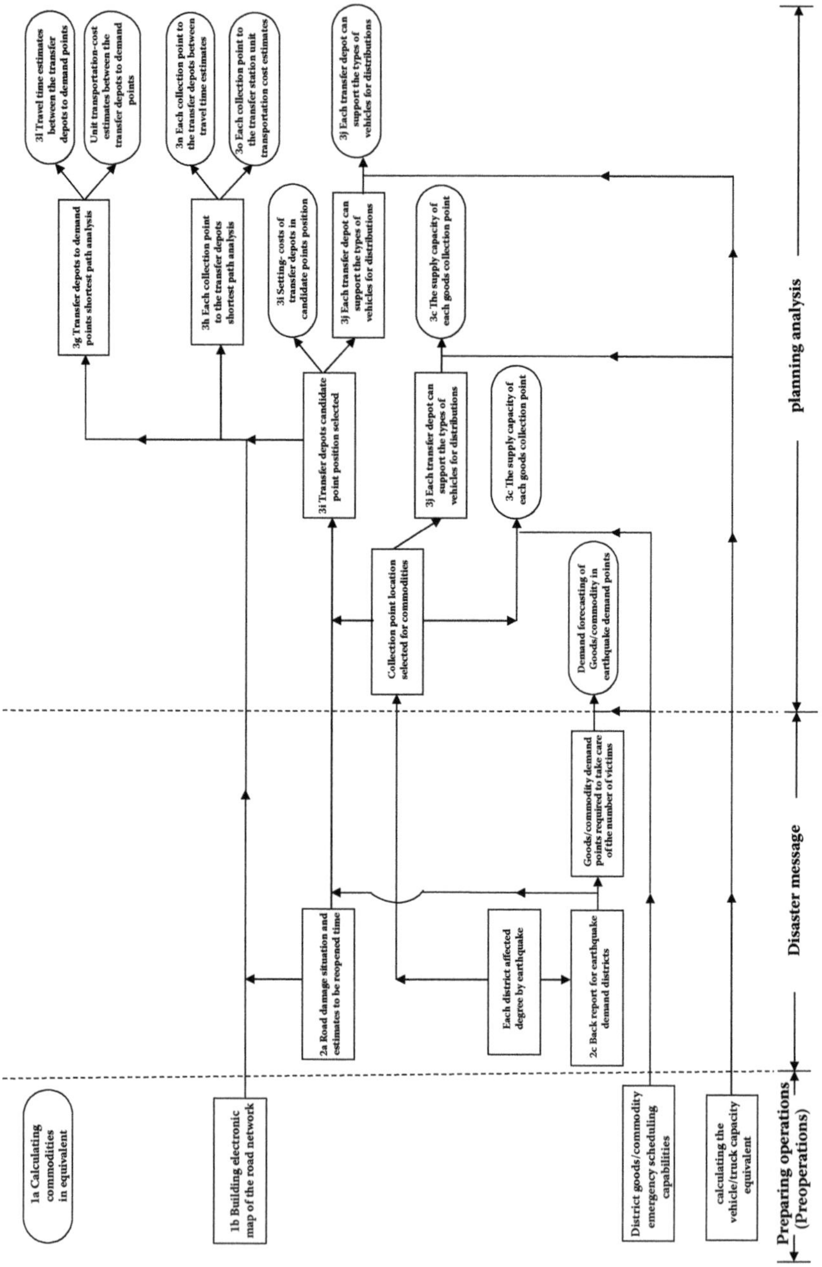

FIGURE 14.2 Procedure of priorities and relationships.

15 Comparative Productivity Efficiency for Global Telecoms

Greater efficiency of an enterprise is the equivalent of being more competitive in the market and more profitable. It is interesting to study the differences between the Forbes Global 2000 rankings of leading companies* and the productivity efficiency ratings for leading global telecom operators. This chapter is the first attempt to compare the operating performances of global telecom operators in the Forbes Global 2000 rankings with the CCR ratings, particularly EBITDA margins, ROA levels, total asset turnover, and net profit ratios.

In recent years, telecommunication industries have encountered fierce competition, the bursting dot.com bubble, 3G high auction license pricing, and rapid overseas development. In response to pressures from technical changes and market competition, telecom operators have worked hard to cut costs to maintain their bottom lines (net incomes). The data envelopment approach (DEA) can combine multiple output and input variables to assess an enterprise's operating performance. One of the incentives for the current study is to understand which geographical areas exhibit better productivity efficiency.

This is the first study to compare relative performance efficiencies of the leading telecom operators by combining the three methods of traditional radical DEA measure, the Andersen and Petersen (A&P) efficiency measure, and the new DEA measure. Critical research focusing on the productivity efficiencies of telecom operators has attracted the attention of academicians, policy regulators, and decision makers the world over.

Several studies have applied DEA methodologies to solve practical problems in the telecom industry. Saunders et al. (1995) discuss some questions about the economics of telecommunications, arranging them around such themes as whether the economic value of the benefits derived from investments in telecommunications can be demonstrated and quantified and which segments of the population derive these benefits. Majumdar (1995) investigates the impact of the adoption of new switching technology on the performances of firms in the United States, telecommunications industry by computing both input-conserving and output-augmenting measures of performance.

* On March 25, 2004, Forbes issued a comprehensive rating of the world's biggest and most powerful companies during year 2003 measured by a composite ranking of sales, profits, assets, and market value. The rating spanned 51 countries and 27 industries (http://www.forbes.com).

Sueyohsi (1998) examines economic effects empirically by comparing the performance of Nippon Telephone & Telegraph (NTT) before and after its privatization in 1985. The move made NTT managerially ineffective under its public–private joint ownership arrangement. Giokas and Pentzaropoulos (2000) studied the technical efficiency and economic benefit of the Hellenic telecommunications organization from 1971 to 1993.

Koskie and Majumdar (2000) examined the efficiency with which countries have been able to develop and provide their telecom infrastructures, and whether disparities in efficiency have diminished or increased over time among different countries. Lien and Peng (2001) explores the production efficiency of telecommunications in 24 Organization for Economic Cooperation and Development (OECD) countries. Pentzaropoulos (2000) compares the operational efficiencies of the main European telecommunications organizations.

Uri (2000) explores whether incentive regulation in the telecommunications industry in the United States resulted in increases of productive efficiency. Thereafter, Uri (2001) consistently implements the DEA approach to measure the changing productive efficiencies arising from incentive regulations for some telecommunications issues. Zhu (2000) develops a multi-factor performance measure model to measure profitability and marketability for Fortune 500 companies. He indicates that the top-ranked companies by revenue do not necessarily exhibit top-ranked performance, and also that reductions in current levels of employees, assets, and equity may actually increase revenue and profit levels.

Karlaftis (2004) measures the efficiency and effectiveness of urban transit systems via the DEA approach over a 5-year period. Luo (2003) evaluates the profitability and management of large banks and shows that profitability and marketability efficiency play key roles in determining a bank's survival. He also indicates that overall technical efficiency (OTE) of profitability performance can predict the likelihood of bank failures.

15.1 GLOBAL TELECOMMUNICATION TRENDS

In the late 20th century, the almost simultaneous arrival of two major innovations—mobile phones and the Internet—changed the face of communications and also gave impetus to dramatic economic growth. Modern communication technologies have been instrumental in reshaping the world's telecommunications market.

The development of the Internet and the progress of information and communication technology (ICT) accelerated the transmission of knowledge and the exchange of information, thereby propelling people all over the world toward an information society. The development of the knowledge economy promoted broadband network construction—a major factor leading to the information society.

15.1.1 Fixed-Mobile Substitution (FMS)

The utilization of mobile instead of fixed-line telephones for calls or access is called fixed-to-mobile substitution. It created a massive threat to fixed-line voice revenues for incumbent telecom carriers. Mobile telephone connections in developing

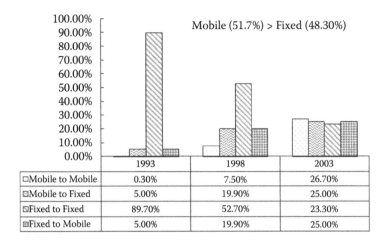

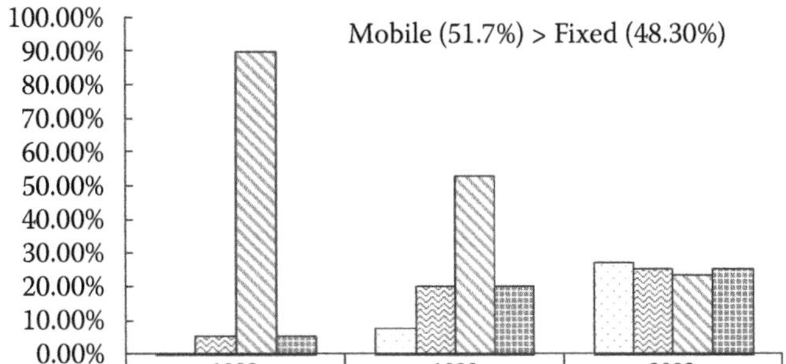

	1993	1998	2003
Mobile to Mobile	0.30%	7.50%	26.70%
Mobile to Fixed	5.00%	19.90%	25.00%
Fixed to Fixed	89.70%	52.70%	23.30%
Fixed to Mobile	5.00%	19.90%	25.00%

FIGURE 15.1 Trend for mobile substitution fixed ratio (1993–2003). (*Source:* United Nations' International Telecommunications Union).

countries have increased considerably because cellular networks can be built faster than fixed-line networks.

Wireless technology makes it unnecessary to run a wire line into every subscriber's home and can also cover geographically challenging areas. During 2003, the number of mobile calls (mobile to mobile and mobile to fixed) comprised 51.70% of total traffic calls, surpassing the number of fixed calls (fixed to fixed, fixed to mobile) that represented 48.30% of all calls. International Telecommunications Union (ITU) data showing the FMS trend from 1993 to 2003 appear in Figure 15.1.

15.1.2 Internet Broadband

Internet broadband now has become fully integrated into a multi-faceted but consolidated information, communications, and entertainment (ICE) marketplace. Along with the diversification of content applications and the increase in broadband penetration, the Internet has become the most common tool creating and exchanging information and also for offering broadcast TV, TV on demand, and videophone platforms worldwide.

Content providers and equipment manufacturers continue to develop new services and products that will allow users to make the most of their broadband connection to facilitate Internet consumer behavior. The telecom industry is looking for value-added services to diversify content and thereby drive innovations to deliver further growth.

15.1.2.1 Broadband Access

Incumbent telecom operators quickly move into Internet broadband fields to stimulate revenue growth and further offset fixed voice revenue losses. Incumbent operators have the resources to dominate the Internet, primarily through ownership of most local and long distance transmission conduits. They also control most of the broadband backbone facilities providing Internet data transport by owning the largest Tier 1 Internet service providers (ISPs) in their local data markets.

According to ITU (2003), Internet broadband utilizes several advanced technologies to provide access, for example, digital subscriber lines (DSLs, copper phone lines), cable modems (copper coaxial), fiber optics (FTTH, FTTB), cable, wireless local area networks (WLANs), satellites, and fixed broadband wireless (IEEE 802.16).

15.1.2.2 Broadband Penetration

Network infrastructure, per capita income, and the degree of openness in a society are the most important determinants for Internet diffusion. External conditions such as legal, economic, political, and social conditions that surround Internet users also affect the level of adoption in a country (Beilock and Dimitrova, 2003).

Among the global broadband penetration rankings as of December 31, 2003, KT Corporation (formerly Korea Telecom) pushed South Korea into having the most high-speed Internet access and greatest broadband penetration. Yoon (1999) examined factors affecting efficiency and demonstrated how the South Korean government's commitment to a market liberalization schedule created a credible competitive threat to the incumbent carriers and enhanced the dynamic efficiency of the telecom industry.

15.1.2.3 Convergence of 4C Services

The convergence of communication, computing, consumer appliance, and content (4C) technologies provides integrated voice, high-speed data, video, and other consumer services. Multimedia services via a broadband network are becoming routine in daily life, and information networks have become essential elements of communication in the information society.

Telecom operators have increasingly bundled various services as a new marketing strategy. This could allow incumbent operators to provide more than plain old telephony service (POTS) and narrowband Internet access, differentiating themselves from their main rivals by offering me-too-but-cheaper service.

As for cable TV operators, broadband access enables them to provide a larger slice of consumer entertainment along with less expensive telephone service. Fastweb Italy deploys 10 Mbps triple-play services including broadcast TV, broadband, and telephony and has become profitable after 3 years of operation. The incumbent Telecom Italia followed by launching video on demand through its Rosso Alice portal and introduced fixed-line videophones to the market in June 2004 according to its annual report.

15.1.3 Mobile Market

Mobile arming is the most important revenue resource when the lack of wireless operation income decreases an incumbent's competitive power. Most developed countries have the greatest levels of mobile penetration. Taiwan has the highest mobile penetration in the world. Mainland China has 17% penetration in a growing market. In 2003, China's mobile phone subscribers overtook fixed-line subscribers as shown in Figure 15.2.

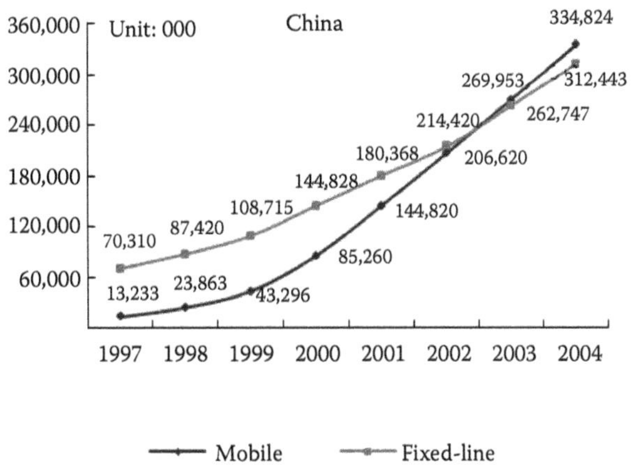

FIGURE 15.2 Growth of mobile and fixed subscribers in China. Source: www.mii.gov.cn

The strategy of mobile operators is to get closer to their high-spending customers to build loyalties and grow their average revenue per user (ARPU). The operating focus is to retain and develop high-value customers while cutting the costs of supporting low-value customers. With the convergence of mobility and the Internet, the industry is witnessing a growing trend from geographic to strategic advantages for the globalization of a wireless value system (Steinbock, 2003).

15.2 DATA AND METHODS

DEA is a non-parametric technique for measuring and evaluating the relative efficiencies of a set of entities with common crisp inputs and outputs (Guo and Tanaka, 2001). The advantages of DEA in this context are its multiple output and input viewpoints and its wide acceptance for performance evaluation. In this study, the application of the traditional radial DEA, A&P efficiency measures, and achievement efficiency measure are used to explore relative productivity efficiency as a way to compare operating performances of 39 leading global telecom operators listed among the Forbes Global 2000 as decision-making units (DMUs). Figure 15.3 illustrates the DEA procedure.

15.2.1 Data Collection

The first step was selecting the top telecom companies from the Forbes global 2000 rankings. Next, data from 40 DMUs were retrieved from the telecom operators' annual reports on their websites* and the related features were checked against the

* Financial data for most operators covered year ending March 31, 2004 and included NTT Corporation, NTT DoCoMo, BT Group, KDDI Group, Singapore Telecommunications, MM2, and Cable and Wireless. Telstra data ended June 30, 2004; data for other telecoms ended December 31, 2003. Data were obtained from company annual reports.

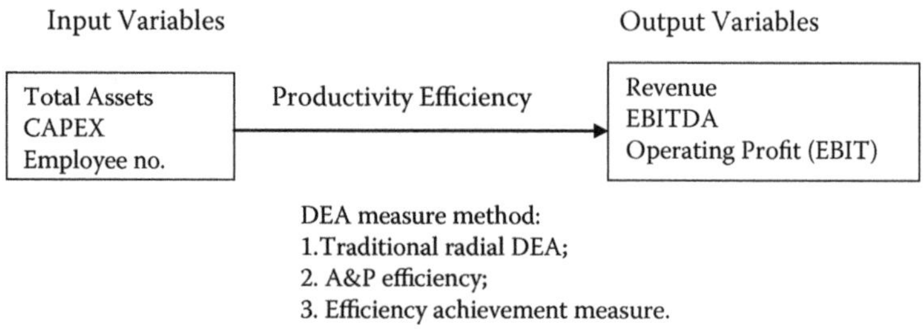

FIGURE 15.3 DEA assessment procedure.

UBS Investment Bank database.* The Vodafone Group ranking 355th among the Forbes 2000 was excluded as an observation DMU to avoid statistical distortion because in fiscal year 2003 it showed a significantly large negative value of –$17,426 million (U.S.) in operating profit (earnings before interest and taxes [EBIT]).

15.2.2 DEA Methods

Three types of efficiency measures can be adopted to assess the relative operational performances of an enterprise: the classical radial efficiency measure, the A&P efficiency measure, and the efficiency achievement measure.

15.2.2.1 Classical Efficiency Measure

The traditional radial DEA method of Charnes, Cooper, and Rhodes (1978) known as the CCR method is based on the pioneering efficiency relative efficiency measurement work of Farrell (1957). The radial system generalizes a multiple output–input performance measure in which the ratio of the weighted outputs to weighted inputs for each observation is maximized.

Two versions of DEA efficiency measures are the CCR measure and the later BCC measure named for Banker, Charnes, and Cooper (1984). The CCR measure is calculated with the constant returns to scale (CRS) assumption whereas the BCC method allows for variable returns to scale (VRS; Lien and Peng, 2001). Every DMU's efficiency evaluation is viewed as one objective function to be maximized (relative efficiency). There are n units or n decision-making units (DMUs) and each has m inputs for producing s outputs.

15.2.2.2 Andersen & Petersen (A&P) Efficiency Measure

The traditional radial DEA measure evaluates the relative efficiency of DMUs but does not allow for a ranking of the units themselves. A modified version of DEA based on a comparison of DMUs is the A&P efficiency measure. Efficient observations are assigned an index value of 1 in the CCR efficiency model and an index

* UBS Investment Bank Research covers more than 3000 companies worldwide and provides data on valuation, strategy, and economics (www.ubs.com: Equity/Research/Sectors/Telecommunication).

equal to or larger than 1 in the A&P model (Andersen and Petersen, 1993). The A&P function is described in Appendix C.

15.2.2.3 Efficiency Achievement Measure

The conventional approach of DEA analysis considers individual DMUs separately, and then employs Equation (15.2) to calculate a set of weights that brings maximal relative efficiency to each group. Such an approach enables most DMUs to be used in efficiency measures. DEA models of efficiency achievement measure can be applied in multi-criteria decision making.

A revised DEA multiple-objective programming approach proposed by Chiang and Tzeng (2000) tries to find a set of common weights by calculating the efficiency ratios of all DMUs. This approach considers the efficiency ratios of all DMUs to calculate and find a set of common identity-based weights so that the efficiency ratios of all DMUs calculated accordingly improve as the ratio gets larger. To achieve this goal, multiple-objective programming can be employed to find a set of consistent weight combinations so that the optimized efficiency value can be calculated for each DMU to determine an overall relative efficiency achievement.

15.2.3 DEA Assessment Procedures

DEA is a linear programming-based technique that converts multiple output and input measures into a single comprehensive measure of operator level performance. The DEA assessment procedure for the input variables associated with output variables to measure productivity efficiency is illustrated in Figure 15.3.

For the DEA, selected telecom companies were chosen from the Forbes global 2000 list rankings. We excluded Vodafone Group and America Telecom because of Vodafone's significant negative operating profit and because America Telecom annual report data are not available on the Internet.

The three types of DEA programming were adopted to capture performance based on output variables of revenue, EBITDA,* and operating profit (EBIT). The input variables utilized were total assets, capital expenditures, and total number of employees† (Karlaftis, 2004). The DEA input-oriented models were chosen because cost minimization or reduction is used in this method (Golany and Roll, 1989). It is now popular to rely on non-GAAP financial measures such as the EBITDA and EBITDA margin (%) to assess the operating performance of a company against that of its counterparts. Finally, the CCR efficiency rankings of the selected companies from the Forbes list were analyzed based on the three performance indicators.

* EBITDA represents operating income plus interest, taxes, depreciation, and amortization. It is a good measure for reconciling free cash flow to investment capital cash earnings distributed to shareholders.

† Employee numbers are added to one of the output variables because all the sample telecom companies cited on the Forbes list are publicly traded in open stock markets. Human resource allocations are assumed to be aligned appropriately to induce better operating performance for facing market challenges.

TABLE 15.1
Descriptive Statistics

Items	Mean	Minimum	Maximum	Std. Dev.	Valid N
Total Assets	38615	4450	186200	42946	39
Capex	3204	636	17900	3443	39
Employee	59848	11717	248153	59608	39
Revenue	21666	2930	100200	21537	39
EBITDA	7868	833	33783	7781	39
EBIT	3838	189	14100	3850	39

15.2.4 Productivity Efficiency Measure

A multiple-objective programming method was applied to improve the discriminating power of the classical DEA method that often results in many relatively efficient DMUs. The efficiency achievement approach achieves more discriminating power than the classical efficiency measure. The descriptive statistics are described below.

15.2.4.1 Descriptive Statistics

The collected observations covered 39 leading global telecom companies as ranked on the Forbes 2000 list to be our DEA programming DMUs. The descriptive statistics cover valid numbers, minimums, maximums, means, and standard deviations as shown in Table 15.1.

15.2.4.2 Isotonicity Test

The variables of input and output for the correlation coefficient matrix should comply with the isotonicity premise. In other words, the increase of an input will not cause a decreasing output of another item. We excluded the net income output variable item because the negative values for Sprint PCS and Cable and Wireless (U.K) revealed a negative correlation with input variables and thus failed to meet the requirements of the DEA isotonicity premise. The test for the Pearson correlation matrix among the input and output variables is isotonicity, i.e., the p-value of 0.0001 ($p < 0.05$) matches the basic assumption of the DEA approach. The correlation coefficients among inputs and outputs are listed in Table 15.2.

15.2.4.3 Common Weight for Efficiency Achievement Measure

The efficiency achievement measure is established by using a common multiplier based on the multiple-objective programming approach. Fuzzy multiple-objective programming utilizes the membership function to convert multiple-objective programming to one-objective programming. Unlike classical DEA, the new approach locates one linear frontier as a common reference to calculate the efficiency measure. By using a common frontier, reducing the number of DMUs to improve the discriminating power of the DEA model is relatively efficient. The common weight employed by efficiency achievement measure is shown in Table 15.3.

TABLE 15.2
Correlation Coefficients among Inputs and Outputs

Items	Revenue	EBITDA	Operating Profit
Total Assets	0.9598	0.9561	0.8907
	p = 0.0001	p = 0.0001	p = 0.0001
Capex	0.9312	0.9379	0.8744
	p = 0.0001	p = 0.0001	p = 0.0001
Employee	0.6500	0.6971	0.6998
	p = 0.0001	p = 0.0001	p = 0.0001

TABLE 15.3
Common Weights Employed by Efficiency Achievement Measure

Efficiency Achievement	Input Weight			Output Weight		
Common Weight	0.00000488	0.00015368	0.00000074	0.00000001	0.00005360	0.00000001

15.3 EMPIRICAL RESULTS AND DISCUSSIONS

The empirical results have five efficiency measures: CCR, BCC, scale, A&P, and efficiency achievement. A comparison of the differences of the best practice frontiers of efficient DMUs was performed for the three DEA methods. Figure 15.4 is a scatter graph of CCR efficiency scores (between 0 and 1) and associated EBITDA

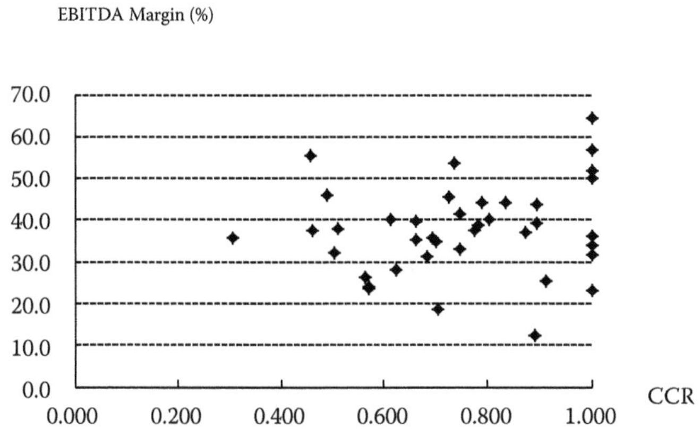

FIGURE 15.4 China telecom performance. Comparison of CCR efficiency scores and EBITDA margins.

margins (%) that displays the efficiency distribution. The efficiency comparison was performed for Asia, Europe, and America. In addition, we discuss representative characteristics of efficient operators and a case study for Chunghwa Telecom to explore operating strategies.

15.3.1 Productivity Efficiency

This study uses the traditional radial DEA measure, A&P efficiency measure, and efficiency achievement measure to compute the various productivity ratings for leading global telecom operators. The efficiency assessment conducted was based on the input and output variables of each DMU.

The CCR model for each DMU is the overall technical efficiency that can be condensed into a measure of pure technical efficiency (BCC) and scale efficiency (Coelli, 1998). All of the efficiency scores are bounded between 0 and 1 (CCR = BCC × scale).

A score equal to 1 indicates relative productivity efficiency while a value less than 1 is regarded as relatively inefficient. The various productivity efficiency estimations by CCR, BCC, scale, A&P, and efficiency achievement measures are listed in Table 15.4. Certain public telecommunications organizations (PTOs) with full efficiency ratings are considered models or benchmarks for less efficient organizations (Pentzaropoulos and Giokas, 2002).

15.3.2 Efficient DMUs

DEA is a mathematical programming technique for identifying efficient frontiers for peer DMUs (Chen and Ali, 2002). Empirical results indicate that the A&P and the efficiency achievement measure can provide clearer DEA efficiency ratings. Efficiency observations for CCR involve observations of eight companies: NTT DoCoMo, Swisscom, KDDI Group, Telstra Corporation, NTT Corporation, Carso Global Telecom, Telkom Indonesia, and China Mobile. NTT DoCoMo shows the best-practice frontier (1.861) for A&P efficiency. The findings of the efficiency achievement measure provide stricter efficiency indicators; the only DMUs that matched them were Swisscom, Telstra. and Carso.

Table 15.5 compares results for the CCR, A&P, and efficiency achievement measures. Telecom companies other than NTT DoCoMo and KDDI are not in identifiable local markets and are thus excluded from a comparison of market share. The market share of NTT DoCoMo (56.3%) surpasses KDDI (20.8%) in the mobile market of Japan and these percentages are therefore consistent with the A&P efficiency values for DoCoMo (1.861) and KDDI (1.543) shown in Table 15.5.

15.3.3 CCR versus EBITDA Margin (%)

Higher EBITDA margin percentages mean that enterprises have free cash flows for making investments and paying returns on shareholders' investments. Generally, an EBITDA margin over 40% implies that an enterprise is in healthy financial condition,

TABLE 15.4
Various Productivity Efficiency Models

	Efficiency Model				
DMUs	**CCR**	**BCC**	**Scale**	**A&P**	**Achievement Measure**
Verizon Communications	0.612	1.000	0.612	0.612	0.525
NTT Corp.	1.000	1.000	1.000	1.185	0.491
NTT DoCoMo	1.000	1.000	1.000	1.861	0.622
SBC Communication	0.662	0.752	0.880	0.662	0.544
BellSouth Corp.	0.727	0.867	0.839	0.727	0.698
BT Group	0.682	1.000	0.682	0.682	0.574
China Mobile	1.000	1.000	1.000	1.021	0.551
AT&T Corp.	0.912	1.000	0.912	0.912	0.619
Telstra Corp.	1.000	1.000	1.000	1.501	1.000
BCE	0.510	0.536	0.951	0.510	0.510
AT&T Wireless	0.564	0.565	0.998	0.564	0.355
Nextel Communications	0.784	0.950	0.826	0.784	0.575
China Telecom	0.458	0.584	0.785	0.458	0.410
Sprint FON	0.704	0.721	0.977	0.704	0.356
KDDI	1.000	1.000	1.000	1.543	0.657
Qwest Communication	0.569	0.582	0.977	0.569	0.372
KT Corp.	0.504	0.539	0.936	0.504	0.461
Alltel	0.661	0.801	0.825	0.661	0.618
Singapore Telecom	0.693	1.000	0.693	0.693	0.588
Swisscom	1.000	1.000	1.000	1.575	1.000
Telenor	0.703	0.968	0.726	0.703	0.703
Chunghwa Telecom	0.735	0.877	0.839	0.735	0.725
Deutsche Telekom	0.746	1.000	0.746	0.746	0.606
France Telecom	0.875	1.000	0.875	0.875	0.774
China Unicom	0.461	0.533	0.865	0.461	0.341
Telefonica	0.836	1.000	0.836	0.836	0.836
Telecom Italia	0.789	0.974	0.811	0.789	0.678
TDC Group	0.625	0.944	0.662	0.625	0.486
Portugal Telecom	0.896	1.000	0.896	0.896	0.861
Carso Global Telecom	1.000	1.000	1.000	1.118	1.000
Royal KPN	0.896	0.896	1.000	0.896	0.799
Hellenic Telecom	0.305	0.878	0.347	0.305	0.295
TeliaSonera Group	0.775	0.797	0.972	0.775	0.694
Sprint PCS	0.746	0.746	1.000	0.746	0.618
mmO2	0.571	1.000	0.571	0.571	0.309
Telkom Indonesia	1.000	1.000	1.000	1.058	0.542
Telkom (S. Africa)	0.804	1.000	0.804	0.804	0.728
Telekom Malaysia	0.489	0.991	0.494	0.489	0.488
Cable & Wireless	0.892	1.000	0.892	0.892	0.309
Mean	0.748	0.885	0.852	0.822	0.598

TABLE 15.5
DMU Efficiency Comparison

Efficient DMU	CCR	A&P	Efficiency Achievement
NTT DoCoMo	1.000	1.861	—
Swisscom	1.000	1.575	1.000
KDDI	1.000	1.543	—
Telstra Corp.	1.000	1.501	1.000
NTT Corp.	1.000	1.185	—
Carso Global Telecom	1.000	1.118	1.000
Telkom Indonesia	1.000	1.058	—
China Mobile (HK)	1.000	1.021	—

and these enterprises comprise 35.9% of all companies studied. DMUs having EBITDA margins between 30 and 40%, represent 41% of the companies studied.

Although KDDI reveals a best-practice frontier, its EBITDA margin of 23.1% is somewhat lower than those of other efficient DMUs. Conversely, China Telecom performs at a higher EBITDA margin of 55.2% but a lower CCR efficiency value of 0.458, revealing its relative non-scale efficiency. The details are shown in Figure 15.4.

15.3.4 Efficiency Ranking Comparison

The construction of alternative objectives can aid our understanding of the complexity of analyzing the performances of telecommunications operations (Giokas and Pentzaropoulos, 2000). Such analyses may have significant implications for foreign investors when they decide whether to invest in certain telecom companies. The establishment of CCR efficiency rankings for a number of DMUs demonstrates the relative advantages of EBITDA margins, returns on assets (ROAs), total asset turnover, and profitability*as analytical tools when compared with the Forbes 2000 rankings of telecom companies. The quantitative performance indicators obtained from four separate measures are significantly different from the Forbes rankings and CCR efficiency ratings. The Forbes rankings along with CCR, EBITDA margin, ROA, total assets turnover, and profitability ratings are listed in Table 15.6.

The profitability rankings and net profit ratios (%) have been added to Table 15.6 to show their relationships to the efficiency measures. The correlation matrix, however, was based on various efficiency rankings. The CCR ranking reveals a high correlation with turnover ranking (0.61478), while the profitability ranking displays a higher correlation with ROA (0.91316) and EBITDA margins (0.56579). Interestingly, the Forbes ranking shows a lower correlation with ROA level (0.20911), profitability (0.19514), and CCR ranking (0.03462) as described in Table 15.7 and Figure 15.5.

* Performance indicator formulae: EBITDA margin (%) = EBITDA/revenue. Return on assets (ROA, %) = net income/total assets. Total asset turnover (%) = revenue/total assets. Profitability (%) = net income/revenue.

TABLE 15.6
Summary of Forbes and Other Rankings of Efficiency Performance

Forbes ranking	Telecom operators	CCR ranking	A&P, CCR	EBITDA ranking	EBITDA Margin (%)	ROA ranking	ROA	Turnover ranking	Total Turnover (%)	Assets Profitability ranking	Net profit ratio (%)
3	NTT DoCoMo	1	1.861	22	36.1	9	9.8	8	76.0	13	12.9
20	Swisscom	2	1.575	30	31.8	2	15.5	1	143.9	18	10.8
15	KDDI	3	1.543	37	23.1	27	4.2	3	101.1	32	4.1
9	Telstra Corp.	4	1.501	6	50.2	7	12.4	15	62.4	8	19.8
2	NTT Corp.	5	1.185	27	33.9	30	3.1	26	53.8	27	5.8
30	Carso Global Telecom	6	1.118	5	51.7	31	3.1	16	60.4	29	5.1
36	Telkom Indonesia	7	1.058	1	64.5	4	14.8	13	65.8	3	22.5
7	China Mobile	8	1.021	2	56.6	1	18.5	7	82.4	4	22.4
8	AT&T Corp.	9	0.912	34	25.5	28	4.1	9	72.9	28	5.6
31	Royal KPN	10	0.896	11	43.7	5	14.1	12	66.6	5	21.2
29	Portugal Telecom	11	0.896	16	39.3	29	3.5	11	69.3	30	5.1
39	Cable & Wireless	12	0.892	39	12.2	39	−7.3	2	113.5	39	−6.5
24	France Telecom	13	0.875	21	37.1	26	4.2	17	60.1	26	7.0
26	Telefonica (Spain)	14	0.836	9	44.4	25	4.3	23	55.8	22	7.8
37	Telkom (S. Africa)	15	0.804	14	40.0	8	10.3	5	93.2	17	11.1
27	Telecom Italia	16	0.789	10	44.3	34	1.7	24	54.9	33	3.1
12	Nextel Commu.	17	0.784	17	39.0	16	7.1	29	52.5	12	13.6
33	TeliaSonera Group	18	0.775	20	37.3	18	6.0	28	52.9	15	11.3
34	Sprint PCS	19	0.746	12	41.6	38	−3.0	21	58.1	38	−5.2
23	Deutsche Telekom	20	0.746	28	32.8	36	1.2	27	53.4	36	2.2
22	Chunghwa Telecom	21	0.735	4	53.5	3	15.2	20	58.8	2	25.9

(continued)

TABLE 15.6
Summary of Forbes and Other Rankings of Efficiency Performance (Continued)

Forbes ranking	Telecom operators	CCR ranking	A&P, CCR	EBITDA ranking	EBITDA Margin (%)	ROA ranking	ROA	Turnover ranking	Total Turnover (%)	Assets Profitability ranking	Net profit ratio (%)
5	BellSouth Corp.	22	0.727	8	45.5	13	7.9	32	45.6	9	17.3
14	Sprint FON	23	0.704	38	18.6	35	1.3	14	64.9	37	2.1
21	Telenor	24	0.703	26	34.6	14	7.6	6	85.4	21	8.9
19	Singapore Telecom	25	0.693	23	35.7	6	14.1	34	41.5	1	33.9
6	BT Group	26	0.682	31	31.4	15	7.4	4	98.2	24	7.5
4	SBC Communication	27	0.662	25	35.1	10	8.6	35	41.0	6	20.9
18	Alltel (U.S.A.)	28	0.661	15	39.7	12	8.0	30	47.9	10	16.7
28	TDC Group	29	0.625	32	28.0	32	2.9	18	60.0	31	4.8
1	Verizon Communi.	30	0.612	13	40.3	24	4.4	36	40.8	19	10.7
35	mmO2	31	0.571	35	24.2	33	1.8	19	60.0	34	2.9
16	Qwest Commu.	32	0.569	36	23.5	17	6.2	25	54.6	16	11.3
11	AT&T Wireless	33	0.564	33	26.3	37	0.9	39	33.0	35	2.7
10	BCE	34	0.510	18	37.7	22	4.6	31	47.1	20	9.7
17	KT Corp.	35	0.504	29	32.0	19	5.2	10	72.6	25	7.2
38	Telekom Malaysia	36	0.489	7	46.2	20	4.9	33	41.9	14	11.8
25	China Unicom	37	0.461	19	37.6	23	4.4	22	57.3	23	7.8
13	China Telecom	38	0.458	3	55.2	11	8.1	37	38.8	7	20.8
32	Hellenic Telecom	39	0.305	24	35.5	21	4.7	38	33.5	11	13.9

TABLE 15.7
Correlations of Various Rankings

Ranking	Forbes	CCR	EBITDA Margin	ROA	Turnover	Profitability
Forbes	1	0.03462	−0.10891	0.20911	−0.15243	0.19514
CCR	0.03462	1	0.11761	0.13482	0.61478	−0.02733
EBITDA Margin	−0.10891	0.11761	1	0.44717	−0.21984	0.56579
ROA	0.20911	0.13482	0.44717	1	0.10466	0.91316
Turnover	−0.15243	0.61478	−0.21984	0.10466	1	−0.23603
Profitability	0.19514	−0.02733	0.56579	0.91316	−0.23603	1

15.3.5 Efficiency Comparison

The application of the Mann-Whitney rank order statistic (Conover, 1980) utilized SPSS software to compare the relative efficiencies of companies operating in the Asia-Pacific, Europe, and Americas regions and the two operating patterns (state-owned and privatized) based on CCR efficiency scores as shown in Table 15.8.

15.3.5.1 Regions

The area mean test proves that the operating performances of Asia-Pacific operations (0.77833) are superior to those of Europe (0.75967), while European performances are superior to those of the Americas (0.70425). However, the differences between regions are not significant because the sig. value is 0.611 ($p > 0.05$). The description of region efficiency comparison is shown in Table 15.9. The results of the F test and significance are listed in Table 15.10.

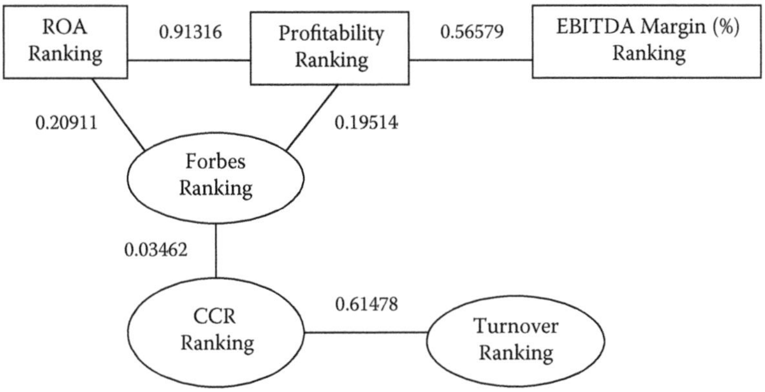

FIGURE 15.5 Correlation comparisons for various rankings.

TABLE 15.8
Operating Patterns and Region Distributions of Telecommunication Companies

Pattern	Region	Companies
State-owned	Asia-Pacific	China Mobile, China Telecom, China Unicom, Telstra, Singapore Telecom, CHT, Telkom Indonesia, Telekom Malaysia
	Europe	Swisscom, Telenor, France Telecom
	Americas	NA
Privatized	Asia-Pacific	NTT, DoCoMo, KDDI, KT, Telkom SA,
	Europe	BT, Deutsche Telekom, Telefonica, Telecom Italia, Portugal Telecom, Royal KPN, TDC, Hellenic Telecom, Telia Sonera Group, MMO2, Cable & Wireless
	Americas	AT&T, AT&T Wireless, Nextel, Verizon, SBC, BellSouth, Qwest (now Century Link), Alltel, BCE, Carso Global Telecom, Sprint PCS, Sprint FON

Notes: NA = not applicable. France Telecom was privatized at the end of 2005.

15.3.5.2 State-Owned and Privatized Patterns

The operating pattern mean test proved that business performance for the state-owned group (0.76491) was superior to performance of the privatized group (0.74186). However, the differences between operating patterns are not significant because the sigma value is 0.734 ($p > 0.05$). Most of the state-controlled telecom companies (Telstra, Singapore Telecom, Chunghwa Telecom, Telkom Indonesia, Telekom Malaysia, Telenor, and others) are full-service operators and maintain mobile and Internet growth segments to offset the revenue declines from fixed-line telephony.

In addition, the state-controlled operations in Chinese markets (China Mobile, China Telecom, and China Union) are all protected by government regulations. The

TABLE 15.9
Area Mean Test Results

Area	Mean	Maximun	Minimum	Std. Deviation	Median	N
Asia-Pacific	0.77833	1.000	0.458	0.24611	0.86750	12
Europe	0.75967	1.000	0.305	0.16998	0.78900	15
America	0.70425	1.000	0.510	0.14369	0.68300	12
Total	0.74836	1.000	0.305	0.18736	0.74600	39

TABLE 15.10
Results of F Test between and within Groups (CRS versus Area)

CRS * Area	Sum of Squares	df	Mean Square	F	Sig.
Between Groups (Combined)	0.036	2	0.018	0.500	0.611
Within Groups	1.298	36	0.036		
Total	1.334	38			

Significance level $\alpha = 0.05$.

comparison of pattern efficiency is shown in Table 15.11. The results of the F test and significance are listed in Table 15.12.

15.3.6 Efficient DMUs

The empirical results reveal that Swisscom, Telstra, and Carso attained the best-practice frontier for CCR efficiency and the efficiency achievement measure. They also indicate that NTT DoCoMo, KDDI Group, NTT Corporation, Telkom Indonesia, and China Mobile (Hong Kong) had the best-practice frontiers for CCR efficiency but were not efficient based on the efficiency achievement measure. In particular, Carso Global Telecom* revealed a relatively better CCR efficiency score and EBITDA margin rating, but ranked 30th among telecom operators on the Forbes 2000 list. Overviews of business strategies of the efficient DMUs are described below.

Swisscom AG — The Swiss government owns 62.7% of the state-controlled incumbent operators. Swisscom's mobile communications business was spun off into a joint venture known as Swisscom Mobile in January 2001. Swisscom owns 75% of Swisscom Mobile and Vodafone owns the remaining 25%. Swisscom controls an extensive network of European mobile communications service providers and resellers, and operates international fixed-line networks in several key European population and business centers.

Telstra — Telstra is 50.1%-owned by the Australian government and provides a full range of telecom services in Australia. Telstra joined forces with PCCW in 2001 to form a Pan-Asia Internet Protocol (IP) backbone network. Telstra has a strong presence in New Zealand through TelstraClear, is a 50:50 partner in a joint

* Carso Global Telecom (CGT) is a Mexican fixed-line communication operator wholly owned by Carlos Slim. CGT controls Telmex with 27% of ownership. It also owns 60% of Prodigy, a U.S. Internet service provider and a small stake in Mcleod USA. Telmex is the incumbent and remains the market leader in Mexico.

TABLE 15.11
Results of Mean Test of Operating Patterns

Operating patterns	Mean	Maximum	Minimum Std.	Deviation	Median	N
State-owned telcos	0.76491	1.000	0.458	0.22489	0.73500	11
Privatized telcos	0.74186	1.000	0.305	0.17469	0.74600	28
Total	0.74836	1.000	0.305	0.18736	0.74600	39

venture with PCCW, and has full ownership of CSL, a mobile operator in Hong Kong. Telstra's market share in Australia continues to erode because of increasing niche competition from companies such as Optus SingTel.

NTT DoCoMo — NTT Corporation owns 61.5% of NTT DoCoMo. NTT DoCoMo is Japan's leading mobile company and the second largest global mobile operator. NTT DoCoMo steadily promotes globalization of both its i-mode services and its 3G systems based on WCDMA technology, as well as promoting overseas operations of mobile multimedia services. NTT DoCoMo had a negative value of net income during the financial year 2002 (2001.3~2002.3) because of its losses in overseas investments and the dot.com bubble effect, but thereafter rapidly recovered financial health in fiscal year 2003.

Telkom Indonesia — The Republic of Indonesia owns 51.19% of this national company. More than a fixed-line telephone provider, Telkom offers a full range of integrated multimedia network services in one package. Telkom has built a hybrid fiber coaxial (HFC) network called Broadband Access Network 2000 to deliver high-speed Internet, video-on-demand, and voice services. Its vision is to become a leading information and communications player in Asia.

15.3.7 Operating Strategy Case Study

In the 1990s some incumbent telecom operators such as NTT, KT, BT, Italia Telecom, and PCCW split their mobile arms from their parent bodies as independent companies for initial public offerings (IPOs) as part of their privatization processes. Nevertheless, CHT, Singapore Telecom, Telstra, and Deutsche Telekom still retain full-service telecom operations.

TABLE 15.12
F Test Results between and within Groups (CRS versus Operating Patterns)

CRS * Operating patterns	Sum of Squares	df	Mean Square	F	Sig.
Between Groups (Combined)	0.004	1	0.004	0.117	0.734
Within Groups	1.330	37	0.036		

Significance level $\alpha = 0.05$.

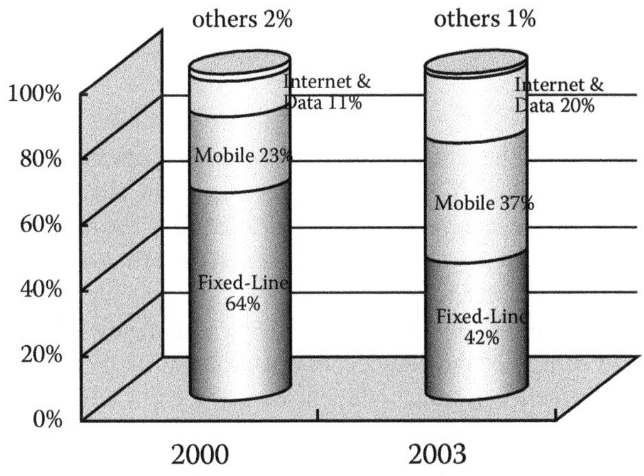

FIGURE 15.6 CHT's fast move to mobile and Internet markets.

CHT in particular has excellent rankings (fourth for EBITDA margin, third for ROA, and second for profitability in comparison to other leading operators. What key factors led to this striking success in the face of strong market competition? Although CHT is 22nd in the Forbes global ranking, consistent with its 21st place in the CCR ranking, it is not as large in economic scale as Forbes measured because its revenue resources are limited to Taiwan. Nevertheless, exploration of the key factors leading to this success is worth exploring. CHT's four strategies created to cope with changes in the telecom industries and the entrance of competitors are: (1) defending the fixed-line market, (2) preempting the mobile market, (3) being the leading broadband supplier, and (4) driving value-added services (VAS).

CHT moved fast to develop Internet broadband subscribers and regain the biggest market share in mobile services. Its dominance in DSL access has made it difficult for competitors to gain scale efficiency and generate enough margin to compete profitably. Although CHT has become more efficient in an intensely competitive market, the slow progress of its privatization has limited its opportunities for international expansion. During 2001, the company widely reduced fixed-line tariffs to cope with competition and quickly moved to mobile markets and Internet broadband segments, with the resulting revenue structure change shown in Figure 15.6.

15.4 CONCLUSIONS

This study produced interesting findings concerning performance ratings and comparisons of productivity efficiency for the leading global telecom operators ranked on the Forbes global 2000 list. The operating performance indicators of EBITDA margin, ROA, total asset turnover, and net profit ratio were assessed by mass investors and relate significantly to market success. However, the Forbes rankings display a low correlation with CCR, ROA, and profitability rankings. The empirical results are summarized below:

1. About 20.5% of the Forbes 2000 telecoms are operating on the best-practice frontier for CCR efficiency measure and only 7.7% match the efficiency achievement measure criteria. Therefore, the efficiency achievement measure provides stricter and clearer DEA efficiency indicators.
2. Although Asia-Pacific telecom operators show relatively higher scoring than area counterparts by the Mann-Whitney test, the differences are insignificant. In sum, 8 of the 39 telecom operators on the Forbes list achieved higher best-practice frontier efficiencies using the CCR model and 6 of those are in the Asia-Pacific region (NTT DoCoMo, NTT Corporation, KDDI, Telkom Indonesia, Telstra, and China Mobile).
3. The state-owned telecoms show somewhat higher scoring than the privatized group although the differences are not significant in light of the present move toward more liberalization in telecommunications. Most state-controlled telecoms are still full-service operators with mobile, fixed-line, and Internet segments that achieve integrated economic efficiency. Privatization also enables a government to sell less than 50% of telecom share holdings to acquire additional income. The privatized operators face more fierce competition in the market. Local government policy protection as a key factor deserves further research efforts.
4. The current focus for incumbent telecom operators is finding new growth opportunities in broadband via DSL or FTTX (FTTB, FTTH) access and mobile data markets. Fixed mobile substitution (FMS) and voiceover Internet (VoIP) caused declines in voice revenue for most of the incumbents. They are now offering fixed networks of SMS, MMS, and other value-added services, but revenue growth from such services cannot easily offset losses in legacy services.

Competition continues to increase in a liberalized market and arises from global and regional alliances formed by telecom operators in fixed-line, Internet, and wireless markets. Telecom operators introduce value-added services to develop value-added content hoping to persuade customers to spend more, allowing the companies to take full advantage of revenue streams from Internet and wireless broadband to stimulate increased use of fixed-line and wireless networks. This chapter's assessment of the relative operational performances of the leading global telecom operators applying traditional radial, A&P, and efficiency achievement measures is useful, and efficiency approaches should be studied further.

APPENDIX A: GLOBAL BROADBAND PENETRATION RANKINGS

Table 15A.1 lists global broadband penetration rankings for 29 countries as of December 31, 2003.

TABLE 15A.1
Broadband Penetration Rankings for 29 Countries

Ranking	Countries	Broadband lines/100 population	Ranking	Countries	Broadband lines/100 population
1	S. Korea	23.48	16	Israel	6.63
2	H.K.	18.18	17	France	5.76
3	Canada	14.81	18	Germany	5.53
4	Switzerland	14.42	19	U.K	5.39
5	Taiwan	13.42	20	Spain	5.30
6	Denmark	12.43	21	Portugal	4.86
7	Belgium	12.35	22	Italy	4.41
8	Netherlands	11.61	23	Australia	3.80
9	Japan	10.70	24	New Zealand	1.74
10	Sweden	10.68	25	China	1.05
11	Singapore	8.98	26	Ireland	0.77
12	U.S.A.	8.71	27	Poland	0.45
13	Finland	8.02	28	Mexico	0.20
14	Norway	7.81	29	Malaysia	0.09
15	Austria	7.20			

Source: International Telecommunicatoins Union World Indicators, 2004.

APPENDIX B: GLOBAL MOBILE PENETRATION RANKINGS

Table 15B.1 lists global mobile penetration rankings for 37 countries as of December 31, 2003.

APPENDIX C: DEA EFFICIENCY MEASURES

C.1 CLASSICAL EFFICIENCY MEASURE

If the a-th DMU uses m-dimension input variables x_{ia} (i = 1,...,m) to produce s-dimension output variables y_{ra} (r = 1,...,s), the efficiency of DMU h_a can be found from the following model:

$$\text{Max } h_a = \frac{\sum_{r=1}^{S} u_r \, y_{ra}}{\sum_{i=1}^{m} v_i x_{ia}} \tag{15C.1}$$

subject to:

$$\frac{\sum_{r=1}^{S} u_r \, y_{rk}}{\sum_{i=1}^{m} v_i x_{ik}} \le 1, \qquad k = 1,\ldots,n$$

$$0 < \varepsilon \le u_r, 0 < \varepsilon \le v_i, \qquad i = 1,\ldots,m; \qquad r = 1,\ldots,s$$

TABLE 15B.1
Global Penetration Rankings for 37 Countries

Countries	Mobile Penetration (%)	Countries	Mobile Penetration (%)
Taiwan, China	110	Australia	72
Hong Kong	106	France	70
Italy	102	S. Korea	69
Iceland	97	Japan	68
Spain	92	New Zealand	65
Norway	91	United States	54
Portugal	90	Poland	45
Finland	90	Malaysia	44
Sweden	90	Canada	42
Denmark	89	South Africa	36
Israel	88	Philippines	32
United Kingdom	88	Russia	30
Austria	88	Mexico	29
Singapore	85	Brazil	26
Ireland	84	China	21
Switzerland	84	Indonesia	9
Germany	79	Viet Nam	3
Greece	78	India	2
Netherlands	77		

Source: International Telecommunication Union World Indicators, 2004.

where x_{ik} denotes the i-th input of the k-th DMU; y_{rk} is the r-th output of the k-th DMU; u_r and v_i indicate the weight of the r-th output and the i-th input, respectively, and h_a is the relative efficiency value. Since Equation (15C.1) involves fractional programming, it is difficult to solve. Charnes et al. (1978) converted it to a linear programming (LP) model to find a solution.

$$\text{Max} \quad h_a = \sum_{r=1}^{s} u_r y_{ra} \tag{15C.2}$$

subject to:

$$\sum_{i=1}^{m} v_i x_{ik} - \sum_{r=1}^{s} u_r y_{rk} \geq 0, \qquad k = 1,\ldots,n$$

$$\sum_{i=1}^{m} v_i x_{ia} = 1$$

$$u_r \geq \varepsilon > 0, \qquad r = 1,\ldots,s$$

$$v_i \geq \varepsilon > 0, \qquad i = 1,\ldots,m$$

When the number of elements in the dual problem in Equation (15C.2) can be reduced in an effort to find an answer, the dual problem after conversion becomes:

$$\text{Min} \left\{ \theta_a - \varepsilon \left[\sum_{i=1}^{m} S_{ia}^{-} + \sum_{r=1}^{s} S_{ra}^{+} \right] \right\} \quad (15C.3)$$

subject to:

$$\theta_a x_{ia} - \sum_{k=1}^{n} \lambda_k x_{ik} - S_{ia}^{-} = 0, \qquad i = 1,\ldots,m$$

$$y_{ra} \quad - \sum_{k=1}^{n} \lambda_k y_{rk} + \quad s_{ra}^{+} = 0, \quad r = 1,\ldots,s$$

$$S_{ia}^{-}, S_{ra}^{+}, \lambda_k \geq 0$$

where S_{ia}^{-} and S_{ra}^{+} are slack variables.

C.2 A&P EFFICIENCY MEASURE

The Andersen and Petersen (A&P) efficiency measure function is proposed as shown below (Chiang and Tzeng, 2000).

Revenue Aspect

max Φ_k

$$s.t. \quad \sum_{j=1}^{n} \lambda_j x_{ij} \leq x_{ik}$$

$$\sum_{r=1}^{n} \lambda_j y_{rj} \geq \theta_k y_{rk}$$

$$\sum_{j=1}^{n} \lambda_j = 1$$

$\lambda_j \geq 0;\ i = 1,\ldots, m;\ j = 1,\ldots ,n;\ r = 1,\ldots, s$

Cost Aspect

min θ_k

$$s.t. \quad \sum_{j=1}^{n} x_{ij} \lambda_j \leq \theta_k x_{ik}$$

$$\sum_{j=1}^{n} y_{rj} \lambda_j \geq y_{rk}$$

$$\sum_{j=1}^{n} \lambda_j = 1$$

$\lambda_j \geq 0;\ i = 1,\ldots,m;\ j = 1,\ldots ,n;\ r = 1,\ldots, s$

where Φ_k = revenue efficiency value ($k = 1,\ldots,n$) and θ_k = cost efficiency value ($k = 1,\ldots,n$). The ratio OC′/OC defines the efficiency measure for evaluating unit C. An index value equal to or larger than 1 may be interpreted as a maximum proportional increase in a vector that contains corresponding input characterization of that observation as an

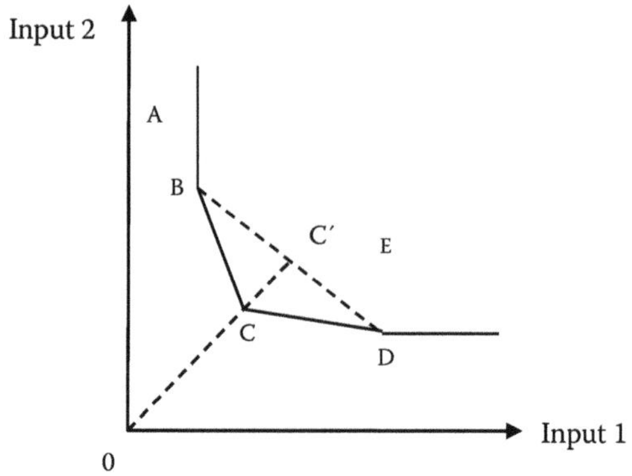

FIGURE 15C.1 Unit isoquant spanned by the A&P model.

efficiency value (Andersen and Petersen, 1993). The unit isoquant spanned by their model is described in Figure 15C.1.

C.3 EFFICIENCY ACHIEVEMENT MEASURE

Multiple objective programming can be employed to find a set of consistent weight combinations so that an optimized efficiency value can be calculated for each DMU to assess relative efficiency achievement (Chiang and Tzeng, 2000). This goal is formulated in Equation (15C.4).

$$\text{Max } h_1 = \frac{\sum_{r=1}^{s} u_r y_{r1}}{\sum_{i=1}^{m} v_i x_{i1}}$$

$$\text{Max } h_2 = \frac{\sum_{r=1}^{s} u_r y_{r2}}{\sum_{i=1}^{m} v_i x_{i2}} \tag{15C.4}$$

$$\vdots$$

$$\text{Max } h_n = \frac{\sum_{r=1}^{s} u_r y_{rn}}{\sum_{i=1}^{m} v_i x_{in}}$$

subject to:

$$\frac{\sum_{r=1}^{s} u_r y_{rk}}{\sum_{i=1}^{m} v_i x_{ik}} \le 1, \qquad k = 1, 2, \cdots, n$$

$$u_r \ge \varepsilon > 0, \qquad r = 1, 2, \cdots, s$$

$$v_i \ge \varepsilon > 0, \qquad i = 1, 2, \cdots, m$$

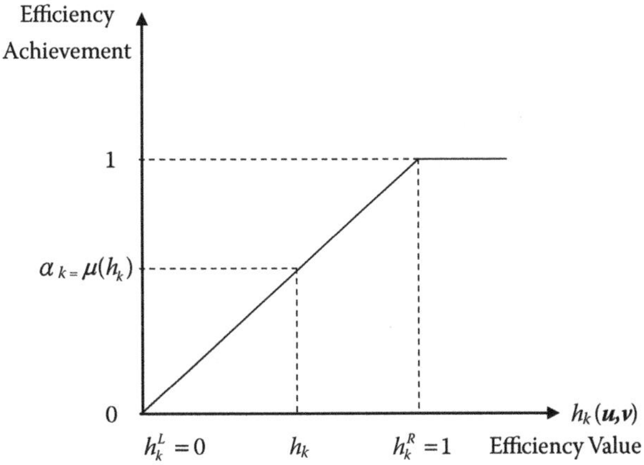

FIGURE 15C.2 Identity function of efficiency achievement.

Given a linear identity function, h_k^L and h_k^R in Figure 15C.2 denote the left and right frontier values of h_k, the k-th objective function value, respectively. The span of h_k^L and h_k^R lies between 0 and 1 because the outcome of the objective equation of Equation (15C.4) is the efficiency ratio, that is $h_k^L = 0$ and $h_k^R = 1$, and the identity value is $\mu(h_k)$, which is the achieved value of efficiency ratio h_k for the DMU. The efficiency value is between 0 and 1 and is called an identity function. Therefore, Equation (15C.4) can be converted to the pattern of fuzzy multiple objective programming (Chiang and Tzeng, 2000).

$$\text{Max} \quad \alpha$$

$$\sum_{r=1}^{s} u_r y_{rk} - \sum_{i=1}^{m} v_i x_{ik} \leq 0, \quad k = 1,\ldots,n \tag{15C.5}$$

$$\sum_{r=1}^{s} u_r y_{rk} - \alpha \sum_{i=1}^{m} v_i x_{ik} \geq 0, \quad k = 1,\ldots,n$$

subject to:

$$0 < \alpha \leq 1$$

$$u_r \geq \varepsilon > 0, \quad r = 1,\ldots,s$$

$$v_i \geq \varepsilon > 0, \quad i = 1,\ldots,m$$

A set of (u^*, v^*) can be calculated according to Equation (15C.5) and the efficiency value hk of a DMU can be calculated with a value of (u^*, v^*). The identity function of efficiency achievement measure is illustrated in Figure 15C.2. Since the efficiency value hk of each DMU actually equals its efficiency achievement αk of the efficiency value hk, it can define the efficiency achievement measure in each DMU as follows:

$$\alpha_k = \sum_{r=1}^{s} u_r^* y_{rk} \Bigg/ \sum_{i=1}^{m} v_i^* x_{ik}, \quad k = 1,\ldots,n \tag{15C.6}$$

APPENDIX D: REFERENCE WEBSITES

Table 15D.1 lists websites for telecom operators cited in the Forbes global 2000 list.

TABLE 15D.1
Websites of Telecom Operators Named in Forbes Global 2000 List

Telecom Ranking	Forbes 2000 Ranking	Country Name	Telecom Operators	Websites
1	26	U.S.A.	Verizon Communication	http://www.verizon.com
2	30	Japan	NTT Corp.	http://www.ntt.co.jp
3	30	Japan	NTT DoCoMo	http://www.nttdocomo.co.jp
4	33	U.S.A.	SBC Communication	http://www.sbc.com
5	84	U.S.A.	BellSouth	http://www.bellsouth.com
6	93	U.K.	BT Group	http://www.btplc.com
7	112	China	China Mobile(HK)	http://www.chinamobile.com.hk
8	116	U.S.A.	AT&T Corp.	http://www.att.com
9	164	Australia	Telstra	http://www.telstra.com.au
10	175	Canada	BCE	http://www.bce.ca
11	187	U.S.A.	AT&T Wireless	http://www.attws.com
12	193	U.S.A.	Nextel Commun.	http://www.nextel.com
13	202	China	China Telecom	http://www.chinatelecom-h.com/chi
14	207	U.S.A.	Sprint FON	http://www.sprint.com
15	211	Japan	KDDI	http://www.kddi.co.jp
16	217	U.S.A.	Qwest Commun.	http://www.qwest.com
17	236	S. Korea	KT Corp.	http://www.kt.co.kr
18	291	U.S.A.	Alltel	http://www.alltel.com
19	318	Singapore	SingTel	http://www.singtel.com.sg
20	320	Switzerland	Swisscom	http://www.swisscom.com
-	355	U.K.	Vodafone	http://www.vodafone.com
21	358	Norway	Telenor	http://www.telenor.nl
22	360	Taiwan, R.O.C.	Chunghwa Telecom	http://www.cht.com.tw
23	364	Germany	Deutsche Telekom	http://www.telekom3.de
24	374	France	France Telecom	http://www.francetelecom.com
25	383	China	China Unicom	http://www.chinaunicom.com.hk

TABLE 15D.1
Websites of Telecom Operators Named in Forbes Global 2000 List (Continued)

Telecom Ranking	Forbes 2000 Ranking	Country Name	Telecom Operators	Websites
26	409	Spain	Telefonica	http://www.telefonica.es
27	412	Italy	Telecom Italia	http://www.telecomitalia.it
28	422	Denmark	TDC Group	http://www.teledanmark.dk
29	442	Portugal	Portugal Telecom	http://www.telecom.pt
30	509	Mexico	Carso Global Telecom	http://www.cqtelecom.com.mx
31	577	Netherlands	Royal KPN	http://www.kpn.com
32	618	Greece	Hellenic Telecom	http://www.ote.gr
33	674	Sweden	Telia Sonera Group	http://www.teliasonera.com
34	677	U.S.A.	Sprint PCS	http://www.sprintpcs.com
35	687	France	mmO2	http://www.o2.com
—	757	Mexico	Am'erica Telecom	http://www.americatelecom.com.mx
36	791	Indonesia	Telekom Indonesia	http://www.telekom.co.id
37	793	S. Africa	Telkom(S. Africa)	http://www.telkom.co.za
38	842	Malaysia	Telekom Malaysia	http://www.telekom.com.my
39	896	U.K.	Cable & Wireless	http://www.cw.com

16 Fuzzy Multiple Objective Programming in Interval Piecewise Regression Model

Tanaka et al. (1982) introduced a fuzzy linear regression model with symmetric triangular fuzzy parameters by using linear programming (LP). Since membership functions of fuzzy sets are often described as possibility distributions, this approach is usually called possibilistic regression analysis. The properties of possibilistic regression formulated by Diamond and Tanaka (1998), Tanaka (1987), Tanaka et al. (1982) and Tanaka and Watada (1988) have been studied further. Kim et al. (1996), Moskowitz and Kim (1993), and Tedden and Woodal (1994) discussed the degree of fit of the fuzzy linear model. The effects of outliers were also examined (Diamond and Tanaka, 1998).

Three shortcomings associated with the fuzzy regression model have been observed (Yu et al., 2005). First, in possibility analysis, Tanaka's methodologies were extremely sensitive to outliers or high variabilities of data and they also ignored certain information contained in the data (Kim et al., 1996; Tedden and Woodal, 1994). Second, in necessity analysis, the necessity area cannot be obtained due to large variations in the data or an inappropriate model (Tanaka et al., 1989; Tanaka and Ishibuchi, 1992).

A necessity model indicates than the assumed model is somewhat reliable. Therefore, Tanaka and Lee (1998) proposed a measure of fitness, which is the ratio of necessity spread divided by possibility spread and averaged over the sample size. However, the approximation model cannot be obtained if the required measure of fitness is set too high. This model must be analyzed with respect to the data property, rather than mere addition of the terms of the polynomial. These issues show the difficulty of managing data with large variations by applying a polynomial or non-linear form. Therefore, the piecewise concept to manage data with large variations was proposed (Yu et al., 1999 and 2001).

Third, when we use LP in possibilistic regression analysis, some coefficients tend to become crisp because of the characteristic of LP. This shortcoming can be alleviated by quadratic programming (QP) proposed by Tanaka and Lee (1998). They devised an interval regression analysis based on QP (Best, 1984; Gill and Murray, 1978; Goldfarb and Indani, 1983) to obtain the possibility and necessity models simultaneously. In their unified approach, they assumed simplicity so that the center coefficients of the possibility regression and the necessity regression model are the same. For a data set with crisp inputs and interval outputs, the possibility and the necessity models can be considered at the same time.

Fuzzy piecewise possibility and necessity regression models (Yu et al., 1999, 2001, and 2005) are employed when a function behaves differently in different parts of the range of crisp input variables. Yu et al.'s 2001 paper requires the analyst to set the number of change points so that the positions of change points and the fuzzy piecewise regression model can be obtained simultaneously. The proper number of change points is still a problem. Hence, we incorporate the concepts of measure of fitness (Tanaka and Lee, 1998), interval piecewise regression (Yu et al., 2005), and multiple-objective technique (Ida and Gen, 1997; Lee and Li, 1993; Li and Lee, 1990; Zimmermann, 1978) to find the measure of fitness and the number of change points, considering all objectives.

The fitness measure should be as high as possible. However, due to the parsimonious rule, the number of change points should be as few as possible. Therefore, the three objectives of minimizing the number of change points, maximizing fitness, and minimizing the objective of obtaining the regression models (Yu et al., 2005) are formulated.

16.1 INTRODUCTION TO MEASURE OF FITNESS AND FUZZY MULTIPLE OBJECTIVE PROGRAMMING

Tanaka and Lee (1998) defined a measure of fitness to gauge the similarities of the obtained possibility and necessity regression models. The larger the fitness, the better the model fits the data. The measure of fitness can be introduced as the overlap of the possibility and necessity models as explained below.

Assume that the input–output data $(x_j; Y_j)$ are $(x_j; Y_j) = (x_{1j}, \ldots, x_{qj}; Y_j), j = 1, \ldots, n$ where x_j is the jth input vector and Y_j is the corresponding interval output. The predicted possibility and necessity models are as follows:

$$\text{Possibility model: } Y^{*}(\mathrm{x}_j) = \left(a_0 + \sum_{i=1}^{q} a_i x_{ij},\ c_0 + \sum_{i=1}^{q} c_i |x_{ij}| + d_0 + \sum_{i=1}^{q} d_i |x_{ij}| \right)$$

$$\text{Necessity model: } Y_{*}(\mathrm{x}_j) = \left(a_0 + \sum_{i=1}^{q} a_i x_{ij},\ c_0 + \sum_{i=1}^{q} c_i |x_{ij}| \right)$$

$a_0 + \sum_{i=1}^{q} a_i x_{ij}$ represents the center of possibility and necessity models and $c_0 + \sum_{i=1}^{q} c_i |x_{ij}| + d_0 + \sum_{i=1}^{q} d_i\, |x_{ij}|$ and $c_0 + \sum_{i=1}^{q} c_i\, |x_{ij}|$ represent the radii of the possibility and necessity models, respectively. The measure of fitness for all data ϕ_Y can be defined as:

$$\phi_Y = \frac{1}{n} \sum_{j=1}^{p} \left(\frac{c_0 + \sum_{i=1}^{q} c_i |x_{ij}|}{c_0 + \sum_{i=1}^{q} c_i |x_{ij}| + d_0 + \sum_{i=1}^{q} d_i |x_{ij}|} \right)$$

where n is the sample size and $0 \le \phi_Y \le 1$. If a necessity model does not exist, $\phi_Y = 0$, which implies the assumed model does not fit the data and must be revised. ϕ_Y is the ratio of necessity spread divided by possibility spread and averaged over the n data. The larger the value of ϕ_Y, the better the model fits the data.

To obtain the possibility and necessity approximation models with the condition of $\phi_Y \ge \omega$, where $0 \le \omega \le 1$, Tanaka and Lee suggested:

1. Take a linear function as an initial regression model.
2. Solve the unified QP problem and calculate ϕ_Y using the obtained possibility and necessity approximation models.
3. Repeat step 2 by increasing the number of terms in the regression polynomial until the assigned condition $\phi_Y \ge \omega$ is satisfied.

The approximation models satisfying $\phi_Y \ge \omega$ may not be obtained if the tolerance limit ω is set too high. Tanaka and Lee mentioned that the problems of determining the reasonable ω and the termination condition of the above procedure are worthy of further study because the above procedure is only conceptual—a trial and error process—and does not help a user decide the number of change points needed in piecewise regression. Therefore, we propose fuzzy multiple-objective programming in interval piecewise regression as a model to determine the fitness and the number of change points.

$$\begin{aligned} &Max \quad \mathbf{Z} = [Z_1, Z_2, \ldots, Z_l]^T, \\ &Min \quad \mathbf{W} = [W_1, W_2, \ldots, W_m]^T, \\ &\text{Subject to constraints} \end{aligned} \tag{16.1}$$

The membership functions for the objective are defined as:

$$\mu_k(Z_k) = \frac{Z_k - Z_k^-}{Z_k^* - Z_k^-}, \quad k = 1, 2, \ldots, l,$$

$$\mu_s(Z_s) = \frac{W_s^- - W_s}{W_s^- - W_s^*}, \quad s = 1, 2, \ldots, m,$$

where Z_k^*, W_s^* and Z_k^-, W_s^- are the ideal and anti-ideal solutions of Equation (16.1).

Max λ

$$\begin{aligned} \text{Subject to } &\lambda \le (Z_k(x) - Z_k^-)/(Z_k^* - Z_k^-), \quad k = 1, 2, \ldots, l, \\ &\lambda \le (W_s^- - W)/(W_s^- - W_s^*), \quad s = 1, 2, \ldots, m, \\ &\text{initial constraints} \end{aligned}$$

where λ is defined as $\lambda = \min_{k,s} (\mu_k(Z)), \mu_s(W_s))$.

Here we adopt this approach to determine the measure of fitness and number of change points of the piecewise regression model simultaneously.

16.2 FUZZY MULTIPLE OBJECTIVE PROGRAMMING IN PIECEWISE REGRESSION MODEL

Our approach involves three objectives in an interval piecewise regression model with automatic change point detection by quadratic programming. As in the following models, the first objective Z_1 is to maximize ϕ_Y, the measure of fitness for all data; the second objective W_1 is to minimize the number of change points; and the third objective W_2 is to minimize the width by quadratic programming and obtain possibility and necessity models (Yu et al., 2005).

In piecewise regression, the initial possibility and necessity models are assumed as follows:

$$Y^*(x_j) = \Bigg(a_0 + \sum_{i=1}^{q} a_i x_{ij} + \sum_{i=1}^{q}\sum_{t=1}^{n-1} b_{it}(|x_{ij} - p_{it}| + x_{ij} - p_{it})/2,$$

$$c_0 + \sum_{i=1}^{q} c_i \, | x_{ij} \, | + \sum_{i=1}^{q}\sum_{t=1}^{n-1} w_{it}(|x_{ij} - p_{it}| + x_{ij} - p_{it})/2 +$$

$$d_0 + \sum_{i=1}^{q} d_i \, | x_{ij} \, |) + \sum_{i=1}^{q}\sum_{t=1}^{n-1} r_{it}(|x_{ij} - p_{it}| + x_{ij} - p_{it})/2)$$

$$Y_*(x_j) = \Bigg(a_0 + \sum_{i=1}^{q} a_i x_{ij} + \sum_{i=1}^{q}\sum_{t=1}^{n=1} b_{it}(|x_{ij} - p_{it}| + x_{ij} - p_{it})/2$$

$$c_0 + \sum_{i=1}^{q} c_i \, |x_{ij}| + \sum_{i=1}^{q}\sum_{t=1}^{n-1} w_{it}(|x_{ij} - p_{it}| + x_{ij} - p_{it})/2)$$

where p_{it} is the ordered ith value of the ith variable and q is the number of independent variables. All data except the last one in each variable are assumed to be potential change points. Detecting the proper number of change points can be done easily by our proposed method.

In the first objective, the measure of fitness ϕ_Y should be as high as possible, so $Z_1^* = 1,\ Z_1^- = 0$ because $0 \le \phi_Y \le 1$. The goal to maximize the measure of fitness is:

$$Z_1 = \phi_Y =$$

$$\frac{1}{n}\sum_{j=1}^{n} \frac{c_0 + \sum_{i=1}^{q} c_i \, |x_{ij}| + \sum_{i=1}^{q}\sum_{t=1}^{n-1} w_{it}(|x_{ij} - p_{it}| + x_{ij} - p_{it})/2}{c_0 + \sum_{i=1}^{q} c_i |x_{ij}| + \sum_{i=1}^{q}\sum_{t=1}^{n-1} w_{it}(|x_{ij} - p_{it}| + x_{ij} - p_{it})/2 + d_0 + \sum_{i=1}^{q} d_i | x_{ij}|) + \sum_{i=1}^{q}\sum_{t=1}^{n-1} r_{it}(|x_{ij} - p_{it}| + x_{ij} - p_{it})/2}$$

In the second objective, the number of change points should be as small as possible, so $W_1^* = 0$, $W_1^- = q(n-1)$ because all data are potential change points except the point with highest value in each variable, assuming no repeated value. The third objective is to obtain the necessity and possibility models by Tanaka and Lee's concept.

$$W_2 = \sum_{j=1}^{n}\left[d_0 + \sum_{i=1}^{q} d_j|x_{ij}| + \sum_{i=1}^{q}\sum_{t=1}^{n-1} r_{it}(|x_{ij} - p_{it}| + x_{ij} - p_{it})/2\right]^2$$
$$+\xi\left(\sum_{i=1}^{q} a_i^2 + \sum_{i=1}^{q} c_i^2 + \sum_{i=1}^{q}\sum_{t=1}^{n-1} b_{it}^2 + \sum_{i=1}^{q}\sum_{t=1}^{n-1} w_{it}^2\right)$$

The membership functions for the three objectives are defined as below:

$$\mu_1(Z_1) = \frac{Z_1 - Z_1^-}{Z_1^* - Z_1^-}$$

$$\mu_1(W_1) = \frac{W_1^- - W_1}{W_1^- - W_1^*}$$

$$\mu_2(W_2) = \frac{W_2^- - W_2}{W_2^- - W_2^*}$$

where Z_1^*, W_1^*, W_2^* and Z_1^-, W_1^-, W_2^- are the ideal and anti-ideal solutions of the problem. The proposed model for interval piecewise regression to consider the above three objectives simultaneously is:

Max λ

Subject to

$$\lambda \le (Z_1 - Z_1^-)/(Z_1^* - Z_1^-)$$

$$\lambda \le (W_1^- - W_1)/(W_1^- - W_1^*)$$

$$\lambda \le (W_2^- - W_2)/(W_2^- - W_2^*)$$

$$a_0 + \sum_{i=1}^{q} a_i x_{ij} + \sum_{i=1}^{q}\sum_{t=1}^{n-1} b_{it}(|x_{ij} - p_{it}| + x_{ij} - p_{it})/2$$
$$-\left\{c_0 + \sum_{i=1}^{q} c_i\,|x_{ij}| + \sum_{i=1}^{q}\sum_{t=1}^{n-1} w_{it}(|x_{ij} - p_{it}| + x_{ij} - p_{it})\,/\,2]\right\}$$
$$-\left\{d_0 + \sum_{i=1}^{q} d_i\,|x_{ij}| + \sum_{i=1}^{q}\sum_{t=1}^{n-1} r_{it}(|x_{ij} - p_{it}| + x_{ij} - p_{it})/2]\right\} \le Y_{jL}$$

$$a_0 + \sum_{i=1}^{q} a_i x_{ij} + \sum_{i=1}^{q}\sum_{t=1}^{n-1} b_{it}(|x_{ij} - p_{it}| + x_{ij} - p_{it})/2$$

$$+\left\{c_0 + \sum_{i=1}^{q} c_i\,|x_{ij}| + \sum_{i=1}^{q}\sum_{t=1}^{n-1} w_{it}(|x_{ij} - p_{it}| + x_{ij} - p_{it})/2]\right\}$$

$$+\left\{d_0 + \sum_{i=1}^{q} d_i\,|x_{ij}| + \sum_{i=1}^{q}\sum_{t=1}^{n-1} r_{it}(|x_{ij} - p_{it}| + x_{ij} - p_{it})/2]\right\} \geq Y_{jR}$$

$$a_0 + \sum_{i=1}^{q} a_i x_{ij} + \sum_{i=1}^{q}\sum_{t=1}^{n-1} b_{it}(|x_{ij} - p_{it}| + x_{ij} - p_{it})/2$$

$$-\left\{c_0 + \sum_{i=1}^{q} c_i\,|x_{ij}| + \sum_{i=1}^{q}\sum_{t=1}^{n-1} w_{it}(|x_{ij} - p_{it}| + x_{ij} - p_{it})/2]\right\} \geq Y_{jL}$$

$$a_0 + \sum_{i=1}^{q} a_i x_{ij} + \sum_{i=1}^{q}\sum_{t=1}^{n-1} b_{it}(|x_{ij} - p_{it}| + x_{ij} - p_{it})/2$$

$$+\left\{c_0 + \sum_{i=1}^{q} c_i\,|x_{ij}| + \sum_{i=1}^{q}\sum_{t=1}^{n-1} w_{it}(|x_{ij} - p_{it}| + x_{ij} - p_{it})/2]\right\} \leq Y_{jR}, \quad j = 1, 2, \ldots, n$$

$$b_{it} < Mu_{it} - 2\delta v_{it} + \delta + \phi v_{it}$$

$$b_{it} > 2\delta u_{it} - Mv_{it} - \delta - \phi u_{it}$$

$$u_{it} + v_{it} \leq I_{it}$$

$$\sum_{i=1}^{q}\sum_{t=1}^{n-1} I_{it} \leq C$$

$$w_{it} \leq MI_{it}$$

$$r_{it} \leq MI_{it}$$

Let $M = 1000$, $\delta = 0.0000000001$, $\phi = 0.000000001$, and $\xi = 0.00001$ for solving this program by LINGO optimization software. Based on piecewise characteristics, our methodology is insensitive to larger variations in the data. A compromise solution (better ϕ_Y and a proper number of change points) can be found automatically.

16.3 NUMERICAL EXAMPLES

Two examples are demonstrated to show the proposed method. The first is from Tanaka and Lee (1998) and the second is from Yu et al. (2005).

16.3.1 Example 1

According to Tanaka and Lee, $\{(x_j; Y_j)\} = \{(1; [15, 30]), (2; [20, 37.5]), (3; [15, 35]), (4; [25, 60]), (5; [25, 55]), (6; [40, 65]), (7; [55, 95]), (8; [70, 100])\}$. The ideal and anti-ideal solutions are $Z_1^* = 1$ and $Z_1^- = 0$, which represent the range of fitness measurement; $W_1^* = 0$ and $W_1^- = 7$, which indicate that the total change points are 7 in the worst case and 0 in the best case; and $W_2^* = 29293.03$ and $W_2^- = 392.01$, which are obtained from W_2 without considering Z_1 and W_1. The goal achievements are:

$$Z_1 = 0.5135$$

$$W_1 = 0.5714$$

$$W_2 = 0.9874$$

The proposed method considers a compromising solution for three objectives. The measure of fitness with different change points is also shown in Table 16.1. As depicted, the best fitness is 0.5135 with three change points. We also show the models with 2, 1, and 0 change points.

TABLE 16.1
Measures of Fitness with Different Change Points

Change Points	Fitness Measure	$Y^*(x)$: Possibility model $Y_*(x)$: Necessity model
3	0.5135	$Y^*(x) = (23.472, 7.4536) + (1.389, 2.407)x + (13.472, 0.417)$ $(\lvert x-3\rvert + x-3)/+ (-20.778, 0)(\lvert x-4\rvert + x-4)/2$ $+ (21.389, 0)(\lvert x-5\rvert + x-5)/2$ $Y_*(x) = (23.472, 4.028) + (1.389, 1.111)x + (13.472, 0.417)$ $(\lvert x-3\rvert + x-3)/+ (-20.778, 0)(\lvert x-4\rvert + x-4)/2$ $+ (21.389, 0)(\lvert x-5\rvert + x-5)/2$
2	0.4225	$Y^*(x) = (16.25, 7.5) + (6.252, 0.625)x + (-7.5, 6.25)$ $(\lvert x-2\rvert + x-2)/+ (11.25, 0)(\lvert x-3\rvert + x-3)/2$ $Y_*(x) = (16.25, 7.5) + (6.252, 0)x + (-7.5, 0)$ $(\lvert x-2\rvert + x-2)/+ (11.25, 0)(\lvert x-3\rvert + x-3)/2$
1	0.3336	$Y^*(x) = (21.667, 5.833) + (2.292, 3.958)x$ $+ (7.5, 0)(\lvert x-3\rvert + x-3)/2$ $Y_*(x) = (21.667, 5.833) + (2.292, 0.208)x$ $+ (7.5, 0)(\lvert x-3\rvert + x-3)/2$
0	0.1258	$Y^*(x) = (7.143, 11.607) + (8.571, 2.679)x$ $Y_*(x) = (7.143, 0) + (8.571, 0.714)x$

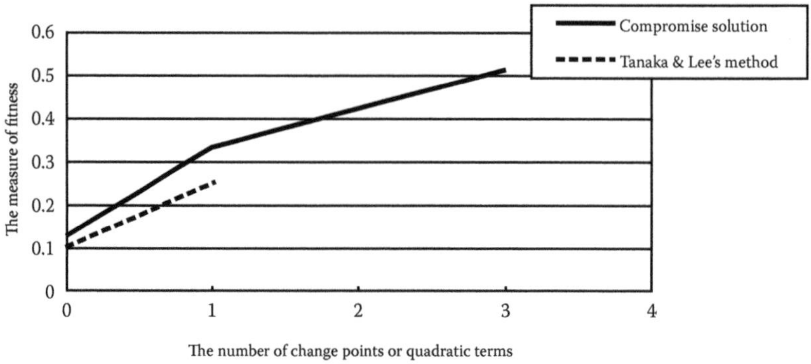

FIGURE 16.1 Comparison of proposed method and Tanaka and Lee's method.

Figure 16.1 illustrates the growth trend with increasing numbers of change points. The solid line represents the fitness by our method and the dotted line represents the fitness with Tanaka and Lee's model. The model with 0 change points is linear. The fitness by the proposed model is still higher than 0.101, which is the measure of fitness by Tanaka and Lee (1998), because we wanted to find the maximum fitness even considering the other two objectives.

Tanaka and Lee use a heuristic method to find a higher ϕ_Y, so after adding a quadratic term, ϕ_Y increases only to 0.252. However, by using our method, ϕ_Y increases to 0.3336 with adding only one change point. Employing Tanaka and Lee's conceptual procedure, the approximation models may not be satisfactory if the tolerance limit ω is set too high. However, by using the fuzzy multiple-objective programming, a better ϕ_Y can be determined automatically.

16.3.2 Example 2

A multivariate example in Yu et al. (2005) is used to explain how our method handles multivariate data. The data are $\{(x_j; Y_j)\}$ = {(1; 2; [22, 32]), (2; 1; [5, 15]), (3; 2; [12, 22]), (5; 5; [43, 53]), (5; 6; [50, 60]), (5; 4; [21, 31]), (5; 3; [4, 14]), (8; 4; [21, 11])}.

The fitness measure ϕ_Y is 0.0702 with three change points. The following is the result of the proposed fuzzy multiple objective program. The ideal and anti-ideal solutions are $Z_1^- = 0$ and $Z_1^* = 1$, which are the range of fitness measurement; $W_1^- = 14$ and $W_1^* = 0$,= 0, indicating that the total numbers of change points are 14 in the worst case and 0 in the best case; and $W_2^- = 720.681$ and $W_2^* = 0.0045$, which are obtained from W_2 without considering Z_1 and W_1. The results are:

$$Z_1 = 0.7143$$

$$W_1 = 0.75$$

$$W_2 = 0.997$$

TABLE 16.2
Comparison of Proposed Method and Yu et al.'s Method

Fitness	Possibility Model	Necessity Model
0.7143	Model 1	Model 2
(Proposed)	$Y^*(x_j) = (-6.167, 5.832) + (17, 0)\,x_2$	$Y_*(x_j) = (-6.167, 4.167) + (17, 0)\,x_2$
	$+ (-11.667, 0)(\lvert x_1 - 2\rvert + x_1 - 2)/2$	$+ (-11.667, 0)(\lvert x_1 - 2\rvert + x_1 - 2)/2$
	$+ (8.056, 0.583)(\lvert x_1 - 5\rvert + x_1 - 5)/2$	$+ (8.056, 0.278)(\lvert x_1 - 5\rvert + x_1 - 5)/2$
	$+ (-13.333, 0)(\lvert x_2 - 5\rvert + x_2 - 5)/2$	$+ (-13.333, 0)(\lvert x_2 - 5\rvert + x_2 - 5)/2$
	$+ (3.75, 0.417)(\lvert x_2 - 4\rvert + x_2 - 4)/2$	$+ (3.75, 0.417)(\lvert x_2 - 4\rvert + x_2 - 4)/2$
0.0702	Model 3	Model 4
(Yu et al.)	$Y^*(x_j) = (-1.949, 3.928)$	$Y_*(x_j) = (-1.949, 0.459)$
	$+ (-2.041, 0)\,x_1 + (16.082, 1.122)\,x_2$	$+ (-2.041, 0)\,x_1 + (16.082, 0)\,x_2$
	$+ (-8.265, 0)(\lvert x_1 - 2\rvert + x_1 - 2)/2$	$+ (-8.265, 0)(\lvert x_1 - 2\rvert + x_1 - 2)/2$
	$+ (5.188, 0.188)(\lvert x_1 - 5\rvert + x_1 - 5)/2$	$+ (5.188, 0.188)(\lvert x_1 - 5\rvert + x_1 - 5)/2$

As described in Table 16.2, the proposed models (1 and 2) and Yu's models (3 and 4) are obtained by considering three objectives via the proposed method. The average goal is 0.7143 with four change points. Models 3 and 4 are obtained by the method of Yu et al. with three pre-specified change points. By increasing one change point ϕ_Y from 0.0702 to 0.7143, we achieve promising results. The fitness difference is 0.6441 by adding only one change point.

16.4 CONCLUSIONS

The proposed method solves the fitness problem of Tanaka and Lee's method. Our contribution is to automatically decide the best fitness measurement rather than use trial and error, and turn Tanaka and Lee's concept into a multiple objective problem. This method can obtain better fitness and a proper number of change points by considering all objectives. An extra objective can be added if required. Fuzzy multiple objective programming is a good technique to solve compromising problems in fuzzy regression and meets practical concerns.

Future research is required to solve the nonlinear issue of fitness and accomplish variable selection in fuzzy regression by employing multiple objective programming. Besides, the multiple objective problem can be solved by the two-phase method to avoid unbalanced solutions (Ida and Gen, 1997; Lee and Li, 1993) if needed.

Bibliography

Abdulaal, M. and LeBlanc, L.J. (1979). Continuous equilibrium network design models. *Transportation Research B* 13(1): 19–32.

Aiyoshi, E. and Shimizu, K. (1981a). Hierarchical decentralized systems and its new solution by a barrier method. *IEEE Transactions on Systems, Man, and Cybernetics* SMC-11: 444–449.

Aiyoshi, E. and Shimizu, K. (1981b). A new computational method for Stackelberg and min-max problems by use of a penalty method. *IEEE Transactions on Automatic Control* 26(2): 460 – 466.

Aiyoshi, E. and Shimizu, K. (1984). A solution method for the static constrained Stackelberg problem via penalty method. *IEEE Transactions on Automatic Control SMC*-29(12): 1111–1114.

Al Harbi, K.M. (2001). Application of the AHP in project management. *International Journal of Project Management* 19(1): 19–27.

Al Khalil, M.I. (2002). Selecting the appropriate project delivery method using AHP. *International Journal of Project Management* 20(6): 469–474.

Al Najjar, B. and Al Syouf, I. (2003). Selecting the most efficient maintenance approach using fuzzy multiple criteria decision making. *International Journal of Production Economics* 84(1): 85–100.

Altrock, C.V. and Krause, B. (1994). Multi-criteria decision-making in German automotive industry using fuzzy logic. *Fuzzy Sets and Systems* 63(3): 375–380.

Andersen, P. and Petersen, N.C. (1993). A procedure for ranking efficient units in data envelopment analysis. *Management Science* 39(10): 1261–1264.

Anderson, E. and Gatignon, H. (1986). Modes of foreign entry: transaction cost and propositions. *Journal of International Business Studies* 17(3): 1–26.

Andreou, A.S., Mateou, N.H., and Zombanakis, G.A. (2005). Soft computing for crisis management and political decision making: the use of genetically evolved fuzzy cognitive maps. *Soft Computing* 9(3): 194–210.

Arino, A. and Abramov, M. (1997). Partner selection and trust building in west European–Russian joint ventures. *International Studies of Management & Organization* 27(1): 19–37.

Armstrong, R.D., Cook, W.D., and Seiford, L.M. (1982). Priority ranking and consensus formation: the case of ties. *Management Science* 28(6): 638–645.

Athanassopoulos, A.D. and Giokas, D. (1998). Technical efficiency and economies of scale in state owned enterprises: the Hellenic telecommunications organization. *European Journal of Operational Research* 107(1): 62–75.

Auster, E. (1994). Macro and strategic perspectives on interorganizational linkages: a comparative analysis and review with suggestions for reorientation. *Advances in Strategic Management* 10(1): 3–40.

Axelrod, R. (1976). *Structure of Decision: The Cognitive Map of Political Elite*. London: Princeton University Press. Azibi, R. and Vanderpooten, D. (2002). Construction of rule-based assignment models. *European Journal of Operational Research* 138(2): 274–293.

Baas, S.M. and Kwakernaak, H. (1997). Rating and ranking of multiple aspect alternative sing fuzzy sets. *Automatica* 13(1): 47–58.

Bakuli, D.L. and Smith, J.M. (1996). Resource allocation in state-dependent emergency evacuation etworks. *European Journal of Operational Research* 89(4): 543–555.

Baky, I.A. (2010). Solving multi-level multi-objective linear programming problems through fuzzy goal programming approach. *Applied Mathematical Modelling*, 34(9): 2377–2387.

Banker, R.D., Charnes, A., and Cooper, W.W. (1984). Some models for estimating technical and scale inefficiencies in data envelopment analysis. *Management Science* 30(9):1078–1092.

Bard, J. F. (1982). A grid search algorithm for the linear bilevel programming problem. In *Proceedings of the 14th Annual Meeting of the American Institute for Decision Science*, pages 256–258.

Bard, J. F. and Falk, J. (1982). An explicit solution to the multilevel programming problem. *Computers and Operations Research*, 9(1):77–100

Bard, J. F. (1983a). An algorithm for the general bilevel programming problem. *Mathematics of Operations Research* 8(2):260–272.

Bard, J. F. (1983b). Coordination of a multidivisional organization through two levels of management. *OMEGA* 11(5):457–468.

Bard, J. F. (1983c). An efficient point algorithm for a linear two-stage optimization problem. Operations Research 31(4):670–684.

Barney, J., Wright, M., and Ketchen, D.J. (2001). The resource-based view of the firm: ten years after 1991. *Journal of Management* 27(6): 625–641.

Barney, J.B. (1991). Firm resources and sustained competitive advantage. *Journal of Management* 17(1): 99–120.

Bedard, J.C. (1985) Use of data envelopment analysis in accounting applications: evaluation and illustration by prospective hospital reimbursement. Ph.D. dissertation, University of WisconsinÐMadison, UMI, Ann Arbor, MI.

Beilock, R. and Dimitrova, D.V. (2003). An exploratory model of inter-country internet diffusion. *Telecommunications Policy* 27(3–4): 237–252.

Beliakov, G. (2002). Monotone approximation of aggregation operators using least squares splines. *International Journal of Uncertainty, Fuzziness and Knowledge-Based Systems* 10(6): 659–676.

Bell, D.E., Raiffa, H., and Tversky, A. (1988). *Decision Making: Descriptive, Normative, and Prescriptive Interactions.* Cambridge: Cambridge University Press.

Bellman, R. (1952). On the theory of dynamic programming. In: Proceedings of the national academy of sciences of the United States of America. National Academy of Sciences, Mathematics 38: 716–719.

Bellman, R. (1953). Bottleneck problems and dynamic programming. In: Proceedings of the national academy of sciences of the United States of America. National Academy of Sciences, Mathematics 39: 747–951.

Bellman, R.E. and Zadeh, L.A. (1970). Decision making in a fuzzy environment. *Management Science* 17(4): 141–164.

Belton, V. and Stewart, T.J. (2002). *Multiple Criteria Decision Analysis: An Integrated Approach.* Boston: Kluwer Academic.

Benayoun, R., Roy, B., and Sussman, N. (1966). *Manual de reference du program ELECTRE. Note de synthese et formation, direction scientifique.* Paris: SEMA.

Bernoulli, D. (1738). Specimen theoriae novae de mensura sortis. *Commentari Academiae Scientiarum Imperialis Petropolitanae* 5: 175–192.

Best, M.J. (1984). Equivalence of some quadratic programming algorithms. *Mathematical Programming* 30(1): 71–87.

Beynon, M.J. and Peel, M.J. (2001). Variable precision rough set theory and data discretisation: an application to corporate failure prediction. *Omega* 29(6): 561–576.

Bialas, W. and Karwan, M. (1978). Multilevel linear programming. Technical Report 78-1, Operations Research Program, State University of New York at Buffalo.

Bialas, W. and Karwan, M. (1982). On two-level optimization. IEEE Transactions on Automatic Control, 27(1):211–214.

Bialas W. and Karwan, M. (1984). Two-level linear programming. *Management Science* 30(8): 1004–1020.

Blair, C. (1992). The computational complexity of multi-level linear programs. Annals of Operations Research, 34(1):13–19.

Bortolotti, B., D'Souza J., Fantini, M. and Megginson, W.L. (2002). Privatization and the source of performance improvement in the global telecommunications industry. *Telecommunications Policy* 26(5–6): 243–268.

Borys, B. and Jemison, D.B. (1989). Hybrid arrangements as strategic alliances: theoretical issues in organizational combinations. *Academy of Management Review* 14(2): 244–249.

Boussofiane, A., Dyson, R.G., and Thanassoulis, E. (1991). Applied data envelopment analysis. *European Journal of Operational Research*, 52(1), 1–15.

Bouyssou, D. and Vansnick, J.C. (1986). Noncompensatory and generalized noncompensatory preference structures. *Theory and Decision* 21(3): 251–266.

Bracken, J. and McGill, J.T. (1973). Mathematical programs with optimization problems in the constraints, *Operations Research* 21(1): 37–44.

Brans, J.P., Mareschal, B., and Vincke, Ph. (1984a). PROMETHEE: a new family of outranking methods in MCDM. In Brans, J.P., Ed., *Operational Research.* Amsterdam: Elsevier, pp. 477–490.

Brans, J.P., Mareschal, B., and Vincke, Ph. (1984b). PROMETHEE: A new family of outranking methods in multi-criteria analysis. In Brans, J.P., Ed., *Operational Research.* Amsterdam, Elsevier, pp. 408–421.

Brans, J.P. and Vincke, Ph. (1985). A preference ranking organization method (the PROMETHEE method for MCDM). *Management Science* 31(6): 647–656.

Bretthauer, K.M. and Shetty, B. (1995). The nonlinear resource allocation problem. *Operations esearch* 43(4): 670–683.

Bretthauer, K.M. and Shetty, B. (1997). Quadratic resource allocation with generalized upper ounds. *Operations Research Letters* 20(1): 51–57.

Brodrick, C.J., Lipman, T.E., Farshchi, M., and Lutsey, N.P. (2002). Evaluation of fuel cell auxiliary ower units for heavy-duty diesel trucks. *Transportation Research D* 7(4): 303–315.

Brouthers, K.D., Brouthers, L.E., and Wilkinson, T.J. (1995). Strategic alliances: choose your partners. *Long Range Planning* 28(3): 18–25.

Buckley, J.J. (1985a). Ranking alternatives using fuzzy numbers. *Fuzzy Sets and Systems* 15(1): 21–31.

Buckley, J.J. (1985b). Fuzzy hierarchical analysis. *Fuzzy Sets and Systems* 17(3): 233–247.

Buckley, J.J. and Eslami, E. (2002). Fuzzy Markov chains: uncertain probabilities. *Mathware and Soft Computing* 9(1): 33–41.

Buyukozkan, G., Ertay, T., and Kahraman, C. (2004). Determining the importance weights for the design requirements in the house of quality using the fuzzy analytic network approach. *International Journal of Intelligent Systems* 19(5): 443–461.

Candler, W. and Norton, R. (1977a). Multilevel programming. Technical Report 20, World Bank Development Research Center, Washington D.C.

Candler, W. and Norton, R. (1977b). Multilevel programming and development policy. Technical Report 258,World Bank Development Research Center, Washington D.C.

Candler, W. and Townsley, R. (1982). A linear two-level programming problem. *Computer and Operations Research* 9(1): 59–76.

Carlsson, C. and Korhonen, P. (1986). A parametric approach to fuzzy linear programming. *Fuzzy Sets and Systems* 20(1): 17–30.

Carlsson, C. and Fuller, R. (1994). Interdependence in fuzzy multiple objective programming. *Fuzzy Sets and Systems* 65(1): 19–29.

Carlsson, C. and Fuller, R. (1995). Multiple criteria decision making: the case for interdependence. *Computers and Operations Research* 22(3): 251–260.

Carlsson, C. and Fuller, R. (1996). Fuzzy multiple criteria decision making: recent developments. *Fuzzy Sets and Systems* 78(2): 139–153.

Carlsson, C. and Fuller, R. (2002). *Fuzzy Reasoning in Decision Making and Optimization*. Heidelberg: Physica.

Chan, C.C. and Tzeng, G.H. (2008). Computing approximations of dominance-based rough sets by bit-vector encodings. In *RSCTC 2008*, Springer Lecture Notes on Artificial Intelligence.

Chan, C.C. and Tzeng, G.H. (2009). Dominance-based rough sets using indexed blocks as granules. *Fundamenta Informaticae* 94(2): 133–146.

Chan, C.C. and Tzeng, G.H. (2011). Bit-vector representation of dominance-based approximation space. In *Transactions on Rough Sets XIII*, Springer Lecture Notes in Computer Science 6499, 1–16. DOI: 10.1007/978-3-642-18302-7_1.

Chandra, P. and Fisher M.L. (1994), Coordination of production and distribution planning. *European Journal of Operational Research* 72 (3): 503–517.

Chang, H.F. and Tzeng, G.H. (2010). A causal decision making model for knowledge management capabilities to innovation performance in Taiwan's high-tech industry. *Journal of Technology Management and Innovation* 5(4): 137–146.

Chang, M.C., Hu, J.L., and Tzeng, G.H. (2009). Decision making on strategic environmental technology licensing: fixed-fee versus royalty licensing methods. *International Journal of Information Technology and Decision Making* 8(3): 609–624.

Chang, P.L. and Chen, Y.C. (1994). A fuzzy multi-criteria decision making method for technology transfer strategy selection in biotechnology. *Fuzzy Sets and Systems* 63(1): 131–139.

Chang, P.T. and Lee, E.S. (1995). The estimation of normalized fuzzy weights. *Computers and Mathematics with Applications* 29(5): 21–42.

Charnes, A. and Cooper, W.W. (1961). *Management models and industrial applications of linear programming*, Wiley, New York.

Charnes, A., Cooper, W.W. and Ferguson, R. (1955). Optimal estimation of executive compensation by linear programming, *Management Science* 1(2): 138–151.

Charnes, A., Cooper, W.W. and Li, S. (1989). Using Data Envelopment Analysis to Evaluate Efficiency in the Economic Performance of Chinese Cities, *Socio-Economic Planning*, 23(6): 325–344.

Charnes, A., Cooper, W.W., and Rhodes, E. (1978). Measuring the efficiency of decision making units. *European Journal of Operations Research* 2(6): 429–446.

Chen, C.H. and Tzeng, G.H. (2011). Creating the aspired intelligent assessment systems for teaching materials. *Expert Systems with Applications* 38(10): 12168–12179.

Chen, F.H., Hsu, T.S., and Tzeng, G.H. (2011). A balanced scorecard approach to establish a performance evaluation and relationship model for hot spring hotels based on a hybrid MCDM model combining DEMATEL and ANP. *International Journal of Hospitality Management* 30(4): 908–932.

Chen, H. and Chen, T.J. (2003). Governance structures in strategic alliances: transaction cost versus resource-based perspective. *Journal of World Business* 38(1): 1–14.

Chen, H.C., Hu, Y.C., Shyu, J.Z. and Tzeng, G.H. (2005). Comparing possibility grey forecasting with neural network-based fuzzy regression by an empirical study. *Journal of Grey Systems* 18(2): 93–106.

Chen, H.C., Hu, Y.C., Shyu, J.Z. and Tzeng, G.H. (2005). Comparison analysis of possibility grey forecasting and fuzzy regression to the stock-market price in Taiwan. *Journal of Information Management* 12(1): 195–214.

Chen, M. and Alfa, A.S. (1991). A network design algorithm using a stochastic incremental traffic assignment approach. *Transportation Science* 25(3): 215–224.

Chen, M.F. and Tzeng, G.H. (2004). Combining grey relation and TOPSIS concepts for selecting an expatriate host country. *Mathematical and Computer Modelling* 40(13): 1473–1490.

Chen, M.F., Tzeng, G.H., and Ding, C.G. (2008). Combing fuzzy AHP with MDS in identifying the preference similarity of alternatives. *Applied Soft Computing* 8(1): 110–117.

Chen, P.T., Lee, Z.Y., Yu, H.C. and Tzeng, G.H. (2003). Analysis of the theoretical basis and practical applicability of a college teacher's achievement evaluation model: the case study of a national university in Hsinchu. *Bulletin of Educational Research* 49(4): 191–218.

Chen, S.H. and Hsieh C.H. (1998). Graded mean integration representation of generalized fuzzy number. In *Proceedings of Sixth Conference on Fuzzy Theory and Its Applications*. CD-ROM, file name: 031.wdl, pp. 1–6. Taipei: Chinese Fuzzy Systems Association.

Chen, S.J. and Hwang, C.L. (1992). *Fuzzy Multiple Attribute Decision Making Methods and Applications*. New York: Springer.

Chen, S.J. and Hwang, C.L. (1993). *Fuzzy Multiple Attribute Decision Making Methods and Applications*. Springer Lecture Notes in Economics and Mathematical Systems 375.

Chen, T.Y. and Tzeng, G.H. (2000). Reducing the amount of information necessary to compute fuzzy measures used in multi-attribute decision making. *Journal of Management* 17(3): 483–514 (Chinese, Taiwan).

Chen, T.Y. and Tzeng, G.H. (2003). Exploring the public attitude using fuzzy integrals. *PanPacific Management Review* 6(1): 45–58.

Chen, T.Y. and Wang, J.C. (2001). Identification of -fuzzy measures using sampling design and genetic algorithms. *Fuzzy Sets and Systems* 123(3): 321–341.

Chen, T.Y., Chang, C.C., and Tzeng, G.H. (2001). Applying fuzzy measures to establish priority-setting procedures for the pavement management system. *PanPacific Management Review* 4(1): 23–33.

Chen, T.Y., Chang, H.L., and Tzeng, G.H. (2001). Using a weight-assessing model to identify route choice criteria and information effects. *Transportation Research A* 35(3): 197–224.

Chen, T.Y., Chang, H.L., and Tzeng, G.H. (2002). Using fuzzy measure and habitual domains to analyze the public attitude and apply to the gas taxi policy. *European Journal of Operational Research* 137(2): 145–161.

Chen, T.Y., Wang, J.C., and Tzeng, G.H. (2000). Identification of general fuzzy measures by genetic algorithms based on partial information. *IEEE Transactions on Systems B*, 30B(4): 517–528.

Chen, Y. and Iqbal, A.I. (2002). Continuous optimization: output–input ratio analysis and DEA frontier. *European Journal of Operational Research* 142(3): 476–479.

Chen, Y.C. (2002). An application of fuzzy set theory to external performance evaluation of distribution centers in logistics. *Soft Computing* 6(1): 64–70.

Chen, Y.C., Lien, H.P., and Tzeng, G.H. (2011). Measures and evaluation for environment watershed plans using a novel hybrid MCDM model. *Expert Systems with Applications* 37(2): 926–938.

Chen, Y.C., Lien, H.P., Liu, C.H., Liou, J.J.H., and Tzeng, G.H. (2011). Fuzzy MCDM approach for selecting the best environment-watershed plan. *Applied Soft Computing* 11(1): 265–275.

Chen, Y.W. and Tzeng, G.H. (1999). A fuzzy multi-objective model for reconstructing post-earthquake road network by genetic algorithm. *International Journal of Fuzzy Systems* 1(2): 85–95.

Chen, Y.W. and Tzeng, G.H. (2000). Fuzzy multi-objective approach to the supply chain model. *International Journal of Fuzzy Systems* 1(3): 220–227.

Chen, Y.W. and Tzeng, G.H. (2001). Using fuzzy integral for evaluating subjectively perceived travel costs in a traffic assignment model. *European Journal of Operational Research* 130(3): 653–664.

Chen, Y.W., Tzeng, G.H. and Lou, P.J. (1997). Fuzzy multi-objectives facility location planning: a case study of Chiang Kai Shek International Airport in Taiwan. *Chinese Public Administration Review* 6(2): 17–42.

Chen, M.F., Tzeng, G.H., and Tang, T.I. (2005). Fuzzy MCDM approach for evaluation of expatriate assignments. *International Journal of Information Technology and Decision Making* 4(2): 277–296.

Cheng, C.C., Shyu, J.Z., and Tzeng, G.H. (2004). The decision/evaluation of Civil Aeronautics Administration public construction plan. *Civil Aviation Journal Quarterly* (Taiwan) 6(1): 37–63.

Cheng, C.C., Tsai, L M., Shyu, J.Z. and Tzeng, G.H. (2004). Fuzzy heuristic algorithm method for transport route choice of low-radiation waste junk. *Journal of Management* (Taiwan) 5(1): 41–56.

Cheng, C.H. and Lin, Y. (2002). Evaluating the best main battle tank using fuzzy decision theory with linguistic criteria evaluation. *European Journal of Operational Research* 142(1): 174–186.

Cheng, H.J. and Tzeng, G.H. (2003). Multi-objective planning for relief distribution systems. *ransportation Planning Journal* 32(3): 561–580.

Cheng, J.W. Chiu, W.L. and Tzeng, G.H. (2013). Do impression management tactics and/or supervisor–subordinate guanxi matter?. Knowledge-Based Systems 40: 123–133.

Cheng, J.Z., Yu, Y.W., Tsai, M.J. and Tzeng, G.H. (2004). Setting a business strategy to weather the telecommunications industry downturn by using fuzzy MCDM. *International Journal of Services Technology and Management* 5(4): 346–361.

Cheng, Y.C. (1993). The theory and characteristics of school-based management. *International Journal of Educational Management* 7(6): 6–17.

Cheung, F.K., Kuen, J.L.F., and Skitmore, M. (2002). Multi-criteria evaluation model for the selection of architecture consultants, *Construction Management and Economics* 20(7): 569–580.

Cheung, S.O., Lam, T.I., Leung, M.Y. and Wan, Y.W. (2001). An analytical hierarchy process based procurement selection method. *Construction Management and Economics* 19(1): 427–437.

Chiang, C.I. and Tzeng, G.H. (2000). A Multiple objective programming approach to data envelopment analysis. In Shi. Y. and Zeleny, M., Eds., *New Frontiers of Decision Making for the Information Technology Era. World Science Publishing Company* 270–285.

Chiang, C.I. and Tzeng, G.H. (2000). A new efficiency measure for DEA: efficiency achievement measure established on fuzzy multiple objectives programming. *Journal of Management* (Taiwan) 17(2): 369–388.

Chiang, C.I., Li, J.M., Tzeng, G.H. and Yu, P.L. (2000). A revised minimum spanning table method for optimal expansion of competence sets. In *Research and Practice in Multiple Criteria Decision Making*, Springer Lecture Notes in Economics and Mathematical Systems, DOI: 10.1007/978-3-642-57311-8–20.

Chiang, C.I. and Tzeng, G.H. (2000). A multiple objective programming approach to data envelopment analysis. In Shi, Y. and Zeleny, M., Eds., *New Frontiers of Decision Making for the Information Technology Era*. Singapore: World Science, pp. 270–285.

Chiang, J.H. (1999). Choquet fuzzy integral-based hierarchical networks for decision analysis. *IEEE Transactions on Fuzzy Systems* 7(1): 63–71.

ChiangLin, C.Y., Lai, C.C., and Yu, P.L. (2007). Programming models with changeable parameters - theoretical analysis on taking loss at the ordering time and making profit at the delivery time. *International Journal of Information Technology and Decision Making*, 6(4): 577–598.

Chin, Y.C., Chang, C.C., Lin, C.S. and Tzeng, G.H. (2010). The impact of recommendation sources on the adoption intention of microblogging based on dominance-based rough sets approach. In Szczuka, M. et al., Eds., *Proceedings of RSCTC*, Springer Lecture Notes on Artificial Intelligence, 6086, 514–523.

Chiou, H.K. and Tzeng, G.H. (2003). An extended approach of multicriteria optimization for MODM problems. In Tanino, T., Tanaka, T., and Inuiguchi, M., Eds., *Multi-Objective Programming and Goal Programming: Theory and Applications*. Heidelberg: Springer, pp. 111–116.

Chiou, H.K., Tzeng, G.H. and Cheng, D.C. (2005). Evaluating sustainable fishing development strategies using a fuzzy MCDM approach. *Omega* 33(3): 223–234.

Chiu, W.Y., Tzeng, H.T., and Li, H. (2013). A new hybrid MCDM model combining DANP with VIKOR to improve e-store business. *Knowledge-Based Systems* 37(1): 48–61.

Chiu, Y.C. and Tzeng, G.H. (1999). The market acceptance of electric motorcycles in Taiwan: experience through a stated preference analysis. *Transportation Research D* 4(2): 127–146.

Chiu, Y.C., Chen, B., Shyu, J.Z. and Tzeng, G.H. (2006), An evaluation model of new product launch strategy. *Technovation* 26(11): 1244–1252.

Chiu, Y.C., Shyu, J.Z., and Tzeng, G.H. (2004). Fuzzy MCDM for evaluating the e-commerce trategy. *International Journal of Computer Applications in Technology* 19(1): 12–22.

Chiu, Y.J., Chen, H.C., Shyu, J.Z. and Tzeng, G.H. (2006). Marketing strategy based on customer behavior for the LCD-TV. *International Journal of Decision Making* 7(2–3): 143–165.

Choi, Y.L. (1984). Land use transport optimization study. *Hong Kong Engineer* 47–57.

Choi, Y.L. (1986), LUTO model for strategic urban development planning. *Asian Geographer* 5(1–2): 155–176.

Choi, Y.L. (1985). The LUTO model and its applications in Hong Kong. *Planning and Development* 1: 21–31.

Choquet, G. (1953). Theory of capacities. *Annales de L'Institut Fourier* 5(1): 131–295.

Chou, C.C. (2003). The canonical representation of multiplication operation on triangular fuzzy numbers. *Computers and Mathematics with Applications* 45(10–11): 1601–1610.

Chou, T.Y., Chou, S.T., and Tzeng, G.H. (2006). Evaluating IT/IS investments: a fuzzy multicriteria decision model approach. *European Journal of Operational Research* 173(3): 026–1046.

Chu, M.T., Shyu, J., and Tzeng, G.H. (2007). Comparison among three analytical methods for knowledge communities group-decision analysis. *Expert Systems with Applications* 33(4): 1011–1024.

Chu, M.T., Shyu, J.Z., and Tzeng, G.H. (2007). Using non-additive fuzzy integral to assess performances of organization transformation via communities of practice. *IEEE Transactions on Engineering Management* 54(2): 327–339.

Chu, P.Y., Tzeng, G.H., and Teng, M.J. (2004). A multivariate analysis of the relationship among market share, growth and profitability: the case of the science-based industrial park. *Yat-Sen Management Review* 12(3): 507–534.

Churchman, C.W. and Ackoff, R.L. (1954). An approximate measure of value. *Journal of Operations Research Society of America* 2(1): 172–187.

Coase, R.H. (1937). The Nature of the firm. *Economica* 4(16): 386–405.

Coelli, T. (1998). Guide to DEAP Version 2.1: A Data Envelopment Analysis. Centre for Efficiency and Productivity Analysis, Department of Econometrics University of New England, Armidale, New South Wales, Australia. CEPA working paper.

Coello, C.A., van Veldhuizen, D.A., and Lamont, B.B. (2002). Evolutionary Algorithms for Solving Multi-Objective Problems. New York: Kluwer.

Cogger, K.O. and Yu, P.L. (1985). Eigenweight vectors and least distance approximation for revealed preference in pairwise weight ratios. *Journal of Optimization Theory and Applications* 46 (4): 483–491.

Cohon, J.L. (1978). Multiobjective programming and planning, Academic Press, New York.

Combarro, E.F. and Miranda, P. (2006). Identification of fuzzy measures from sample data with genetic algorithms. *Computers and Operations Research* 33(10): 3046–3066.

Conover, W.J. (1980). *Practical Nonparametric Statistics*, 2nd ed. Lubbock: Texas Tech University.

Cook, W.D. and Seiford, L.M. (1978). Priority ranking and consensus formation. *Management Science* 24(16): 1721–1732.

Cooper, J., Heron, T., and Heward, W. (2007). *Applied Behaviour Analysis*. New Jersey: Pearson Education.

Cooper, W.W., Park, K.S., and Yu, G. (1999). IDEA and ARIDEA: Models for dealing with imprecise data in data envelopment analysis, *Management Science*, 45(4): 597–607.

Cooper, W.W. and Tone, K. (1997). Measures of inefficiency in data envelopment analysis and stochastic frontier estimation, *European Journal of Operational Research* 99(1): 72–88.

Csutora, R. and Buckley, J. (2001). Fuzzy hierarchical analysis: the lambda-max method. *Fuzzy Sets and Systems* 120(2): 181–195.

Current, J.R. and Min, H. (1986). Multiobjective design of transportation networks: taxonomy and annotation. *European Journal of Operational Research* 26(2): 187–201.

Current, J.R., Revelle, C.S., and Cohon, J.L. (1987). The median shortest path problem: a multiobjective approach to analyze cost vs. accessibility in the design of transportation networks. *Transportation Science* 21(3): 188–197.

Dacin, M.T., Hitt, M.A., and Levitas, E. (1997). Selecting partners for successful international alliances: examination of U.S. and Korean firms. *Journal of World Business* 32(1): 3–16.

Dantzig GB (1951) A proof of the equivalence of the programming problem and the game problem. In: Koopmans TC (ed) Activity analysis of production and allocation. Wiley, New York, 330–335.

D'Ambra, J. and Rice, R.E. (2001). Emerging factors in user evaluation of the World Wide Web. *Information and Management* 38(6): 373–384.

Das, T.K., and Teng, B.S. (2000). A resource-based theory of strategic alliances. *Journal of Management* 26(1): 31–61.

David, J.L. (1989). Synthesis of research on school-based management. *Educational Leadership* 46(8): 45–53.

Deb, K. (2001). *Multi-Objective Optimization Using Evolutionary Algorithms*. New York: John Wiley & Sons.

Delgado, M., Verdegay, J.L., and Vila, M.A. (1992). Linguistic decision-making models. *International Journal of Intelligent Systems* 7(5): 479–492.

DeLucchi, M.A. (1989). Hydrogen vehicles: an evaluation of fuel storage, performance, safety, environmental impacts, and cost. *International Journal of Hydrogen Energy* 4(2): 81–130.

DeLucchi, M.A. and Lipman, T.E. (2001). An analysis of the retail and lifecycle cost of battery-powered electric vehicles. *Transportation Research D* 6(6): 371–404.

DeLucchi, M.A., Murphy, J.J., and McCubbin, D.R. (2002). The health and visibility cost of air pollution: a comparison of estimation methods. *Journal of Environmental Management* 64(2): 139–152.

Dempe, S., Kalashnikov, V.V., Pérez-Valdés, G.A. (2006). Mixed-integer bilevel programming: Application to an extended gas cash-out problem. In: Clute, R.P. (Ed.), International Business and Economics Research Conference. The Clute Institute for Academic Research.

Dempe, S., V.V. Kalashnikov, V.V., and Kalashnykova, N. (2006). Optimality conditions for bilevel programming problems. In S. Dempe and N. Kalashnykova, editors, Optimization with Multivalued Mappings: Theory, Applications and Algorithms. Springer-Verlag New York Inc. (New York 2006) pp. 11–36.

Dempe, S., Dutta, J., and Lohse, S. (2006). Optimality conditions for bilevel programming problems, *Optimization*, 55(6): 505–524.

Deng, J. (1982). Control problem of grey systems. *Systems and Control Letters* 1(5): 288–294.

Deng, J. (1985). *Grey System Fundamental Method.* Wuhan: Huazhong University of Science and Technology.

Deng, J. (1988). *Grey System Book.* Windsor: Sci-Tech.

Deng, J. (1989). Introduction to grey system theory. *Journal of Grey Systems* 1(1): 1–24.

DeRoche, E.F. (1987). *An Administrator's Guide for Evaluation Programs and Personnel: An Effective School Approach,* 2nd ed. Newton, MA: Allyn & Bacon.

Dhaenens-Flipo, C. (2000). Spatial decomposition for a multi-facility production and distribution problem. *International Journal of Production Economics* 64(1–3): 177–186.

Diamond, P, and Tanaka, H. (1998), Fuzzy regression analysis. In Slowinski, R., Ed., *Fuzzy ets in Decision Analysis: Operations Research and Statistics.* Dordrecht, Kluwer, pp. 49–387.

Dimitras, A.I., Slowinski, R., Susmaga, R. and Zopounidis, C. (1999). Business failure prediction using rough sets. *European Journal of Operational Research* 114(2): 263–280.

Doumpos, M. and Zopounidis, C. (2002). *Multi-Criteria Decision Aid Classification Methods.* Dordrecht: Kluwer.

Dubois, D. and Prade, H. (1987). Fuzzy numbers: an overview. In Bezdek, J., Ed., *Analysis of Fuzzy Information,* Vol. I. Boca Raton, FL: CRC Press, pp. 3–39.

Dubois, D. and Prade, H. (1980). *Fuzzy Sets and Systems.* New York: Academic Press.

Dubois, D. and Prade, H. (1988). *Possibility Theory. An Approach to Computerized Processing of Uncertainty.* New York: Plenum.

Dubois, D. and Prade, H. (1992). Putting rough sets and fuzzy sets together. In Slowinski, R., Ed., *Intelligent Decision Support: Handbook of Applications and Advances of the Rough Sets Theory.* Dordrecht: Kluwer, pp. 203–232.

Dubois, D., Kerre, E., Mesiar, R. et al. and Prade, H. (2000). Fuzzy interval analysis. In Dubois, D. and Prade, H., Eds., *Fundamentals of Fuzzy Sets.* Boston: Kluwer, p. 483–581.

Dubois, D., Prade, H., and Yager, R.R. (1993). *Readings in Fuzzy Sets for Intelligent Systems.* San Mateo, CA: Morgan & Kaufmann.

Dyer, J.H. and Singh, H. (1998). The relational view: Cooperative strategy and sources of interorganizational competitive advantage. *Academy of Management Review* 23(4): 660–679.

Elton, E.J. and Gruber, M.J. (1995). *Modern Portfolio Theory and Investment Analysis.* New York: John Wiley & Sons.

Elton, E.J., Gruber, M.J., and Urich, T.J. (1978). Are betas best? *Journal of Finance* 33(5): 1357–1384.

Entani, T., Maeda, Y., and Tanaka, H. (2002). Dual models of interval DEA and its extension to interval data, *European Journal of Operational Research,* 136(1): 32–45.

Ertay, T., Buyukozkan, G., Kahraman, C., and Ruan, D. (2005). Quality function deployment implementation based on analytic network process with linguistic data: an application in automotive industry. *Journal of Intelligent and Fuzzy Systems* 16(3): 221–232.

Fan, T. F., Liu, D. R., and Tzeng, G. H. (2005). Arrow decision logic. In *Rough Sets, Fuzzy Sets, Data Mining, and Granular Computing,* Springer Lecture Notes in Computer Science 3641, Part I, 651–659.

Fan, T. F., Liu, D.R., and Tzeng, G.H. (2006). Arrow decision logic for relational information systems. Springer Lecture Notes in Computer Science 4100, 240–262.

Fan, T.F., Liu, D.R., and Tzeng, G.H. (2007). Rough set-based logics for multi-criteria decision analysis. *European Journal of Operational Research* 182(1): 340–355.

Fang, S.K., Shyng, J.Y., Lee, W.S., and Tzeng, G.H. (2012). Exploring the preferences of customers between financial companies and agents based on TCA. *Knowledge-Based Systems* 27(2): 137–151.

Fare, R. and Grosskopf, S. (1996). Intertemporal Production Frontiers: With Dynamic DEA, Boston: Kluwer Academic Publishers.

Farrell, M.J. (1957). The measurement of productive efficiency. *Journal of the Royal Statistical Society* 120A(3): 253–290.

Fidler, B. and Bowles, G. (1989). *Effective Local Management of Schools*. London: Longman.

Fishburn, P.C. (1970). *Utility Theory for Decision Making*. New York: John Wiley & Sons.

Flavell, J. H. (1976). Metacognitive aspects of problem solving. In L. B. Resnick (Ed.), *The nature of intelligence* (pp. 231–235). Hillsdale, NJ: Erlbaum.

Fliege, J. and Vicente, L.N. (2006). Multicriteria approach to bilevel optimization. *Journal of Optimization Theory and Applications*, 131(2): 209–225.

Fong, S.W. and Choi, S.K.Y. (2000). Final contractor selection using the analytical hierarchy process. *Construction Management and Economics* 18(5): 547–557.

Fortuny-Amat, J. and McCarl, B. (1981). A representation and economic interpretation of a two-level programming problem. *Journal of Operational Research Society* 32(9): 783–792.

Freimer, M. and Yu, P.L. (1976). Some new results on compromise solutions for group decision problems. *Management Science* 22(6): 688–693.

Friesz, T.L. and Harker, P.T. (1983). Multi-criteria spatial price equilibrium network design: theory and computational results. *Transportation Research B* 17(5): 411–426.

Funk, K. and Rabl, A. (1999). Electric versus conventional vehicles: social costs and benefits in France. *Transportation Research D* 4(6): 397–411.

Gabus, A. and Fontela, E. (1973). *Perceptions of World Problems*. Report 1: Communication Procedure: Communicating with Those Bearing Collective Responsibility. Geneva: Battelle Research Institute. K15327.indb 234 14/02/13 7:38 PM Bibliography 235.

Gabus, A. and Fontela, E. (1972). *World Problems: An Invitation to Further Thought within the Framework of DEMATEL*. Geneva: Battelle Research Institute.

Gaur, A., Arora, S.R. (2008). Multi-level multi-objective integer linear programming problem, *Advanced Modelling and Optimization*, 2: 297–322.

Gen, M. and Cheng, R. (2000). Genetic algorithm and engineering optimization, John Wily and Sons, New York.

Geoffrion, A.M. (1987). An introduction to structured modeling. *Management Science* 33(5): 547–588.

Gerlinger, J.M. (1991). Strategic determinants of partner selection criteria in international joint ventures. *Journal of International Business Studies* 22(1): 43–62.

Gill, P.E. and Murray, W. (1978). Numerically stable methods for quadratic programming. Mathematical Programming 14(1): 349–372.

Giokas, D.I. and Pentzaropoulos G.C. (2000). Evaluating productive efficiency in telecommunications: evidence from Greece. *Telecommunications Policy* 24(8–9): 781–794.

Glover, F. (1975). Improved linear integer programming formulations of nonlinear integer problems. *Management Science* 22(4): 455–460.

Goh, C. and Law, R. (2003). Incorporating the rough sets theory into travel demand analysis. *Tourism Management* 24(5): 511–517.

Golany, B., Roll, Y., and Rybak, D. (1994). Measuring Efficiency of Power Plants in israel by Data Envelopment Analysis, *IEEE Transactions on Engineering Management*, 41(3): 291–300.

Goldfarb, D. and Indani, A. (1983). A numerically stable dual method for solving strictly convex quadratic programs. Mathematical Programming 27(1): 1–33.

Gorman, W.M. (1968). Conditions for additive separability. *Econometrica* 36(3–4): 605–609.

Gorman, W.M. (1968). The structure of utility function. *Review of Economic Studies* 35(4): 367–390.

Grabisch, M. (1995a). A new algorithm for identifying fuzzy measures and its application to pattern recognition. In Proceedings of Fourth Joint International IEEE Conference on Fuzzy Systems and Second International Fuzzy Engineering Symposium. Yokohama, pp. 145–150.

Grabisch, M. (1995b). Fuzzy integral in multicriteria decision making. *Fuzzy Sets and Systems* 69(3): 279–298.

Grabisch, M. (1996). The application of fuzzy integrals in multicriteria decision making. *European Journal of Operational Research* 89(3): 445–456.

Grabisch, M. (1997). k-Order additive discrete fuzzy measures and their representation integral in multi-criteria decision making. *Fuzzy Sets and Systems* 92(2): 167–189.

Grant, R.M. (1991). The resource-based theory of competitive advantage. California Management Review 33(3): 114–135.

Greco, S., Matarazzo, B., and Slowinski, R. (2001). Rough sets theory for multicriteria decision analysis. *European Journal of Operational Research* 129(1): 1–47.

Greene, W. (1993). *Econometric Analysis* (second edition). Macmillan Publishing Company, New York.

Greffenstette, J.J. (1991). Lamarckian learning in multi-agent environments. *Proc. Fourth International Conference of Genetic Algorithms.* San Mateo, CA: Morgan Kaufmann, 303–310.

Griffith P. and Gleason G. (1996). Six years of battery-electric bus operation at the Santa Barbara Metropolitan Transit District. In Proceedings of 13th Electric Vehicle Conference.

Gulati, R. (1998). Alliances and networks. *Strategic Management Journal* 19(4): 293–317.

Guo, P. and Tanaka, H. (2001). Fuzzy DEA: a perceptual evaluation method. *Fuzzy Sets and Systems* 119(1):149–160.

Hackman, S.T. and Platzman, L.K. (1990). Near-optimal solution of generalized resource allocation problems with large capacities. *Operations Research* 38(5): 902–910.

Hagedoorn, J. (1993). Understanding the rationale of strategic technology partnering: interorganizational modes of cooperation and sectoral differences. *Strategic Management Journal* 14(5): 371–385.

Haimes,Y.Y., Hall, W.A. and Freedman, H.T. (1975). *Multiobjective Optimization in Water Resources Systems: The Surrogate Worth Trade-off Method.* Elsevier Scientific Publishing Company, Amsterdam.

Hao, X.R. and Shi Y. (1996). Large-Scale Program: A C++ Program Run on PC or Unix. Omaha: University of Nebraska College of Information Science and Technology.

Harary, F., Norman, R., and Cartwright, D. (1965). Structural Models: An Introduction to the Theory of Directed Graphs. New York: John Wiley & Sons.

Hardie, B.G.S., Johnson, E.J., and Fader, P.S. (1993). Modeling loss aversion and reference dependence effects on brand choice. *Marketing Science* 12(4): 378–394.

Harding, G.G. (1999). Electric vehicles in the next millennium. *Journal of Power Sources* 78(2): 193–198.

Harrigan, K.R. (1985). *Strategies for Joint Ventures*. Lanham, MD: Lexington Books.

Harrigan, K.R. and Newman, W.H. (1990). Bases of inter-organizational cooperation: propensity, power, persistence. *Journal of Management Studies* 27(4): 417–434.

Harrison, J.S., Hitt, M.A., Hoskisson, R.E. and Ireland, R.D. (1991). Synergies and post acquisition performance: differences versus similarities in resource allocations. *Journal of Management* 17(1): 173–190.

Hassan, Y. and Tazaki, E. (2001). Rule extraction based on rough set theory combined with genetic programming and its application to medical data analysis. Seventh Australian and New Zealand Intelligent Information System Conference, Perth.

Hastak, M. (1998). Advanced automation or conventional construction process. *Automation in Construction* 7(4): 299–314.

Hennart, J.F. and Reddy, S. (1997). The choice between mergers/acquisitions and joint ventures: Japanese investors in the United States. *Strategic Management Journal* 18(1): 1–12.

Hennig-Thurau, T. and Klee, A. (1997). The impact of customer satisfaction and relationship quality on customer retention: a critical reassessment and model development. *Psychology and Marketing* 14(8): 737–764.

Herrera, F., Herrera-Viedma, E., and Verdegay, J.L. (1996). A model of consensus in group decision making under linguistic assessments. *Fuzzy Sets and Systems* 78(1): 73–87.

Hess, P. and Siciliano, J. (1996). *Management Responsibility for Performance*, New York; Irwin McGraw-Hill.

Highhouse, S. and Johnson, M.A. (1996). Gain/loss asymmetry and riskless choice: loss aversion in choices among job finalists. *Organizational Behavior and Human Decision Processes* 68(3): 225–233.

Hillier, F.S. (2001). *Evaluation and Decision Models*: A Critical Perspective. Boston: Kluwer.

Hitt, M.A., Ireland, R.D., and Hosskison, R.E. (1997). Strategic Management: Competitiveness and Globalization. St. Paul: West Publishing.

Ho, W.R.J., Tsai, C.L., Tzeng, G.H., and Fang, S.K. (2011). Combined DEMATEL technique with a novel MCDM model for exploring portfolio selection based on CAPM. *Expert Systems with Applications* 38(1): 16–25.

Hochbaum, D.S. (1995) A nonlinear knapsack problem. *Operations Research Letters* 17(3): 103–110.

Holland, J.M. (1975). Adaptation in Natural and Artificial Systems. Ann Arbor: University of Michigan Press.

Hori, S. and Shimizu, Y. (1999). Designing methods of human interface for supervisory control systems. *Control Engineering Practice* 7(11): 1413–1419.

Hougaard, J.L. and Keiding, H. (1996). Representation of preferences on fuzzy measures by a fuzzy integral. *Mathematical Social Sciences* 31(1): 1–17.

Hsieh, T.Y., Lu, S.T., and Tzeng, G.H. (2004). Fuzzy MCDM approach for planning and design tenders selection in public office buildings. *International Journal of Project Management* 22(7): 573–584.

Hsieh, Y.L., Tzeng, G.H., Lin, T.R., and Yu, H.C. (2010). Wafer sort bit map data analysis using the PCA-based approach for yield analysis and optimization. *IEEE Transactions on Semiconductor Manufacturing* 24(4): 493–502.

Hsu, C.F. and Tzeng, G.H. (2002). Evaluation analysis for Taiwan agriculture productivity: an application of DEA method. *Journal of the Land Bank of Taiwan* 39(2): 139–157.

Hsu, C.H., Wang, F.K., and Tzeng, G.H. (2012). The best vendor selection for conducting the recycled material based on a hybrid MCDM model combining DANP with VIKOR. *Resources, Conservation and Recycling* 66: 95–111.

Hsu, C.S., Lee, Z.Y., Hung, C.Y., and Tzeng, G.H. (2004). Key factors in performance appraisal for R&D organizations: the case of the industrial technology research institute in Taiwan. *Journal of Biomedical Fuzzy Systems* 10(1–2): 19–29.

Hsu, C.S., Lee, Z.Y., Shi, C.S., and Tzeng, G.H. (2003). The clustering and performance efficiency of ITRI's R&D units. *Journal of Technology Management (Taiwan)* 8(1): 33–60.

Hsu, C.S., Lee, Z.Y., Shi, C.S., and Tzeng, G.H. (2003). Application of the DEA approach to evaluate management performance of R&D organizations in the Industrial Technology Research Institute of Taiwan. *Management Review* 22(2): 25–53.

Hsu, C.Y., Chen, K.T., and Tzeng, G.H. (2007). FMCDM with fuzzy DEMATEL approach for customers' choice behavior model. *International Journal of Fuzzy Systems* 9(4): 236–246.

Hsu, Y.G., Shyu, J.Z., and Tzeng, G.H. (2005). Policy tools on the formation of new biotechnology firms in Taiwan. *Technovation* 25(3): 281–292.

Hsu, Y.G., Tzeng, G.H., and Shyu, J.Z. (2003). Fuzzy multiple criteria selection of government-sponsored frontier technology R&D projects. *R&D Management* 33(5): 539–551.

Hu, Y.C. and Tzeng, G. H. (2003). Elicitation Of classification rules by fuzzy data mining. *Engineering Applications of Artificial Intelligence* 16(7–8): 709–716.

Hu, Y.C., Chen, R.S., Hsu, Y.T., and Tzeng, G.H. (2002). Grey self-organizing feature maps. *Neurocomputing* 48(4): 863–877.

Hu, Y.C., and Tzeng, G.H. (2002). Mining fuzzy association rule for classification problems. *Computers and Industrial Engineering* 43(4): 735–750.

Hu, Y.C., Chen, R.S., and Tzeng, G.H. (2002). Generating learning sequences for decision makers through data mining and competence set expansion. *IEEE Transactions on Systems, Man and Cybernetics B* 32(5): 679–686.

Hu, Y.C., Chen, R.S., and Tzeng, G.H. (2003). An effective learning algorithm for discovering fuzzy sequential patterns. International Journal of Uncertainty, *Fuzziness, and Knowledge-Based Systems* 11(2): 173–193.

Hu, Y.C., Chen, R.S., and Tzeng, G.H. (2003). Finding fuzzy classification rules using data mining techniques. *Pattern Recognition Letters* 24(1–3): 509–519.

Hu, Y.C., Chen, R.S., and, Tzeng, G.H. (2003). Discovering fuzzy association rules using fuzzy partition methods. *Knowledge-Based Systems* 16(3): 137–147.

Hu, Y.C., Chen, R.S., Tzeng, G.H., and Chiu, Y.J. (2003). Acquisition of compound skills and learning costs for expanding competence sets. *Computers and Mathematics with Applications* 46(5–6): 831–848.

Hu, Y.C., Chen, R.S., Tzeng, G.H., and Shieh, J.H. (2003). A fuzzy data mining algorithm for finding sequential patterns. International Journal of Uncertainty, *Fuzziness, and Knowledge-Based Systems* 11(2): 173–193.

Hu, Y.C., Chiu, Y.J., and Tzeng, G.H. (2003). Grey theory and competence sets for multiple criteria project scheduling. *Management Review* 1(2): 257–273.

Hu, Y.C., Chiu, Y.J., Chen, C.M., and Tzeng, G.H. (2003). Competence set expansion for obtaining scheduling plans in intelligent transportation security systems. In Tanino, T. et al., Eds., Multi-Objective Programming and Goal Programming: Theory and Applications. Heidelberg: Springer, pp. 347–352.

Hu, Y.C., Chiu, Y.J., Hu, J.S., and Tzeng, G.H. (2002). Acquisitions of learning costs for expanding competence sets using grey relations and neural networks. *Journal of Chinese Grey System Association* 5(2): 75–82.

Hu, Y.C., Hu, J.S., Chen, R.S., and Tzeng, G.H. (2004). Assessing weights of product attributes from fuzzy knowledge in a dynamic environment. *European Journal of Operational Research* 154(1): 125–143.

Hu, Y.C., Tzeng, G.H., and Chen, C.M. (2004). Deriving two stage learning sequence from knowledge in fuzzy sequential pattern mining. *Information Sciences* 159(1–2): 69–86.

Hu, S.K., Chuang, Y.C., Yeh, Y.F., and Tzeng, G.H. (2012). Hybrid MADM with fuzzy integral for exploring the smart phone improvement in M generation. *International Journal of Fuzzy Systems* 14(2): 204–214.

Huang, C.Y., Chang, S.Y., Yang, Y.H., and Tzeng, G.H. (2010). Next generation passive optical networking technology predictions by using hybrid MCDM methods. *Journal of Advanced Computational Intelligence and Intelligent Informatics* 15(4): 400–405.

Huang, C.Y., Hung, Y.H., and Tzeng, G.H. (2010). Using hybrid MCDM methods to assess fuel cell technology for the next generation of hybrid power automobiles. *Journal of Advanced Computational Intelligence and Intelligent Informatics* 15(4): 406–417.

Huang, C.Y., Shyu, J.Z., and Tzeng, G.H. (2007). Reconfiguring the innovation policy portfolios for Taiwan's SIP mall industry. *Technovation* 27(12): 744–765.

Huang, C.Y. and Tzeng, G.H. (2007). Post-merger high technology R&D human resources optimization through the de novo perspective. In Shi, Y. et al., Eds., Advances in Multiple Criteria Decision Making and Human Systems Management. Fairfax, VA: IOS Press, pp. 47–64.

Huang, C.Y. and Tzeng, G.H. (2008). Multiple generation product life cycle predictions using a novel two-stage fuzzy piecewise regression analysis method. *Technological Forecasting and Social Change* 75(1): 12–31.

Huang, C.Y., Tzeng, G.H., Chan, C.C., and Wu, H,C, (2008). Semiconductor market fluctuation indicators and rules derivations by using the rough set theory. *International Journal of Innovative Computing, Information and Control* 5(6): 1485–1503.

Huang, C.Y., Tzeng, G.H., Chen, Y.T., and Chen, H. (2012). Performance evaluation of leading integrated circuit design houses by using a multiple objective programming based data envelopment analysis approach. *International Journal of Innovative Computing Information and Control.* 8(8): 5899–5916.

Huang, C.Y., Tzeng, G.H., and Ho, W.R.J. (2011). System on chip design service e-business value maximization through a novel MCDM framework. Expert Systems with Applications 38(7): 7947–7962.

Huang, C.Y., Wang, P.Y., and Tzeng, G.H. (2012). Evaluating top information technology firms in Standard & Poor's 500 index by using a multiple objective programming based data envelopment analysis. In Jiang, H. et al., Eds., Advanced Research in Applied Artificial Intelligence, Springer Lecture Notes in Computer Science, 7345, 720–730. DOI: 10.1007/978-3-642-31087-4_73.

Huang, C.Y., Wu, M.J., Liu, Y.W., and Tzeng, G.H. (2012). Using the DEMATEL based network process and structural equation modeling methods for deriving factors influencing the acceptance of smart phone operation systems. In Jiang, H. et al., Eds., Advanced Research in Applied Artificial Intelligence, Springer Lecture Notes in Computer Science, 7345, 731–741. DOI: 10.1007/978-3-642-31087-4_74.

Huang, J.J., Chen, C.Y., Liu, H.H., and Tzeng, G.H. (2011). A multiobjective programming model for partner selection-perspectives of objective synergies and resource allocations. *Expert Systems with Applications* 37(5): 3530–3536.

Huang, J.J., Ong, C.S., and Tzeng, G.H. (2004). Using rough set theory for detecting the interaction terms in a generalized logit model. In Rough Sets and Current Trends in Computing, Springer Lecture Notes in Artificial Intelligence 3066, pp. 624–629.

Huang, J.J., Ong, C.S., and Tzeng, G.H. (2005a). A novel hybrid model for portfolio selection. *Applied Mathematics and Computation* 169(2): 1195–1210.

Huang, J.J., Ong, C.S., and Tzeng, G.H. (2005b). Building credit scoring models using genetic programming. *Expert Systems with Applications* 29(1): 41–47.

Huang, J.J., Ong, C.S., and Tzeng, G.H. (2005c). Model identification of ARIMA family using genetic algorithms. *Applied Mathematics and Computation* 164(3): 885–912.

Huang, J.J., Ong, C.S., and Tzeng, G.H. (2005d). Motivation and resource allocation for strategic alliance through the de novo perspective. *Mathematical and Computer Modelling* 41(6–7): 711–721.

Huang, J.J., Ong, C.S., and Tzeng, G.H. (2006). Fuzzy principal component regression (FPCR) for fuzzy input and output data. *International Journal of Uncertainty and Knowledgebased Systems* 14(1): 87–100.

Huang, J.J., Ong, C.S., and Tzeng, G.H. (2006). Optimal fuzzy multi-criteria expansion of competence sets using multi-objectives evolutionary algorithms. *Expert Systems with Applications* 30(4): 739–745.

Huang, J.J. and Tzeng, G.H. (2007). Marketing segmentation using support vector clustering. *Expert Systems with Applications* 32(2): 313–317.

Huang, J.J., Tzeng, G.H., and Ong C.S. (2005). Multidimensional data in multidimensional scaling using the analytic network process. *Pattern Recognition Letters* 26(6): 755–767.

Huang, J.J., Tzeng, G.H., and Ong, C. S. (2006). A novel algorithm for dynamic factor analysis. *Applied Mathematics and Computation* 175(2): 1288–1297.

Huang, J.J., Tzeng, G.H., and Ong, C.S. (2006a). A novel algorithm for uncertain portfolio Selection. *Applied Mathematics and Computation* 173(1): 350–359.

Huang, J.J., Tzeng, G.H., and Ong, C.S. (2006b). Choosing best alliance partners and allocating optimal alliance resources using the fuzzy multi-objective dummy programming model. *Journal of Operational Research Society* 57(10): 1216–1223.

Huang, J.J., Tzeng, G.H., and Ong, C.S. (2006c). Interval multidimensional scaling for group decision using rough set concept. *Expert Systems with Applications* 31(3): 525–530.
Huang, J.J., Tzeng, G.H., and Ong, C.S. (2006d). Two-stage genetic programming (2SGP) for the credit scoring model. *Applied Mathematics and Computation* 174(2): 1039–1053.
Hung, Y.H., Chou, S.C.T., and Tzeng, G.H. (2011). Knowledge management adoption and assessment for SMEs by a novel MCDM approach, *Decision Support Systems* 51(2): 270–291.
Hung, Y.H., Huang, T.L., Hsieh, J.C., Tsuei, H.J., Cheng, C.C., and Tzeng, G.H. (2012). Online reputation management for improving marketing by using a hybrid MCDM model. *Knowledge-Based Systems* 35: 87–93.
Hwang, M.J., Chiang, C.I., Chiu, I.C., and Tzeng, G.H. (2001). Multi-stage optimal expansion of competence sets in fuzzy environment. *International Journal of Fuzzy Sets* 3(3): 486–492.
Hwang, M.J., Tzeng, G.H., and Liu, Y.H. (1996). A dynamic model for optimal expansion of multistage competence sets. *Journal of Management* 14(1): 115–133.
Hwang, C.L. and Yoon, K. (1981). Multiple-Attribute Decision Making Methods and Applications, Springer Lecture Notes in Economics and Mathematical Systems 186, Springer-Verlar, Berlin.
Ida, K. and Gen, M. (1997). Improvement of two-phase approach for solving fuzzy multiple objective linear programming. *Journal of Japan Society for Fuzzy Theory and Systems* 9(1): 115–121.
Ignizio, J.P. (1976). *Goal Programming and Extensions*. Lexington Books, Lexington, MA.
Ijiri, Y. (1965). *Management Goals and Accounting for Control*, North-Holland, Amsterdam.
Inuiguchi, M. (2004). Generalizations of rough sets: from crisp to fuzzy cases. In Tsumoto, S. et al., Eds., Rough Sets and Current Trends in Computing, Springer Lecture Notes in Artificial Intelligence 3066(1): 26–37.
Ishii, H., Shiode, S., Hwang, H. et al. (2012), Preface. *In Soft Computing for Management Systems*. In press.
Ishii, K. and Sugeno, M. (1985). A model of human evaluation process using fuzzy measure. *International Journal of Man–Machine Studies* 22(1): 19–38.
Jackson, A.G., Leclair, S.R., Ohmer, M.C., Ziarko, W., and AL-Kamhwi, H. (1996). Rough sets applied to materials data. *Acta Materia* 44(11): 4475–4484.
Jeng, J.F. and Tzeng, G.H. (2012). Social influence on the use of clinical decision support systems: Revisiting the unified theory of acceptance and use of technology by the fuzzy DEMATEL technique. *Computers and Industrial Engineering* 62(3): 819–828.
Johnsson, B. and Ahman, M. (2002). A comparison of technologies for carbon-neutral passenger transport. *Transportation Research part D* 7(3): 175–196.
Kahneman, D., Knetsch, J.L., and Thaler, R.H. (1991). Anomalies: the endowment effect, loss aversion, and status quo bias. *Journal of Economic Perspectives* 5(1): 193–206.
Kahneman, D. and Tversky, A. (2000). Choices, Values, and Frames. *Cambridge: Cambridge University Press.*
Kahneman, D. and Tversky, A. (1984). Choices, Values, and Frames. *American Psychologist* 39(4): 341–350.
Kahneman, D. and Tversky, A. (1979). Prospect theory: an analysis of decision under risk. *Econometrica* 47(2): 263–291.
Kahraman, C., Ertay, T., and Ertay, G. (2006). A fuzzy optimization model for QFD planning processes using analytic network approach. *European Journal of Operational Research* 171(22): 390–411.
Kaiser, H.F. (1958). The Varimax criteria for analysis rotation in factor analysis. *Psychometrics* 23(2): 187–200.
Kantorovich, L.V. and Koopmans, T.C. (1976). Problems of application of optimization methods in industry. Stockholm: Federation of Swedish Industries.

Karlaftis, M.G. (2004). A DEA approach for evaluating the efficiency and effectiveness of urban transit systems. *European Journal of Operational Research* 152(2): 354–364.

Karsak, E.E., Sozer, S., and Alptekin, S.E. (2002). Product planning in quality function deployment using a combined analytic network process and goal programming approach. *Computers and Industrial Engineering* 44(1): 171–190.

Kartam, N., Tzeng, G.H., and Teng, J.Y. (1993). Robust contingency plans for transportation investment planning, *IEEE Transactions on Systems, Man, and Cybernetics* 23(1): 5–13.

Kaufmann, A. and Gupta, M.M. (1985). Introduction to Fuzzy Arithmetic: Theory and Applications. New York: Van Nostrand Reinhold.

Kaufmann, A. and Gupta, M.M. (1988). Fuzzy Mathematical Models in Engineering and Management Science. Amsterdam: North Holland.

Kazimi, C. (1997). Evaluating the environmental impact of alternative-fuel vehicles. *Journal of Environmental Economics and Management* 33(2), 163–185.

Keeney, R.L. and Raiffa, H. (1976). Decisions with Multiple Objectives: Preferences and Value Trade-Offs. New York: John Wiley & Sons.

Kempton, W. and Kubo, T. (2000). Electric-drive vehicles for peak power in Japan. *Energy Policy* 28(1): 9–18.

Kim, H., Ida, K., and Gen, M. (1993). A de novo approach for bicriteria 0–1 linear programming with interval coefficients under GUB structure. *Computers and Industrial Engineering* 25(1–4) : 17–20.

Kim, K.J., Moskowitz, H., and Koksalan, M. (1996). Fuzzy versus statistical linear regression. *European Journal of Operations Research* 92(2): 417–434.

Ko, Y.C., Fujita, H. and Tzeng, G.H. (2013), An extended fuzzy measure on competitiveness correlation based on WCY 2011, *Knowledge-Based Systems* 37(1): 86–93.

Kleindorfer, P.R., Kunreuther, H.C., and Schoemaker, P.J.H. (1993). Decision Sciences: An Integrative Perspective. Cambridge: Cambridge University Press.

Klir, G.J. and Folger, T.A. (1988). Fuzzy Sets, Uncertainty, and Information. Englewood Cliffs, NJ: Prentice Hall.

Klir, G. J. and Pan, Y. (1998). Constrained fuzzy arithmetic: basic questions and some answers. *Soft Computing* 2(2): 100–108.

Klir, G.J. and Pan, Y. (1997). Fuzzy arithmetic with requisite constraints. *Fuzzy Sets and Systems* 91(2): 165–175.

Ko, Y.C., Fujita, H., and Tzeng, G.H. (2012). Using DRSA and fuzzy measure to enlighten policy making for enhancing national competitiveness by W CY 2011. In Jiang, H. et al., Eds., Springer Lecture Notes in Science 7345, pp. 709–719.

Ko, Y.C., Fujita, H., and Tzeng, G.H. (2012). An extended fuzzy measure on competitiveness correlation based on WCY 2011, Knowledge-Based Systems.

Kogut, B. (1988). Joint ventures: Theoretical and empirical perspectives. *Strategic Management Journal* 9(4): 319–332.

Koskie, H.A. and Majumdar, S.K. (2000). Convergence in telecommunications infrastructure development in OECD counties. *Information Economics and Policy* 12(2): 111–131.

Kosko, B. (1988). Hidden patterns in combined and adaptive knowledge networks. *International Journal of Approximate Reasoning* 2(4): 377–393.

Kough, B. (1988). Joint ventures: theoretical and empirical perspective. *Strategic Management Journal* 9(4): 319–332.

Kough, B. (1991). Joint ventures and the option to expand and acquire. *Management Science* 37(1): 19–33.

Koza, J. (1992). Genetic Programming: On the Programming of Computers by Natural Selection. Cambridge, MA: MIT Press.

Kristense, K., Martensen, A., and Gronholdt, L. (1999). Measuring the impact of buying behaviour on customer satisfaction. *Total Quality Management* 10(4): 602–614.

Krusinska, E., Slowinski, R., and Stefanowski, J. (1992). Discriminant versus rough set approach to vague data analysis. *Applied Stochastic Models and Data Analysis* 8(1):43–56.

Kuan, M.J., Tzeng, G.H. and Hsiang, C.C. (2012). Exploring the quality assessment system for new product development process by combining DANP with MCDM model. International Journal of Innovative Computing, Information, and Control 8(8): 5745–5762.

Kuhn, H.W. and Tucker, A.W. (1951). Nonlinear programming. In Proceedings of Second Symposium on Mathematical Statistics and Probability. Berkeley: University of California Press, pp. 481–491.

Kuo, M.S., Tzeng, G.H., and Huang, W.C. (2007). Group decision making based on concepts of ideal and anti-ideal points in fuzzy environment. *Mathematical and Computer Modeling* 45(3–4): 324–339.

Laarhoven, P.J.M. and Pedrycz, W. (1983). A fuzzy extension of Saaty's priority theory. *Fuzzy Sets and Systems* 11(3): 229–241.

Lai, K.K. and Li, L. (1999). A dynamic approach to multiple-objective resource allocation problem. *European Journal of Operational Research* 117(2): 293–309.

Lai, Y.J., Liu, T.Y. and Hwang, C.L. (1994). TOPSIS for MODM, *European Journal of Operational Research*, 76(3): 486–500.

Larbani, M., Huang, C.Y., and Tzeng, G.H. (2011). A novel method for fuzzy measure identification. *International Journal of Fuzzy Systems* 13(1): 24–34.

Lebesgue, H. (1966). Measurement and the Integral. San Francisco: Holden Day.

LeBlanc, L.J, and Boyce, D.E. (1986). A bilevel programming algorithm for exact solution of the network design problem with user-optimal flows. *Transportation Research B* 20(3): 259–265.

LeBlanc, L.J. (1975). An algorithm for the discrete network design problem. *Transportation Science* 9(3): 183–199.

LeBlanc, L.J., Morlok, E., and Pierskalla, W. (1975). An efficient approach to solving road-network equilibrium traffic assignment problems. *Transportation Research B* 9(5): 309–318.

Lee, C., Liu, L.C., and Tzeng, G.H. (2000). Multi-criteria evaluation of the schooling efficiency for promoting outstanding junior colleges to technological institutes. *Journal of Chinese Grey Society* 3(2): 95–113.

Lee, C., Liu, L.C., and Tzeng, G.H. (2001). Hierarchical fuzzy integral evaluation approach for vocational education performance: case of junior colleges in Taiwan. *International Journal of Fuzzy Systems* 3(3): 476–485.

Lee, C.C., Tzeng, G.H., and Chiang, C. (2011). Determining key service quality measurement indicators in a travel website using a fuzzy analytic hierarchy process. *International Journal of Electronic Business Management* 9(4): 322–333.

Lee, C.F., Tzeng, G.H., and Wang, S.Y. (2005). A fuzzy set approach to generalize CRR model: an empirical analysis of S&P 500 index option. *Review of Quantitative Finance and Accounting* 25(3): 255–275.

Lee, C.F., Tzeng, G.H., and Wang, S.Y. (2005). A new application of fuzzy set theory to the Black-Scholes option pricing model. *Expert Systems with Applications* 29(2): 330–342.

Lee, E.S. and Li, R.J. (1993). Fuzzy multiple objective programming and compromise programming with Pareto optimum. *Fuzzy Sets and Systems* 53(2): 275–288.

Lee, J.W. and Kim, S.H. (2000). Using analytic network process and goal programming for interdependent information system project selection. *Computers and Operations Research* 27(4): 367–382.

Lee, K.M. and Leekwang, H. (1995). Identification of λ-fuzzy measure by genetic algorithms. *Fuzzy Sets and Systems* 75(3): 301–309.

Lee, S.M. (1972). *Goal Programming for Decision Analysis*, Auerbach Publishers, Philadelphia, Pennsylvania.

Lee, W.S., Tzeng, G.H., Guan, J.L., Kuo-Ting Chien. K.T., and Huang, J.M. (2009). Combined MCDM techniques for exploring stock selection based on Gordon model. *Expert Systems with Applications* 36 (3): 6421–6430.

Leszczyńsk, K., Penczek, P., and Grochulski, W. (1985). Sugeno's fuzzy measure and fuzzy clustering. *Fuzzy Sets and Systems* 15(2): 147–158.

Li, C.W. and Tzeng, G.H. (2009a). Identification of a threshold value for the DEMATEL method using the maximum mean de-entropy algorithm to find critical services provided by a semiconductor intellectual property mall. *Expert Systems with Applications* 36(6): 9891–9898.

Li, C.W. and Tzeng, G.H. (2009b). Identification of interrelationship of key customers' needs based on structural model for services and capabilities provided by a semiconductor intellectual property mall. *Applied Mathematics and Computation* 215(6): 2001–2010.

Li, H.L. (1982). The use of hierarchical multiobjective programming in urban transportation network design. Ph. D. Dissertation, University of Pennsylvania, Philadelphia.

Li, H.L. (1988). Expert system for evaluating multicriteria transportation networks. *Microcomputers in Civil Engineering* 3(3): 259–265.

Li, R. J. (1999). Fuzzy method in group decision making, *Computer and Mathematical with Applications*, 38(1), 91–101.

Li, R. and Wang, Z.O. (2004). Mining classification rules using rough sets and neural networks. *European Journal of Operational Research* 157(2): 439–448.

Li, R.J. and Lee E.S. (1990a). Multicriteria de novo programming with fuzzy parameters. *Computer and Mathematical with Applications* 19(1): 13–20.

Li, R.J. and Lee, E.S. (1990b). Fuzzy approaches to multicriteria de novo programs. *Journal of Mathematical Analysis and Applications* 153(1): 97–111.

Li, Y.Z., Ida, K., and Gen, M. (1997). Improved genetic algorithm for solving multi-objective solid transportation problem with fuzzy numbers. *Computers and Industrial Engineering* 33(3/4): 589–592.

Lien, D. and Peng Y. (2001). Competition and production efficiency telecommunications in OECD countries. *Information Economics and Policy* 13(1): 51–76.

Lin, C.L. and Tzeng, G.H. (2010). A value-created system of science (technology) park by using DEMATEL. *Expert Systems with Applications* 36(6): 9683–9697.

Lin, C.L., Chen, C.W., and Tzeng, G.H. (2010). Planning the development strategy for the mobile communication package based on consumers' choice preferences. *Expert Systems with Applications* 37(7): 4749–4760.

Lin, C.L., Hsieh, M.S., and Tzeng, G.H. (2010). Evaluating vehicle telematics system by using a novel MCDM techniques with dependence and feedback. *Expert Systems with Applications* 37(10): 6723–6736.

Lin, C.M., Huang, J.J., Gen, M., and Tzeng, G. H. (2006). Recurrent neural network for dynamic portfolio selection. *Applied Mathematics and Computation* 175(2): 1139–1146.

Lin, C.S., Tzeng, G.H., and Chin, Y.C. (2011). Combined rough set theory and flow network graph to predict customer churn in credit card accounts. *Expert Systems with Applications* 38(1): 8–15.

Lin, C.S., Tzeng, G.H., Chin, Y.C., and Chang, C.C. (2010). Recommendation sources on the intention to use e-books in academic digital libraries. *Electronic Library* 28(6): 844–857.

Lin, Y.T., Lin, C.L., Yu, H.C., and Tzeng, G.H. (2011), Utilisation of interpretive structural modeling method in the analysis of interrelationship of vendor performance factors. *International Journal of Business Performance Management* 12(3): 260–275.

Lin, Y.T., Lin, C.L., Yu, H.C., and Tzeng, G.H. (2010). A novel hybrid MCDM approach for outsourcing vendor selection: a case study for a semiconductor company in Taiwan. *Expert Systems with Applications* 37(7): 4796–4804.

Liou, J.J.H. and Tzeng, G.H. (2007). A non-additive model for evaluating airline service quality. *Journal of Air Transport Management* 13(3): 131–138.

Liou, J.J.H. and Tzeng, G.H. (2010). A dominance-based rough set approach to customer behavior in the airline market. *Information Sciences* 180(11): 2230–2238.

Liou, J.J.H. and Tzeng, G.H. (2012). Comments on "Multiple criteria decision making (MCDM) methods in economics: An overview. *Technological and Economic Development of Economy*, 18(4): 672–695.

Liou, J.J.H., Tsai, C.Y., Lin, R.H., Tzeng, G.H. (2011). A modified VIKOR multiple-criteria decision method for improving domestic airlines service quality. *Journal of Air Transport Management* 17(2): 57–61.

Liou, J.J.H., Tzeng, G.H. Tsai, C.Y., and Hsu, C.C. (2011). A hybrid ANP model in fuzzy environments for strategic alliance partner selection in the airline industry. *Applied Soft Computing*, 11(4): 3515–3524.

Liou, J.J.H., Tzeng, G.H., and Chang, H.C. (2007). Airline safety measurement using a hybrid model. *Journal of Air Transport Management* 13(4): 243–249.

Liou, J.J.H., Tzeng, G.H., Hsu, C.C., and Yeh, W.C. (2012). Reply to comment on using a modified grey relation method for improving airline service quality, *Tourism Management* 33(3): 719–720.

Liou, J.J.H., Yen, L., and Tzeng, G.H. (2008). Building an effective safety management system for airlines. *Journal of Air Transport Management* 14(1): 20–26.

Liou, J.J.H., Yen, L., and Tzeng, G.H. (2010). Using decision rules to achieve mass customization of airline services. *European Journal of Operational Research* 205(3): 680–686.

Liu, L.C., Lee, C., and Tzeng, G.H. (2001). Hierarchical fuzzy integral evaluation approach for vocational education performance: case of junior colleges in Taiwan. *International Journal of Fuzzy Sets* 3(3): 476–485.

Liu, C.H., Tzeng, G.H., Lee, M.H., and Lee, P.Y. (2013). Improving metro–airport connection service for tourism development: Using hybrid MCDM models. *Tourism Management Perspectives* 6: 95–107.

Liu, L.C., Lee, C., and Tzeng, G.H. (2003). Using DEA of REM and EAM for efficiency assessment of technology institutes upgraded from junior colleges: the case in Taiwan. In Tanino, T. et al., Eds., Multi-Objective Programming and Goal-Programming: Theory and Applications. Heidelberg: Springer, pp.361–366.

Liu, L.C., Lee, C., and Tzeng, G.H. (2004). DEA approach of the current-period and cross-period efficiency for evaluating the vocational education. *International Journal of Information Technology and Decision Making* 3(2): 353–374.

Liu, B.C., Tzeng, G.H., and Hsieh C.T. (1992). Energy planning and environment quality management: a decision support system approach. Energy Economics 14(4): 302–307.

Liu, C.H., Tzeng, G.H., and Lee, M.H. (2012). Improving tourism policy implementation: the use of hybrid MCDM models. *Tourism Management* 33(2): 239–488.

Liu, C.H., Tzeng, G.H., and Lee, M.H. (2013), Strategies for improving cruise product sales in the travel agency: using hybrid MCDM models. *Service Industry Journal* (Forthcoming, accepted: 21 Jul 2011). DOI: 10.1080/02642069.2011.614342.

Liu, Y.H., Tzeng, G.H., and Park, D.H. (2004). Set covering problem and the reliability of thecovers. International Journal of Reliability and Applications 5(4): 145–154.

Lockett, A. and Thompson S. (2001). The resource-based view and economics. *Journal of Management* 27(6): 723–754.

Lorange, P. and Roos, J. (1992). Strategic Alliances. Cambridge, MA: Oxford University Press.

Lu, M.T., Tzeng, G.H., and Tang, L.L. (2012). Environmental strategic orientations for improving green innovation performance in the electronics industry using fuzzy hybrid MCDM model. *International Journal of Fuzzy Systems*. In press.

Lu, S.T., Hsieh, T.Y., and Tzeng, G.H. (2004). Fuzzy MCDM approach for planning and design tenders selection in public office buildings. *International Journal of Project Management* 22(7): 574–584.

Luo, X. (2003). Evaluating the profitability and marketability efficiency of large banks an application of data envelopment analysis. *Journal of Business Research* 56(8): 626–635.

Mahdi, I.M., Riley, M.J., Fereig, S.M. et al. (2002). A multi-criteria approach to contractor selection engineering. *Construction and Architectural Management* 9(1): 29–37.

Majumdar, S.K. (1995). Does new technology adoption pay? Electronic switching patterns and firm-level performance in U.S. telecommunications. Research Policy 24(8): 803–822.

Mandal, A. and Deshmukh, S.G. (1994). Vendor selection using interpretive structural modeling. *International Journal of Operations and Production Management* 14(6): 52–59.

Mares, M. (1994). Computation of Fuzzy Quantities. Boca Raton. FL: CRC Press.

Marglin, S. (1967). Public Investment Criteria. Cambridge, MA: MIT Press, 103 pages.

Marichal, J.L. and Roubens, M. (2000). Determination of weights of interacting criteria from a reference set. *European Journal of Operational Research* 124(3): 641–650.

Markowitz, H. (1952). Portfolio selection. Journal of Finance 7(1): 77–91.

Markowitz, H. (1959). Portfolio Selection: Efficient Diversification of Investments. New York: John Wiley & Sons.

Markowitz, H. (1987). Mean Variance Analysis in Portfolio Choice and Capital Markets. New York: Blackwell.

Martinson, F.K. (1993). Fuzzy versus min–max weighted multiobjective linear programming illustrative comparisons. *Decision Sciences* 24(4): 809–824.

Matheny, M.S., Erickson, P.A., Niezrecki, C. et al. (2002). Interior and exterior noise emitted by a fuel cell transit bus. *Journal of Sound and Vibration* 251(5): 937–943.

Mathur, K., Salkin, H.M., and Mohanty, B.B. (1986). A note on a general non-linear knapsack problem. *Operations Research Letters* 4(6): 339–356.

McIntyre, C. and Parfitt, M.K. (1998). Decision support system for residential land development site selection process. *Journal of Architectural Engineering ASCE* 4(4): 125–131.

McNicol, B.D., Rand, D.A.J., and Williams, K.R. (2001). Fuel cells for road transportation purposes: yes or no? *Journal of Power Sources* 100(1): 47–59.

Meade, L.M. and Presley, A. (2002). R&D project selection using the analytic network process. *IEEE Transactions on Engineering Management* 49(1): 59–66.

Mi, J., Wu, W., and Zhang, W. (2004). Approaches to knowledge reduction based on variable precision rough set model. *Information Sciences* 159(3–4): 255–272.

Mikhailov, L. (2000). A fuzzy programming method for deriving priorities in the analytic hierarchy process. *Journal of Operational Research Society* 51(3): 341–349.

Mikhailov, L. (2003). Deriving priorities form fuzzy pairwise comparison judgments. *Fuzzy Sets and Systems* 134(3): 365–385.

Mikhailov, L. and Singh, M.G. (2003). Fuzzy analytic network process and its application to the development of decision support systems. *IEEE Transactions on Systems, Man, and Cybernetics* 33(1): 33–41.

Miller, D. and Shamsie, J. (1996). The resource-based view of the firm in two environments: the Hollywood film studios from 1936 to 1965. *Academy of Management Journal* 39(3): 519–543.

Miranda, P., Grabisch, M., and Gil, P. (2002). P-symmetric fuzzy measures. International Journal of Uncertainty, Fuzziness, and Knowledge-Based Systems 10 (Suppl.): 105–123.

Mohanty, R.P., Agarwal, R., Choudhury, A.K. et al. (2005). A fuzzy ANP-based approach to R&D project selection: a case study. *International Journal of Production Research* 43(24): 5199–5216.

Momoh, J.A. and Zhu, J. (2003). Optimal generation scheduling based on AHP/ANP. *IEEE Transactions on Systems, Man, and Cybernetics* 33(3): 531–535.

Mon, D.L., Cheng, C.H., and Lin, J.C. (1994). Evaluating weapon system using fuzzy analytic hierarchy process based on entropy weight. *Fuzzy Sets and Systems* 62(2): 127–134.

Mon, D.L., Tzeng, G.H., and Lu, H.C. (1995). Grey decision making in weapon system evaluation. Journal of Chung Cheng Institute of Technology 24(1): 73–84.

Mori, T. and Murofushi, T. (1989). An analysis of evaluation model using fuzzy measure and the Choquet integral. In Proceedings of Fifth Fuzzy Systems Symposium, pp. 207–212 (in Japanese).

Morita, K. (2003). Automotive power source in the 21st century. *Journal of Society of Automotive Engineers of Japan* 24(1): 3–7.

Morris, V.C. and Young, P. (1976). Philosophy and the American School: An Introduction to Philosophy of Education. Boston: Houghton Mifflin.

Moseley, P.T. (1999). High-rate, valve-regulated lead–acid batteries: suitable for hybrid electric vehicles? *Journal of Power Sources* 84(2): 237–242.

Moskowitz, H. and Kim, K. (1993). On accessing the H value in fuzzy linear regression. *Fuzzy Sets and Systems* 58(3): 303–327.

Murchland, J. (1970). Braess' paradox of traffic flow. *Transportation Research, Part B* 16(1): 45–55.

Murofushi, T. and Sugeno, M. (1989). An interpretation of fuzzy measure and the Choquet integral as an integral with respect to a fuzzy measure. *Fuzzy Sets and Systems* 29(2): 201–227.

Murofushi, T. and Sugeno, M. (1991). A theory of fuzzy measures representations, the Choquet integral, and null sets. *Journal of Mathematical Analysis and Applications* 159(2): 532–549.

Murofushi, T. and Sugeno, M. (1993). Some quantities represented by the Choquet integral. Fuzzy Sets and Systems 56(2): 229–235.

Nagata, M., Yamaguch, T., and Komo, Y. (1995), An Interactive method for multi-period multi-objective production–transportation programming problems with fuzzy coefficients. *Journal of Japan Society for Fuzzy Theory and Systems* 7(1): 153–163.

Nelson, R.R. and Winter, S.G. (1982). An Evolutionary Theory of Economic Change. Cambridge, MA: Harvard University Press.

Nesbitt, K. and Sperling, D. (1998). Myths regarding alternative fuel vehicle demand by light-duty vehicle fleets. *Transportation Research, Part D* 3(4): 259–269.

Nohria, N. and Garcia-Pont, C. (1991). Global strategic linkages and industry structure. *Strategic Management Journal* 12(1): 105–124.

Ohuchi, A., and Kaji, I. (1989). Correction procedures for flexible interpretive structural modeling. *IEEE Transactions on Systems, Man, and Cybernetics* 19(1): 85–94.

Ohuchi, A., Kase, S., and Kaji, I. (1988). MINDS: a flexible interpretive structural modeling system. *In Proceedings of IEEE International Conference on Systems, Man, and Cybernetics* 2: 1326–1329.

Ong, C.S., Huang, J.J., and Tzeng, G.H. (2005). Motivation and resource allocation for strategic alliance through de novo perspective. *Mathematical and Computer Modeling* 41(6–7): 711–721.

Opricovic, S. (1998). Multicriteria optimization of civil engineering systems. Faculty of Civil Engineering, University of Belgrade.

Opricovic, S. and Tzeng, G.H. (2002). Multicriteria planning of post-earthquake sustainable reconstruction. *Computer-Aided Civil and Infrastructure Engineering* 17(3): 211–220.

Opricovic, S. and Tzeng, G.H. (2003). Comparing DEA and MCDM methods. In Tanino, T. et al., Eds., Multi-Objective Programming and Goal Programming: Theory and Applications. Heidelberg, Springer, pp. 227–232.

Opricovic, S. and Tzeng, G.H. (2003). Defuzzification within a fuzzy multicriteria decision model. International Journal of Uncertainty. *Fuzziness and Knowledge-Based Systems* 11(5): 635–652.

Opricovic, S. and Tzeng, G.H. (2003). Fuzzy multicriteria model for post-earthquake land-use planning. *Natural Hazards Review* 4(2): 59–64.

Opricovic, S. and Tzeng, G.H. (2003). Multicriteria expansion of a competence set using genetic algorithm. In Tanino, T. et al., Eds., Multi-Objective Programming and Goal Programming: Theory and Applications. Heidelberg: Springer 221–226.

Opricovic, S. and Tzeng, G.H. (2004). Compromise solution by MCDM methods: a comparative analysis of VIKOR and TOPSIS. *European Journal of Operational Research* 156(2): 445–455.

Opricovic, S. and Tzeng, G.H. (2007). Extended VIKOR method in comparison with outranking methods. *European Journal of Operational Research* 178(2): 514–529.

Ostermark, R. (1997). Temporal interdependence in fuzzy MCDM problems. *Fuzzy Sets and Systems* 88(1): 19–29.

Ou Yang, Y.P., Shieh, H.M., and Tzeng, G.H. (2009). A VIKOR-based multiple criteria decision method for improving information security risk. *International Journal of Information Technology and Decision Making* 8(2): 267–287.

Ou Yang, Y.P., Shieh, H.M., and Tzeng, G.H. (2012). A VIKOR technique based on DEMATEL and ANP for information security risk control assessment. Information Sciences. In press. Available online 2011.

Ou Yang, Y.P., Shieh, H.M., Leu, J.D., and Tzeng, G.H. (2008). A novel hybrid MCDM model combined with DEMATEL and ANP with applications. *International Journal of Operations Research* 5(3): 1–9.

Ou Yang, Y.P., Shieh, H.M., Tzeng, G.H., Yen, L., and Chan, C.C. (2008). Business aviation decision-making using rough sets. In Proceedings of RSCTC, Springer Lecture Notes on Artificial Intelligence 5306, 329–338. DOI: 10.1007/978-3-540-88425-5_34.

Ou Yang, Y.P., Shieh, H.M., Tzeng, G.H. (2011). Combined rough sets with flow graph and formal concept analysis for business aviation decision-making. *Journal of Intelligent Information Systems* 36(3): 347–366.

Owen, S.H. and Daskin, M.S. (1998). Strategic facility location: a review. *European Journal of Operational Research* 111(3): 423–447.

Ozaki, T., Lo, M.C., Kinoshita, E., and Tzeng, G.H. (2011). Decision-making for the best selection of suppliers by using minor ANP. *Journal of Intelligent Manufacturing*. DOI: 10.1007/s10845-011-0563-z.

Ozaki, T., Sugiura, S., Kinoshita, E., and Tzeng, G.H. (2010). Dissolution of the dilemma problems by defining the criteria matrix in ANP. Nagoya Gakuin Daigaku 23: 10–12.

Papageorgiou, E.I. and Groumpos, P.P. (2005). A new hybrid method using evolutionary algorithms to train fuzzy cognitive maps. *Applied Soft Computing* 5(4): 409–431.

Parkhe, A. (1993). Strategic alliance structuring: a game theoretic and transaction cost examination of interfirm cooperation. *Academy of Management Journal* 36(4) :794–829.

Pawlak, Z. (1982). Rough sets. *International Journal of Computer and Information Science* 11(5): 341–356.

Pawlak, Z. (1984). Rough classification. *International Journal of Man–Machine Studies* 20(5): 469–483.

Pawlak, Z. (1997). Vagueness: a rough set view. In Structures in Logic and Computer Science, Springer Lecture Notes in Computer Science, 1261, 106–117, DOI: 10.1007/3-540-63246-8_7.

Pawlak, Z. (2002). Rough sets, decision algorithms and Bayes' theorem. European Journal of Operational Research 136(1): 181–189.

Pawlak, Z. (2004). Decision networks. In Tsumoto, S. et al., Eds., Rough Sets and Current Trends in Computing, Springer Lecture Notes in Artificial Intelligence 3066(1), 1–7.

Pawlak, Z.(2002). Rough set theory and its applications. *Journal of Telecommunications and Information Technology* 3(1): 7–10.

Penrose, E. (1959). The Theory of the Growth of the Firm. New York: John Wiley & Sons.

Pentzaropoulos, G.C. and Giokas, D.I. (2002). Comparing the operational efficiency of the main European telecommunications organizations: A quantitative analysis. *Telecommunications Policy* 26(11): 595–606.

Petrovic, D., Roy, R., and Petrovic, R. (1998). Modelling and simulation of a supply chain in an uncertain environment. *European Journal of Operational Research* 109(2): 299–309.

Petrovic, D., Roy, R., and Petrovic, R. (1999). Supply chain modeling using fuzzy sets. *International Journal of Production Economics* 59(3): 443–453.

Poh, K.L. and Ang, B.W. (1999). Transportation fuels and policy for Singapore: an AHP planning approach. *Computers and Industrial Engineering* 37(4): 507–525.

Polkowski, L. (2004). Toward rough set foundations: a mereological approach. In Tsumoto, S. et al., Eds., Rough Sets and Current Trends in Computing, Springer Lecture Notes in Artificial Intelligence 3066(1), 8–25.

Porter, M.E. (1980). *Competitive Strategies*. New York: *Free Press*.

Porter, M.E. and Fuller, M.B. (1986). Coalitions and global strategy. In Porter, M.E., Ed., Competition in Global Industries. Boston: Harvard Business School Press, pp. 315–343.

Preffer, J. and Nowak, P. (1976). Joint ventures and interorganizational interdependence. *Administrative Science Quarterly* 21(3): 398–418.

Preffer, J. and Salancik, G. (1978). *The External Control of Organizations: A Resource Dependence Perspective. New York: Harper & Row.*

Quafafou, M. (2000). α-RST: a generalization of rough set theory. *Information Sciences* 124(4): 301–316.

Quattrone, G.A. and Tversky, A (1988). Contrasting rational and psychological analyses of political choice. *American Political Science Review* 82(3): 719–736.

Ralescu, D.A. and Adams, G. (1980). Fuzzy integral. *Journal of Mathematical Analysis and Applications* 75(2): 562–570.

Ramanathan, K., Seth, A. and Thomas H. (1997). Explaining joint ventures: alternative theoretical perspectives. In Beamish, P.W. and Killing, J.P., Eds., Cooperative Strategies, Vol. 1. San Francisco: New Lexington Press pp. 51–85.

Rao, S.S. (2004). *Optimization: Theory and Applications* (2nd). John Wiley, New York.

Rao, S.S. (2009). *Engineering Optimization: Theory and Practice*. John Wiley, New Jersey.

Rechenberg, I. (1973). *Evolutions Strategie: Optimierung Technischer Systeme und Prinzipien der Biologischene Evolution*. Stuttgart: Frommann-Holzboog.

Reklaitis, G. V., Ravindran, A. and Ragsdell, K. M. (1983). *Engineering optimization methods and applications*. New York: Wiley.

Reynolds, D. and Cuttance, P. (1992). *School Effectiveness: Research, Policy, and Practice.* London: Villiers House.

Rindfleisch, A. and Heide, J.B. (1997). Transaction cost analysis: past, present, and future applications. *Journal of Marketing* 61(4): 30–54.

Robinson, A.G., Jiang, N., and Lerme, C.S. (1992). On the continuous quadratic knapsack problem. *Mathematical Programming* 55(1): 99–108.

Romero, C. (1991). Carlos Romero (1991). *Handbook of Critical Issues in Goal Programming*. Pergamon Press, Oxford, 124 pp.

Romero, C. (2004). A general structure of achievement function for a goal programming model. *European Journal of Operational Research,* 153(3): 675–686.

Roy, B. (1989). The outranking approach and the foundations of ELECTRE methods. University of Paris Dauphine, Document du Lamsade Laboratoire.

Roy, B. (1990). The outranking approach and the foundation of the ELECTRE method. In Bana-Costa, C., Ed., Readings in Multiple Criteria Decision Aids. Berlin: Springer: pp. 155–183.

Roy, B. and Bertier, P. (1973). La methode electre II: une application au media planning. In Ross, M., Ed., Proceedings of OR '73. Amsterdam: North Holland, pp. 291–302.

Roy, B., Present, M., and Silhol, D. (1986). A programming method for determining which Paris Metro stations should be renovated. *European Journal of Operational Research* 24(2): 318–334.

Romero, C. (2004). A general structure of achievement function for a goal programming model. European Journal of Operational Research, 153(3): 675–686.

Saaty, T.L. (1977). A scaling method for priorities in hierarchical structures. *Journal of Mathematical Psychology* 15(3): 234–281.

Saaty, T.L. (1996). Decision Making with Dependence and Feedback: The Analytic Network Process. Pittsburgh: RWS Publications, p. 481.

Saaty, T.L. (1999). Fundamentals of the analytic network process. International Symposium on Analytic Hierarchy Processes, Kobe.

Saaty, T.L. (2003). Decision-making with the AHP: why is the principal eigenvector necessary? *European Journal of Operational Research* 145(1): 85–91.

Saaty, T.L. and Keams, K.P. (1985). *Analytical Planning*. New York: Pergamon.

Saaty, T.L. and Vargas, L.G. (1982). *The Logic of Priorities*. Boston: Kluwer.

Saaty, T.L. and Vargas, L.G. (1998). Diagnosis with dependent symptoms: Bayes theorem and the analytic hierarchy process. *Operations Research* 46(4): 491–502.

Sakawa, M. (1983). Interactive fuzzy decision making for multiobjective linear programming and its application, Proceedings of IFAC Symposium on Fuzzy Information, Knowledge Representation and Decision Analysis, pp. 295–300.

Sakawa, M. (1984a). Interactive fuzzy decision making for multiobjective nonlinear programming problems. In Grauer, M. and Wierzbicki, A.P., Eds., *Interactive Decision Analysis*, Berlin: Springer, pp. 105–112.

Sakawa, M. (1984b). Interactive fuzzy goal programming for multiobjective nonlinear programming problems and its application to water quality management. *Control and Cybernetics* 13(3): 217–228.

Sakawa, M. (1993). Fuzzy Sets and Interactive Multi-Objective Optimization. New York: Plenum.

Sakawa, M., Kato, K., Sundad, H., Enda, Y. (1995). An interactive fuzzy satisfying method for multiobjective 0–1 programming problems through revised genetic algorithms. *Journal of Japan Society for Fuzzy Theory and Systems* 17(2): 361–370.

Salomon, V.A.P. and Montevechi, J.A.B. (2001). A compilation of comparisons on the analytic hierarchy process and others multiple criteria decision making methods: some cases developed in Brazil. Bern: ISAHP.

Salukvadze, M.E. (1971), Optimization of vector functionals I. The programming of optimal trajectories. *Avtomatika i Telemekhanika* 8(1), 5–15.

Salukvadze, M.E. (1974). On the existence of vector functions in problems of optimization under vector-valued criteria. *Journal of Optimization Theory and Applications* 13(2): 203–217.

Salukvadze, M.E. (1979). Vector-Valued Optimization Problems in Control Theory. New York: Academic Press.

Sarkis, J. (2003). A strategic decision framework for green supply chain management. *Journal of Cleaner Production* 11(4): 397–409.

Sato, T. and Ichii, K. (1996), Optimization of post-earthquake restoration of lifeline networks using genetic algorithms. Japan Society of Civil Engineers 537/I–25: 245–256.

Saunders, R.J., Warford, J.J., and Wellenius, B.A. (1995). Telecommunications and Economic Development: A World Bank Publication. Baltimore: Johns Hopkins University Press.

Schmitt, G. (1985). Second order design. In E.W. Grafarend et al. (Eds), *Optimization and Design of Geodetic Networks*. Springer-Verlag: Berlin, Hidelberg, pp. 74–121.

Sekitani, K. and Takahashi, I. (2001). A unified model and analysis for AHP and ANP. *Journal of Operations Research Society of Japan* 44(1): 67–89.

Sharma, H.D., Gupta, A.D., and Sushil (1995). The objectives of waste management in India: a future inquiry. *Technological Forecasting and Social Change* 48(3): 285–309.

Shee, D.Y. and Tzeng, G.H. (2002). The key dimensions of criteria for the evaluation of ISPs: an exploratory study. *Journal of Computer and Information Systems* 42(4): 112–121.

Shee, D.Y., Tang, T., and Tzeng, G.H. (2000). Modeling the supply–demand interaction in electronic commerce: a bi-level programming approach. *Journal of Electronic Commerce Research* 1(2): 79–93.

Shee, D.Y., Tzeng, G.H., and Tang, T. (2003). AHP, fuzzy measure, and fuzzy integral approaches for the appraisal of information service providers in Taiwan. *Journal of Global Information Technology Management* 6(1): 8–30.

Shen, Y.C., Lin, Grace T.R., Tzeng, G.H. (2012). A novel multi-criteria decision-making combining trial and evaluation laboratory technique for technology evaluation, *Foresight*, 14(2): 139–153.

Shen, Y.C., Lin, Grace T.R., and Tzeng, G.H. (2011). Combined DEMATEL techniques with novel MCDM for the organic light emitting diode technology selection. *Expert Systems with Applications* 38(3): 1468–1481.

Shi, K. (1990). Grey relation theory and its applications. ISUMA.

Shi, Y. (1995) Studies on optimum-path ratios in multi-criteria de novo programming problems. *Computers and Mathematics with Applications* 29(5): 43–50.

Shi, Y. (1999). Optimal system design with multi-decision makers and possible debt: a multi-criteria de novo approach. *Operations Research* 47(5): 723–729.

Shi, Y. (2001). Multiple Criteria Multiple Constraint-Level Linear Programming: Concepts, Techniques, and Applications. Singapore: World Scientific, Chap. 14.

Shi, D.S. and Yu, P.L. (1996). Optimal expansion and design of competence sets with asymmetric acquiring costs, *Journal of Optimization Theory and Applications*, 88(3): 643–658.

Shibano, T., Sakawa, M., and Obata, H. (1996). Interactive decision making for multiobjective 0–1 programming problems with fuzzy parameters through genetic algorithms. *Journal of Japan Society for Fuzzy Theory and Systems* 8(6): 1144–1153.

Shih, H.S. and Lee, E.S. (1999). Fuzzy multi-level minimum cost flow problems. *Fuzzy Sets and Systems* 107(2): 159–176.

Shih, H.S., Lai, Y.J., and Lee, E.S. (1996). Fuzzy approach for multiple-level programming problems. Computers and Operations Research 23(1): 73–91.

Shin, H.W. and Sohn, S.Y. (2004). Multi-attribute scoring method for mobile telecommunication subscribers. *Expert Systems with Applications* 26(3): 363–368.

Shu, J.Z., Tzeng, G.H., and Chen, R.S. (2004). ANP MCDM approach for site selection of biotechnology park. *National Policy Quarterly (Taiwan)* 3(4): 185–200.

Shuai, J.J., Tzeng, G.H., and Li, H.L. (2004). The multi-source partnership selection model. *International Journal of Manufacturing Technology and Management* 6(1–2): 137–154.

Shyng, J.Y., Shieh, H.M., and Tzeng, G.H. (2011a). An integration method combining rough set theory with formal concept analysis for personal investment portfolios. *Knowledge-Based Systems* 23(6), 586–597.

Shyng, J.Y., Shieh, H.M., and Tzeng, G.H. (2011b). Compactness rate as a rule selection index based on rough set theory to improve data analysis for personal investment portfolios. *Applied Soft Computing* 11(4): 3671–3679.

Shyng, J.Y., Shieh, H.M., and Tzeng, G.H. (2011c). Using FSBT technique with rough set theory for personal investment portfolio analysis. *European Journal of Operational Research* 201(2): 601–607.

Shyng, J.Y., Tzeng, G.H., and Wang, F K. (2007). Rough set theory in analyzing the attributes of combination values for insurance market. *Expert Systems with Applications* 32(1): 56–64.

Simon, H.A. (1977). The New Science of Management Decision. New York: Prentice Hall.

Skowron, A. and Grzymala-Busse, J.W. (1993). From the rough set theory to the evidence theory. In Fedrizzi M., et al., Eds., Advances in the Dempster-Shafer Theory of Evidence. New York: John Wiley & Sons, pp. 295–305.

Spencer, L.M. and Spencer, S.M. (1993). Competence at Work: Model for Superior Performance. New York: John Wiley & Sons.

Sperling, D., Setiawan, W., and Hungerford, D. (1995). The target market for methanol fuel. *Transportation Research A* 29(1): 33–45.

Steenbrink, P.A. (1974). Optimization of Transportation Networks. New York: John Wiley & Sons.

Steinbock, D. (2003). Globalization of wireless value system: From geographic to strategic advantages. *Telecommunications Policy* 27(3–4): 207–235.

Stylios, C.D. and Groumpos, P.P. (2004). Modeling complex systems using fuzzy cognitive maps. *IEEE Transactions on Systems, Man, and Cybernetics A* 34(1): 155–162.

Sueyoshi T. (1998). Theory and methodology: privatization of Nippon Telegraph & Telephone. Was it a good policy decision? European Journal of Operational Research 107(1): 45–61.

Sugeno, M. (1974). Theory of fuzzy integrals and its applications, Ph. D. dissertation. Tokyo Institute of Technology, Tokyo, Japan.

Sugeno, M. (1977). Fuzzy measures and fuzzy integrals: a survey in: Gupta, M. M. et al., Eds., Fuzzy Automata and Decision Processes. Amsterdam: North Holland, pp. 89–102.

Sugeno, M. and Terano, T. (1977). A model of learning based on fuzzy information. *Kybernetes* 16(3), 157–166.

Sugeno, M. and Kwon, S.H. (1995). A clusterwise regression-type model for subjective evaluation. *Journal of Japan Society for Fuzzy Theory and Systems* 7(2): 291–310.

Sugeno, M., Fujimoto, K., and Murofushi, T. (1995). A hierarchical decomposition of Choquet integral model. International Journal of Uncertainty, *Fuzziness, and Knowledge-Based Systems* 3(1): 213–222.

Sun, C.C., Lin, Grace T.R., and Tzeng, G.H. (2010). The evaluation of cluster policy by fuzzy MCDM: empirical evidence from Hsinchu Science Park. *Expert Systems with Applications* 36(9): 11895–11906.

Suwansirikul, C., Friesz, T.L., and Tobin, R.L. (1987). Equilibrium decomposed optimization: a heuristic for the continuous equilibrium network design problem. *Transportation Science* 21(4): 254–263.

Swiniarski, R.W. and Skowron, A. (2003). Rough set methods in feature selection and recognition. Pattern Recognition Letters 24(6): 833–849.

Tamiz, M., Jones, D.F., El-Darzi, E. (1995). A review of goal programming and its applications, *Annals of Operations Research*, 58(1): 39–53.

Tamura, M., Nagata, H., and Akazawa, K. (2002). Extraction and systems analysis of factors that prevent safety and security by structural models. In Proceedings of 41st Annual SICE Conference, Osaka.

Tanaka, H. (1987). Fuzzy data analysis by possibilistic linear models. *Fuzzy Sets and Systems* 24(3): 363–375.

Tanaka, H. and Guo, P. (2001). Possibilistic Data Analysis for Operations Research, New York: Physica Verlag.

Tanaka, H., Hayashi, I., and Watada, J. (1989). Possibilistic linear regression analysis for fuzzy data. *European Journal of Operational Research* 40(3): 389–396.

Tanaka, H. and Ishibuchi, H. (1992). Possibilistic regression analysis based on linear programming. In Kacprzyk, J. and Fedrizzi, M., Eds., Studies in Fuzziness: Fuzzy Regression Analysis. Warsaw: Omnitech Press, 47–60.

Tanaka, H. and Lee, H. (1998). Interval regression analysis by quadratic programming approach. IEEE Transactions on Fuzzy Systems 6(4): 473–481.

Tanaka, H., Okuda, T., and Asai, K. (1974). On fuzzy mathematical programming. *Journal of Cybernetics* 3(1): 37–46.

Tanaka, H. and Watada, J. (1988). Possibilistic linear systems and their application to the linear regression model. *Fuzzy Sets and Systems* 27(3): 275–289.

Tanaka, K. and Sugeno, M. (1991). A study on subjective evaluation of color printing images. *International Journal of Approximate Reasoning* 5(3): 213–222.

Tang, T.I., Shee, D.Y., and Tzeng, G.H. (2002). An MCDM framework for assessing ISPs: the fuzzy synthesis decisions of additive and non-additive measurements. J*ournal of Information Management* 8(2): 175–192.

Tang, M.T., Tzeng, G.H., and Wang, S.W. (1999). A hierarchy fuzzy MCDM method for studying electronic marketing strategies in the information service industry. *Journal of International Information Management* 8(1): 1–22.

Tang, T.I. and Tzeng, G.H. (1998). Fuzzy MCDM model for pricing strategies in the internet environment. *Journal of Commercial Modernization* 1(1): 19–34.

Tang, T.I., Tzeng, G.H., and Wang, S.W. (1999). A hierarchy fuzzy MCDM method for studying electronic marketing strategies in the information service industry. *Journal of International Information Management* 8(1): 1–22.

Tay, E.H. and Shen, L. (2002). A modified chi2 algorithm for discretization. *IEEE Transactions on Knowledge and Data Engineering* 14(3): 666–670.

Redden, D.T. and Woodal, W.H. (1994). Properties of certain fuzzy linear regression methods. *Fuzzy Sets and Systems* 64(3): 361–375.

Teece, D.J., Pisano, G., and Shuen, A. (1997). Dynamic capabilities and strategic management. *Strategic Management Journal* 18(7): 509–533.

Teng, J.Y. and Tzeng, G.H. (1996). Multiobjective programming approach for selecting non-independent transportation investment alternatives. *Transportation Research* 30(4): 291–307.

Teng, J.Y. and Tzeng, G.H. (1996). Fuzzy multicriteria ranking of urban transportation investment alternatives. *Transportation Planning and Technology* 20(1): 15–31.

Ting, S.C. and Tzeng, G.H. (2003). Ship scheduling and cost analysis for route planning in liner shipping. *Maritime Economics and Logistics* 5(4): 378–392.

Ting, S.C. and Tzeng, G.H. (2003). Ship scheduling and service network integration for liner shipping companies and strategic alliances. *Journal of Eastern Asia Society for Transportation Studies* 5: 765–777.

Ting, S.C. and Tzeng, G.H. (2004). An optimal container ship slot allocation for liner shipping revenue management. *Maritime Policy and Management* 31(3): 199–211.

Tsai, H. C., Chen, C.M., and Tzeng, G.H. (2006). The comparative productivity efficiency for global telecoms. *International Journal of Production Economics* 103(2): 509–526.

Tsang, E.W.K. (1998). Motives for strategic alliance: a resource-based perspective. *Scandinavian Journal of Management* 14(3): 207–221.

Tsaur S.H., Tzeng, G.H., and Wang, G.C. (1997). Evaluating tourist risks from fuzzy perspectives. *Annals of Tourism Research* 24(4): 796–812.

Tsaur, S.H. and Tzeng, G.H. (1996). Multiattribute decision making analysis for customer preference of tourist hotels. *Journal of Travel and Tourism Marketing* 4(4): 55–69.

Tsaur, S.H., Tzeng, G.H., and Chang, T.Y. (1997). Travel agency organization buying behavior: an application of logit model. *Journal of Management and Systems* 4(2): 127–146.

Tsaur, S.H., Tzeng, G.H., and Chiang, C.I (1996). The comparison of four kinds of prediction methods: traditional econometric regression, fuzzy linear regression, GMDH, and artificial neural network. *Journal of Chinese Statistical Association* 34(2): 132–161.

Tseng, F.M. and Tzeng, G.H. (1999). The comparison of four kinds of prediction methods: ARIMA, fuzzy time series, fuzzy regression time series and grey forecasting. *Journal of Chinese Grey System Association* 2(2): 83–98.

Tseng, F.M. and Tzeng, G.H. (1999). Forecast seasonal time series by comparing five kinds of hybrid grey models. *Journal of the Chinese Fuzzy Systems Association* 5(2): 45–55.

Tseng, F.M. and Tzeng, G.H. (2002). A fuzzy seasonal ARIMA model for forecasting. *Fuzzy Sets and Systems* 126(3): 367–376.

Tseng, F.M., Tzeng, G.H., and Yu, H.C. (1999). Fuzzy seasonal time series for forecasting the production value of mechanical industry in Taiwan. *Technological Forecasting and Social Change, An International Journal* 60(3): 263–273.

Tseng, F.M., Tzeng, G.H., and Yu, H.C. (2000). Combining neural network with seasonal time series ARIMA model to forecast the total production value of Taiwan mechanical industry. *Technological Forecasting and Social Change* 65: 1–17.

Tseng, F.M., Tzeng, G.H., Yuan, B.J.C. et al. (2001). Fuzzy ARIMA model for forecasting the foreign exchange market. *Fuzzy Sets and Systems* 118(1): 9–19.

Tseng, F.M., Yu, H.C., and Tzeng, G.H. (2001). Applied hybrid grey model to forecast seasonal time series. *Technological Forecasting and Social Change* 67(2): 291–302.

Tseng, F.M., Yu, H.C., and Tzeng, G.H. (2002). Combing neural network with seasonal time series ARIMA model. *Technological Forecasting and Social Change* 69(1): 71–87.

Tzeng, G.H. (1990). Modeling energy demand and socioeconomic development of Taiwan. *The Energy Journal* 10(2): 133–152.

Tzeng, G.H. (1990). Energy policy in Taiwan. *Energy Systems and Policy* (*An International Journal*) 13(4): 267–284.

Tzeng, G.H., Chang, J.R., Lin, J.D., and Hung, C.T. (2002). Non-additive grey relation model for the evaluation of flexible pavement condition. *International Journal of Fuzzy Systems* 4(2): 715–724.

Tzeng, G.H., Chang, C.Y., and Lo, M.C. (2005). MADM approach for effecting information quality of knowledge management. *International Journal of Information Systems for Logistics and Management* 1(1): 55–67.

Tzeng, G.H., Chang, C.Y., and Lo, M.C. (2005). The simulation and forecast model for human resources of semiconductor wafer Fab operation. *International Journal of Industrial Engineering and Management Systems* 4(1): 47–53.

Tzeng, G.H., Chang, S.L., Wang, J.C., Hwang, M.J., Yu, G.C., and Juang, M.C. (1998). Application of fuzzy multi-objective programming to economic-energy-environment model. *Journal of Management (Taiwan)* 15(4): 683–707.

Tzeng, G.H. and Chen, C.H. (1993). Multiobjective decision making for traffic assignment. *IEEE Transactions on Engineering Management* 40(2): 180–187.

Tzeng, G.H. and Chen, J.J. (1997). Developing Taipei automobile driving cycles for emissions. *Energy and Environment* 8(3): 227–238.

Tzeng, G.H. and Chen, J.J. (1998). Developing Taipei motorcycle driving cycle for emissions and fuel economy. *Transportation Research, Part D* 3(1): 19–27.

Tzeng, G.H., Chen, J.J., and Lan, C.J. (1991). The influence of modal choice on energy conservation: Application of logit model. *Energy Economics* (*An International Journal*) 13(4): 290–299.

Tzeng, G.H., Chen, J.J., and Teng, J.D. (1997). Evaluation and selection of suitable battery for electrics motorcycle in Taiwan: application of fuzzy multiple attribute decision making. *Journal of Chinese Institute of Industrial Engineers* 14(3): 319–331.

Tzeng, G.H., Chen, J.J., and Yen, Y.K. (1996). The strategic model of multicriteria decision making for managing the quality of the environment in metropolitan Taipei. *Asian Journal of Environmental Management* 4(1): 41–52.

Tzeng, G.H., Chen, T.Y., and Wang, J.C. (1998). A weight assessing method with habitual domains. *European Journal of Operational Research* 110(2): 342–367.

Tzeng, G.H., Chen, W.H., Yu, R. et al. (2010). Fuzzy decision maps: a generalization of the DEMATEL methods. *Soft Computing* 14(11): 1141–1150.

Tzeng, G.H. and Chen, Y.W. (1998). Implementing an effective schedule for reconstructing post-earthquake road network based on asymmetric traffic assignment: An application of genetic algorithm. *International Journal of Operations and Quantitative Management* 4(3): 229–246.

Tzeng, G.H. and Chen, Y.W. (1999). The optimal location of airport fire stations: a fuzzy multiobjective programming through revised genetic algorithm. *Transportation Planning and Technology* 23(1): 37–55.

Tzeng, G.H., Chen, Y.W., and Lin C.Y. (2000). Fuzzy multi-objective reconstruction plan for post-earthquake road network by genetic algorithm. In Research and Practice in Multiple Criteria Decision Making, Springer Lecture Notes in Economics and Mathematical Systems 487, Part III, pp. 510–528, DOI: 10.1007/978-3-642-57311-8_43.

Tzeng, G.H., Cheng, H.J., and Huang, T.D. (2007). Multi-objective optimal planning for designing relief delivery systems. *Transportation Research E* 43(6): 673–686.

Tzeng, G.H. and Chiang, C.I. (1998). Applying possibility regression to grey model. *Journal of Chinese Grey System Association* 1(1): 19–31.

Tzeng, G.H., Chiang, C.H., and Li, C.W. (2007). Evaluating intertwined effects in e-learning programs: a novel hybrid MCDM model based on factor analysis and DEMATEL. *Expert Systems with Applications* 32(4): 1028–1044.

Tzeng, G.H., Chiang, C.I., and Hwang, M.J. (1996). Multiobjective programming approach to the allocation of air pollution monitoring station. *Journal of Chinese Institute of Environmental Engineering* 6(1): 99–105.

Tzeng, G.H., Chiou, Y.C., and Sheu, S.K. (1997). A comparison of genetic and stepwise algorithms for optimal sites of mainline barrier-type toll station. *Journal of Chinese Institute of Civil and Hydraulic Engineering* 9(1): 171–178.

Tseng, Y.H., Durbin, P., and Tzeng, G.H. (2001). Using a fuzzy piecewise regression to predict the nonlinear time-series of turbulent flows with automatic change-point detection. *Flow, Turbulence and Combustion* 67(2): 81–106.

Tzeng, G.H., Feng, G.M., and Kang, C.C. (2001). The fuzzy set theory and DEA model for forecasting production efficiency: case study for Taipei City Bus Company. *Journal of Advanced Computational Intelligence*, 5(3), 128–138.

Tzeng, G.H., Hu, C.P. and Junn-Yuan Teng, J.Y. (1991). Urban environmental evaluation and improvement: Applicaiton of multiattribute utility and compromise programming. *Behaviormetrika* 29:71–87.

Tzeng, G.H. and Hu, Y.C. (1996). The selection of bus system operation and service performance indicators: application of grey relation analysis. *Journal of the Chinese Fuzzy Systems Association* 2(1): 73–82.

Tzeng, G.H. and Huang, C.Y. (2012). Combined DEMATEL technique with hybrid MCDM methods for creating the aspired intelligent global manufacturing and logistics systems. *Annals of Operations Research* 197(1): 159–190.

Tzeng, G.H. and Huang, J.J. (2011). Multiple Attribute Decision Making: Methods and Applications. Boca Raton, FL: Taylor & Francis.

Tzeng, G.H. and Huang, W.C. (1997). The spatial and temporal bi-criteria parallel savings-based heuristic algorithm for vehicle routing problem with time windows. *Transportation Planning and Technology* 20(2): 163–181.

Tzeng, G.H., Hwang, M.J., and Liu, Y.H. (1997). Dynamic optimal expansion for fuzzy competence sets of multistage-multi-objective planning. *Pan-Pacific Management Review* 1(1): 55–70.

Tzeng, G.H., Hwang, M. J., and Yeh, W.C. (1998). Multi-objective planning for integrated land use and outside allied transportation systems in recreation area. *Journal of Management (Taiwan)* 15(1): 133–161.

Tzeng, G.H., Jen, W., and Hu, K.C. (2002). Fuzzy factor analysis for selecting service quality factors: a case of the service quality of city bus service. *International Journal of Fuzzy Systems* 4(4): 911–921.

Tzeng, G.H. and Kuo, J.S. (1996). Fuzzy multi-objective-double sampling plans with genetic algorithms based on Bayesian model. *Journal of the Chinese Fuzzy Systems Association* 2(2): 57–74.

Tzeng, G.H. and Lee, M.Y. (2001). Intellectual capital in the information industry. In Chang, C.Y. and Yu, P.L., Eds., Made by Taiwan: Booming in the Information Technology Era. Singapore: World Science, pp. 298–344.

Tzeng, G.H. and Lin, C.W. (2000). Evaluation on new substitute fuel mode of bus vehicle for suitable urban public transportation. *Transportation Planning Journal* 29(3): 665–692, in Chinese.

Tzeng, G.H., Lin, C.W., and Opricovic, S. (2005). Multi-criteria analysis of alternative fuel buses for public transportation. *Energy Policy* 33(11): 1373–1383.

Tzeng, G.H., Ouyang, P., Lin, C.T., and Chen, C.B. (2005). Hierarchical MADM with fuzzy integral for evaluating enterprise intranet Web sites. *Information Sciences* 169(3–4): 409–426.

Tzeng, G.H., Shiah, J.Y., and Chiang, C.I. (1998). Application of de novo programming to land use plans of theory of Chiao Tung University. *Journal of City Planning* 25(1): 93–105.

Tzeng, G.H., Shiau, T.A. (1987). Energy conservation strategies in transportation - Application of multiple criteria decision-making. *Energy Systems and Policy (An International Journal)* 11(1): 1–19.

Tzeng, G.H., Shiau, T.A. (1988). Multiple objective programming for bus operation: A case study for Taipei city, *Transportation Research, Part B: Methodological* 22(3): 195–206.

Tzeng, G.H., Shiau, T.A., and Lin, C.Y. (1992). Application of multi-criteria decision making to the evaluation of a new energy system development in Taiwan. *Energy* 17(10): 983–992.

Tzeng, G.H., Shiau, T.A., and Teng, J.Y. (1994). Multi-objective decision making approach to energy supply mix decisions in Taiwan. *Energy Sources* 16(3): 301–316.

Tzeng, G.H., Shieh, H.M., and Shiau, T.A. (1989). Route choice behavior in transportation: an application of the multi-attribute utility theorem. *Transportation Planning and Technology* 13(4): 289–301.

Tzeng, G.H., Tang T.I., Hung, Y.M., and Chang, M.L. (2006). Multiple-objective planning for a production and distribution model of the supply chain: case of a bicycle manufacturer. *Journal of Scientific and Industrial Research* 65(4): 309–320.

Tzeng, G.H. and Teng, J.Y. (1994). Multicriteria evaluation for strategies of improving and controlling air-quality in the super city: A case of Taipei City. *Journal of Environmental Management*, 40(3): 213–229.

Tzeng, G.H. and Teng, J.Y. (1998). Transportation investment project selection using fuzzy multi-objective programming. *Fuzzy Sets and Systems* 96(3): 259–280.

Tzeng, G.H., Teng, M.H., and Chen, J.J., (2002). Multi-criteria selection for a restaurant location in Taipei. *International Journal of Hospitality Management* 21(2): 171–187.

Tzeng, G.H., Teodorovic, D., and Hwang, M.J. (1996), Fuzzy bi-criteria multi-index transportation problem for coal allocation planning of Taipower. *European Journal of Operational Research* 95(1): 62–72.

Tzeng, G.H. and Tsaur, S.H. (1994). The multiple criteria evaluation of grey relation model. *Journal of Grey Systems* 6(2): 87–108.

Tzeng, G.H. and Tsaur, S.H. (1993). Application of multicriteria decision making to old vehicle elimination in Taiwan. *Energy and Environment* 4(2): 268–283.

Tzeng, G.H. and Tsaur, S.H. (1995). Energy demand Forecast for motor freight transportation in Taiwan: Application of a dynamic interregional input-output model. Journal of Applied Input-Output Analysis 2(2): 41–53.

Tzeng, G.H. and Tsaur, S.H. (1997). Application of multiple criteria decision making for network improvement. *Journal of Advanced Transportation* 31(1): 49–74.

Tzeng, G.H., Tsaur, S.H., Law, Y.D., and Opricovic, S. (2002). Multi-criteria analysis of environmental quality in Taipei: public preferences and improvement strategies. *Journal of Environmental Management* 65(2): 109–120.

Tzeng, G.H., Wang, J.C., and Hwang, M.J. (1996). Using genetic algorithms and the template path concept to solve the traveling salesman problem. *Transportation Planning Journal* 25(3): 493–516.

Tzeng, G.H., Wang, H.F., Wen, U.P., Yu, P.L. (1994). Multiple Criteria Decision Making. Heidelberg: Springer.

Uri, N.D. (2000). Measuring productivity change in telecommunications. *Telecommunications Policy* 24(5): 439–452.

Uri, N.D. (2001a). Changing productive efficiency in telecommunications in the United States. *International Journal of Production Economics* 72(2): 121–137.

Uri, N.D. (2001b). Incentive regulation and the change in productive efficiency of local exchange carriers. *Applied Mathematical Modeling* 25(5): 335–345.

Uri, N.D. (2001c). Telecommunications in the United States and changing productive efficiency. *Journal of Industry, Competition, and Trade* 1(3): 321.

Uri, N.D. (2001d). The effect of incentive regulation on productive efficiency in telecommunications. Journal of Policy Modeling 23(8): 825–846.

van der Vorst, J.G.A.J., Beulens, A.J.M., De Wit, W., and Beek, P.B. (1998). Supply chain management in food chains: improving performance by reducing uncertainty. *International Transactions in Operational Research* 5(6): 487–499.

van Laarhoven, P.J. and Pedrycz, W. (1983). A fuzzy extension of Saaty's priority method. Fuzzy Sets and Systems 11(1–3): 229–241.

von Altrock, C. (1996). Fuzzy Logic and Neurofuzzy Applications in Business and Finance. New York: Prentice Hall.

von Stackelberg, H. (1952). The theory of market economy. Oxford University Press, Oxford.

von Neumann, J. and Morgenstern, O. (1947). Theory of Games and Economic Behavior, 2nd ed. Princeton, NJ: Princeton University Press.

Wagenknecht, M. and Hartmann, K. (1983). On fuzzy rank ordering in polyoptimization. Fuzzy Sets and Systems 11: 243–251.

Wakabayashi, T., Itoh, L. and Ohuchi, A. (1995). A method for constructing system models by fuzzy flexible interpretive structural modeling fuzzy systems. In Proceedings of Joint Fourth International IEEE Conference on Fuzzy Systems and Second International Fuzzy Engineering Symposium, pp. 913–918.

Walczak, B. and Massart, D.L. (1999). Tutorial rough sets theory. Chemometrics and Intelligent Laboratory Systems 47(1): 1–16. Walters and Smith (1995) Chap. 2.

Wang, J.C., Chen, T.Y., and Shen, H.M. (2001). Using fuzzy densities to determine the value for fuzzy measures. In Proceedings of Ninth National Conference on Fuzzy Theory and Its Applications, pp. 54–59.Wang, B., Shyu, J. Z., and Tzeng, G.H. (2007). Appraisal model for admitting new tenants to the incubation center at ITRT. International Journal of Innovative Computing Information and Control 3(1): 119–130.

Wang, C.H., Chin, Y.C., and Tzeng, G.H. (2010). Mining the R&D innovation performance processes for high-tech firms based on rough set theory. Technovation 30(7–8): 447–458.

Wang, Z. and Klir, G.J. (1992). Fuzzy Measure Theory. New York: Plenum. Wang, Y.L. and Tzeng, G.H. (2012). Brand marketing for creating brand value based on a MCDM model combining DEMATEL with ANP and VIKOR methods. Expert System with Applications 39(5): 5600–5615.

Warfield, J.N. (1973). On arranging elements of a hierarchy in graphic form. IEEE Transactions on Systems, Man, and Cybernetics 3(2): 121–132.

Warfield, J.N. (1974). Developing interconnection matrices in structural modeling. IEEE Transactions on Systems, Man, and Cybernetics 4(1): 81–87.

Warfield, J.N. (1974). Developing subsystem matrices in structural modeling. IEEE Transactions on Systems, Man, and Cybernetics 4(1): 74–80.

Warfield, J.N. (1974). Toward interpretation of complex structural models. IEEE Transactions on Systems, Man, and Cybernetics 4(5): 405–417.

Warfield, J.N. (1976). Societal Systems: Planning, Policy, and Complexity. New York: John Wiley & Sons.

Warfield, J.N. (1977). Crossing theory and hierarchy mapping. IEEE Transactions on Systems, Man, and Cybernetics 7(7): 502–523.

Wei, P.L., Huang, J.H., Tzeng, G.H., and Wu, S.I. (2010). Causal modeling of web advertising effects by improving SEM based on DEMATEL technique. International Journal of Information Technology and Decision Making 9(5): 799–829.

Wernerfelt, B. (1984). A resource-based view of the firm. Strategic Management Journal 5(2): 171–180.

Werners, B. (1987). Interactive multiple objective programming subject to flexible constraints. European Journal of Operational Research 31(2): 342–349.

White, P.A. (1989). An overview of school-based management: what does the research say? NASSP Bulletin 73(518): 1–8.

Whitelock, J. and Rees, M. (1993). Trends in mergers, acquisitions and joint ventures in the single European market. European Business Review. 93(4): 26–32.

Williamson, O.E. (1975). Markets and Hierarchies: Analysis and Antitrust Implications, New York: Basic Books.

Williamson, O.E. (1985). The Economic Institutions of Capitalism, New York: Free Press.

Williamson, O.E. (1991a). Comparative economic organization: The analysis of discrete structural alternatives. Administrative Science Quarterly 36(2): 269–296.

Williamson, O.E. (1991b). Strategizing, economizing, and economic organization. Strategic Management Journal 12(8): 75–94.

Winebrake, J.J. and Creswick, B.P. (2003). The future of hydrogen fueling systems for transportation: an application of perspective-based scenario analysis using the analytic hierarchy process. Technological Forecasting and Social Change 70(2): 359–384.

Winston, W.L. (1994). *Operations Research: Applications and Algorithms* (third ed). Duxbury Press, Belmont, California

Witlox, F. and Tindemans, H. (2004). The application of rough sets analysis in activity-based modeling, opportunities and constraints. Expert Systems with Applications 27(2): 171–180.

Wu, C.H., Tzeng, G.H., Goo, Y.J., and Fang, W.C. (2007). A real-valued genetic algorithm to optimize the parameters of support vector machine for predicting bankruptcy. Expert Systems with Applications 32(2): 397–408.

Wu, C.H., Tzeng, G.H., and Lin, R.H. (2009). A novel hybrid genetic algorithm for kernel function and parameter optimization in support vector regression. Expert Systems with Applications 36 (3): 4725–4735.

Wu, D.S., Chen, D.S., and Zhao, C.G. (1990). Plane Design of Architecture. Taipei: Scientific & Technical Publishing.

Wu, H.Y., Tzeng, G.H. and Chen, Y.H. (2009). A fuzzy MCDM approach for evaluating banking performance based on Balanced Scorecard. Expert Systems with Applications 36(6): 10135–10147.

Wu, I.C., Liu, D.R., and Tzeng, G.H. (2002). Application of AHP method and grey relation model to evaluation of portal site performance. Journal of Chinese Grey System Association 5(1): 41–54.

Wu, K.H. and Tzeng, G.H. (2002). The application of possibility grey forecasting to stock market prices in Taiwan. Journal of Chinese Grey System Association 5(1): 7–16.

Wu, W.W., Lee, Y.T., and Tzeng, G.H. (2005). Simplifying the manager competency model by using the rough set approach, rough sets, fuzzy sets, data mining, and granular computing, 3642 (Part II): 484–495.

Yang, C.A., Fu, G.L., and Tzeng, G.H. (2007). A multicriteria analysis of the strategies to open Taiwan's mobile virtual network operators services. International Journal of Information Technology and Decision Making 6(1): 85–112.

Yang, C.A., Fu, G.L., and Tzeng, G.H. (2005). Creating a win–win in the telecommunications industry: the relationship between MVNOs and MNOs in Taiwan. Canadian Journal of Administration Sciences 22(4): 316–328.

Yang, J.L., Chiu, H.N., and Tzeng, G.H. (2008). Vendor selection by integrated fuzzy MCDM techniques with independent and interdependent relationships. Information Sciences 178(21): 4166–4183.

Yang, J.L. and Tzeng, G.H. (2011). An integrated MCDM technique combined with DEMATEL for a novel cluster-weighted with ANP method Original Research. Expert Systems with Applications 38(3): 1417–1424.

Yang, Y.H., Lee, Z.Y., Hung, C.Y., and Tzeng, G.H. (2003). Fuzzy MCDM approach for analyzing the best mobilization timing of Taiwan. Journal of Defense Management 24(1): 27–44 (Chinese, Taiwan).

Yen, K.C., Lin, J.J., and Tzeng, G.H. (2005). Applying eco-planning and Fuzzy AHP on the evaluation of the development potentiality of land use in the waterfront environment. Journal of Architecture and Planning (Taiwan) 6(1): 21–42.

Yen, K.C., Tzeng, G.H., and Lin, J.J. (2005). Performance standards control rules for sustainable management of coastal wetland by using grey relation. Journal of City and Planning 32(4): 421–441.

Yoon, C.H. (1999). Liberalization policy, industry structure and productivity changes in Korea's telecommunications industry. Telecommunications Policy 23(3–4): 289–306.

Yoon, K.P. (1996). A probabilistic approach to rank complex fuzzy numbers. Fuzzy Sets and Systems 80(2): 167–176.

Yoshino, M.Y. and Rangan, U.S. (1995). Strategic Alliances: An Entrepreneurial Approach to Globalization. Cambridge: Harvard Business School Press.

Yu, J.R. and Tzeng, G.H. (2009). A fuzzy multiple objective programming in interval piecewise regression model. International Journal of Uncertainty, Fuzziness, and Knowledge-Based Systems 17(3): 365–376.

Yu, J.R., Tzeng, G.H., Chiang, C.I. et al. (2007). Raw material supplier ratings in the semiconductor manufacturing industry through fuzzy multiple objectives programming to DEA.
International Journal of Operations and Quantitative Management 13(4): 101–111.

Yu, J.R., Tzeng, G.H., and Li, H.L. (1999). A general piecewise necessity regression analysis based on linear programming. Fuzzy Sets and Systems 105(3): 429–436.

Yu, J.R., Tzeng, G.H. and Li, H.L. (2001). General fuzzy piecewise regression analysis with automatic change-point detection. Fuzzy Sets and Systems 119(2): 247–257.

Yu, J.R., Tzeng, G.H., and Li, H.L. (2005). Interval piecewise regression model with automatic change-point detection by quadratic programming. International Journal of Uncertainty, *Fuzziness, and Knowledge-Based Systems* 13(3): 347–361.

Yu, J.R., Tzeng, Y.C., Tzeng, G.H. et al. (2004). A fuzzy multiple objective programming approach to DEA with imprecise Data. International Journal of Uncertainty, *Fuzziness, and Knowledge-Based Systems* 12(5): 591–600.

Yu, P.L. (1973). A class of solutions for group decision problems. *Management Science* 19(8): 936–946.

Yu, P. L. (1990). Forming winning strategies: An integrated theory of habitual domains. Springer-Verlag, Berlin, Heidelberg, New York, London, Paris, Tokyo, 1990 (392 pages).

Yu, P.L. and Zeleny, M. (1972). The set of all nondominated solutions in linear cases and a multicriteria simplex method, CSS 72–03. The University of Rochester, Rochester, NY.

Yu, P.L. and Zeleny, M. (1973). On some linear-parametric programs, Center for Systems Science, CSS 73–05. The University of Rochester, Rochester, NY.

Yu, P.L. and Zeleny, M. (1975). The Set of All Nondominated Solutions in Linear Cases and Multicriteria Simplex Method, *Journal of Math. Analysis and Applications*, 49(2): 430–468.

Yu, P.L. (1985). *Multiple-Criteria Decision Making*. New York: Plenum.

Yu, P.L. and Leitmann, G. (1974). Compromise solutions, domination structures, and Salukvadze's solution. *Journal of Optimization Theory and Applications* 13(3): 362–378.

Yu, P.L. and Li, H.L. (1994). Optimal Competence Set Expansion Using Deduction Graph, *Journal of Optimization Theory and Applications*. 80(1): 75–91.

Yu, P.L. and Zhang, D. (1992). Optimal expansion of competence sets and decision support, *Information Systems and Operational Research* 30(2): 68–85.

Yu, R.C. and Tzeng, G.H. (2006). A soft computing method for multi-criteria decision making with dependence and feedback. *Applied Mathematics and Computation* 180(1): 63–75.

Yuan, B., Tzeng, G.H., Kang, T.H., and Cheng, C.C. (2005). MADM approach for selecting the consignment performance of the information service computer room of De-Lin Institute of Technology. *Journal of Information Management* 12(1): 131–148 (Chinese, Taiwan).

Yuan, B.J.C., Wang, C.P., and Tzeng, G.H. (2005). An emerging approach for strategy evaluation in fuel cell development. *International Journal of Technology Management* 32(3–4): 302–338.

Zadeh, L. (1965). Fuzzy sets. *Information and Control* 8(3): 338–353.

Zadeh, L.A. (1975a). The concept of a linguistic variable and its application to approximate reasoning I. *Information Science* 8(3): 199–249.

Zadeh, L.A. (1975b). The concept of a linguistic variable and its application to approximate reasoning II. *Information Science* 8(4): 301–357.

Zadeh, L.A. (1978). Fuzzy sets as a basis for a theory of possibility. *Fuzzy Sets and Systems* 1(1): 3–28.

Zajac, E.J. and Olsen, C.P. (1993). From transaction cost to transactional value analysis: implications for the study of interorganizational strategies. *Journal of Management Studies* 30(1): 131–45.

Zanakis, S.H., Solomon, A., Wishart, N. et al. (1998). Multi-attribute decision making: a simulation comparison of select methods. *European Journal of Operational Research* 107(3): 507–529.

Zeleny, M. (1981). A case study in multiobjective design: De Novo programming. In P. Nijkamp and J. Spronk (eds). Multiple Criteria Analysis: Operational Methods, Grower Publisher Co., Hampshire, pp.37–52.

Zeleny, M. (1982). *Multiple Criteria Decision Making*. New York: McGraw-Hill.

Zeleny, M. (1986). Optimal system design with multiple criteria: de novo programming approach. *Engineering Cost Production Economics* 10(1): 89–94.

Zeleny, M. (1990). Optimal given system versus designing optimal system: the de novo programming approach. *International Journal of General System* 17(3): 295–307.

Zeleny, M. (1995). Trade-offs: free management via de novo programming. International Journal of Operations and Quantitative Management 1(1): 3–13.

Zhai, L.Y., Khoo, L.P., and Fok, S.C. (2002). Feature extraction using rough set theory and genetic algorithms: an application for the simplification of product quality evaluation. *Computers and Industrial Engineering* 43(4): 661–676.

Zhou, H. and Sperling, D. (2001). Traffic emission pollution sampling and analysis on urban streets with high-rising buildings. *Transportation Research D* 6(4): 269–281.

Zhu, J. (2000). Multi-factor performance measure model with an application to Fortune 500 companies. *European Journal of Operational Research* 123(1): 105–124.

Zilla, S.S. and Friedman, L. (1998). Theory and methodology: DEA and the discriminant analysis of ratios for ranking units. *European Journal of Operational Research* 111(3): 470–478.

Zimmerman, H.J. (1978). Fuzzy programming and linear programming with several objective functions. *Fuzzy Sets and Systems* 1(1): 45–55.

Notes*

CHAPTER 1 INTRODUCTION

Basic Concepts of Multiple Objective Programming (MOP) Problem

Most multi-objective programming (MOP) problems can mathematically be represented as:

$$\begin{aligned} \max \quad & \left[f_1(\boldsymbol{x}), f_2(\boldsymbol{x}), \dots, f_k(\boldsymbol{x})\right] \\ s.t. \quad & \boldsymbol{A}\boldsymbol{x} \le \boldsymbol{b} \\ & \boldsymbol{x} \ge \boldsymbol{0} \end{aligned} \tag{N1.1}$$

Example: Two objectives and two variables $\boldsymbol{x}' = (x_1, x_2)$

$$\begin{aligned} &\max f_1(\boldsymbol{x}) = x_1 + x_2 \\ &\max f_2(\boldsymbol{x}) = x_2 - x_1 \\ &s.t. \\ &0 \le x_1, x_2 \le 3 \end{aligned} \quad \text{or} \quad \begin{aligned} &\max f(\boldsymbol{x}) = x_1 + x_2 \\ &\max f(\boldsymbol{x}) = x_2 - x_1 \\ &s.t. \\ &0 \le x_1 \quad \le 3 \\ &0 \le \quad x_2 \le 3 \end{aligned} \quad \text{or} \quad \begin{aligned} &\max f(\boldsymbol{x}) = x_1 + x_2 \\ &\max f(\boldsymbol{x}) = x_2 - x_1 \\ &s.t. \\ &\begin{bmatrix} 1 & 0 \\ 0 & 1 \end{bmatrix} \begin{bmatrix} x_1 \\ x_2 \end{bmatrix} \le \begin{bmatrix} 3 \\ 3 \end{bmatrix} \\ &x_1 \ge 0, x_2 \ge 0. \end{aligned}$$

Note that the Pareto optimal solution is also called the non-inferior or non-dominated solution.

Developing the criteria and designing the fuzzy linguistic scale — The first step is defining the decision goals and developing criteria for the specific research question. Linguistic variables take on values defined in the term set (set of linguistic terms). Figure N1.1 displays a triangular fuzzy number (TFN). Linguistic terms are subjective categories for linguistic variables. The values of a linguistic variable are words or sentences in a natural or artificial language. A triangular fuzzy number $x \in \tilde{A}$ and $\tilde{A} = (l, m, u)$ on R may be a TFN if its membership function $\mu_{\tilde{A}}(x) : \Re \rightarrow [0,1]$ is equal to the following equation:

$$\mu_{\tilde{A}}(x) = \begin{cases} (x-l)/(m-l), & l \le x \le m \\ (u-x)/(u-m), & m \le x \le u \\ 0 & \text{otherwise} \end{cases} \tag{N1.2}$$

* The contents are a result of Professor Tzeng taking part of his related teaching courses "Research Methods for Problem-Solving" outline in each part.

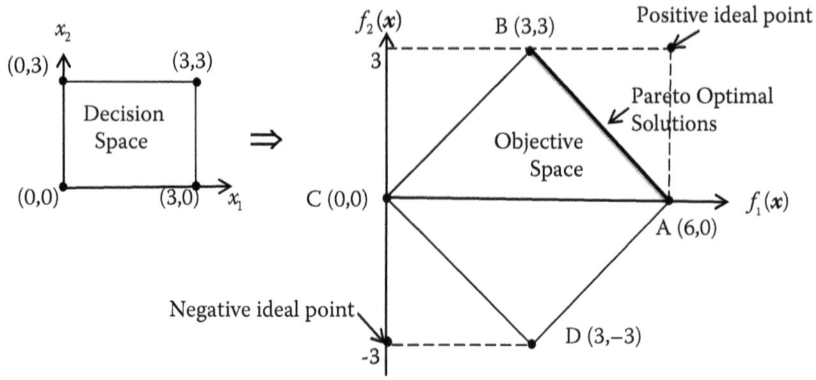

FIGURE N1.1 Basic concept of decision space and decision space in MODM.

(**Note:** "Pareto Optimal Solutions" is also named "Non-inferior Solutions" or Non-dominated solutions" or called "Efficiency Solutions.")

From Equation (N1.2), the diagonal l and u denote the lower and upper bounds of the fuzzy number $\tilde{A}$, and m is the modal value for $\tilde{A}$. The TFN can be denoted by $\tilde{A} = (l,m,u)$. The operational laws of TFNs $\tilde{A}_1 = (l_1, m_1, u_1)$ and $\tilde{A}_2 = (l_2, m_2, u_2)$ are displayed as Equations (N1.3) through (N1.7).

Addition of fuzzy number ⊕:

$$\tilde{A}_1 \oplus \tilde{A}_2 = (l_1, m_1, u_1) \oplus (l_2, m_2, u_2) = (l_1 + l_2, m_1 + m_2, u_1 + u_2) \qquad \text{(N1.3)}$$

Multiplication of fuzzy number ⊗:

$$\begin{aligned} \tilde{A}_1 \otimes \tilde{A}_2 &= (l_1, m_1, u_1) \otimes (l_2, m_2, u_2) \\ &= (l_1 l_2, m_1 m_2, u_1 u_2) \text{ for } l_1, l_2 > 0;\ m_1, m_2 > 0;\ u_1, u_2 > 0 \end{aligned} \qquad \text{(N1.4)}$$

Subtraction of fuzzy number Θ:

$$\tilde{A}_1 \Theta \tilde{A}_2 = (l_1, m_1, u_1) \Theta (l_2, m_2, u_2) = (l_1 - u_2, m_1 - m_2, u_1 - l_2) \qquad \text{(N1.5)}$$

Division of fuzzy number ∅:

$$\begin{aligned} \tilde{A}_1 \varnothing \tilde{A}_2 &= (l_1, m_1, u_1) \varnothing (l_2, m_2, u_2) \\ &= (l_1/u_2, m_1/m_2, u_1/l_2) \text{ for } l_1, l_2 > 0;\ m_1, m_2 > 0;\ u_1, u_2 > 0 \end{aligned} \qquad \text{(N1.6)}$$

Reciprocal of fuzzy number:

$$\tilde{A}^{-1} = (l_1, m_1, u_1)^{-1} = (1/u_1, 1/m_1, 1/l_1) \text{ for } l_1, l_2 > 0;\ m_1, m_2 > 0;\ u_1, u_2 > 0 \qquad \text{(N1.7)}$$

We use this kind of expression to evaluate two shopping websites by nine basic linguistic terms (natural language) for measuring perceptions and feelings. Examples are beautiful, good, perfect, very high influence, high influence, low influence, very low influence, and no influence on a fuzzy level scale as shown in Table N1.1 and Figure N1.2.

TABLE N1.1
Linguistic Scales for Importance (Example)

Linguistic Influence	Linguistic Value
Perfect	(1, 1, 1)
Very high (VH)	(0.5, 0.75, 1)
High (H)	(0.25, 0.5, 0.75)
Low (L)	(0, 0.25, 0.5)
Very low (VL)	(0, 0, 0.25)
None (No)	(0, 0, 0)

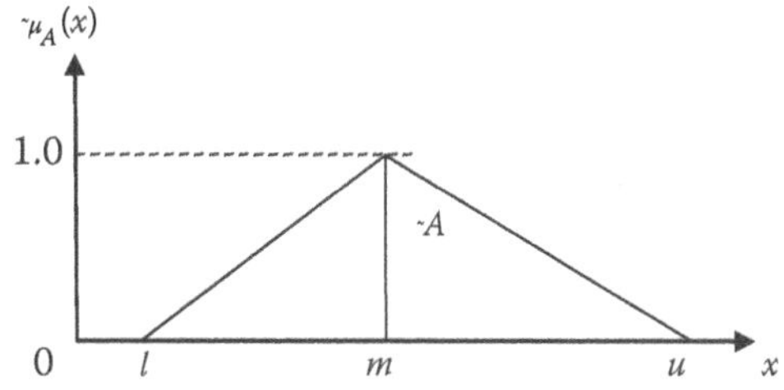

FIGURE N1.2 Membership function of triangular fuzzy number.

Extension Principle for Fuzzy Arithmetic Operations

Let $\tilde{m}$ and $\tilde{n}$ be two fuzzy numbers and z denote a specific event. The membership functions of the four basic arithmetic operations for $\tilde{m}$ and $\tilde{n}$ can be defined by

$$\mu_{\tilde{m}+\tilde{n}}(z) = \sup_{x,y}\{\min(\tilde{m}(x), \tilde{n}(y)) \mid x + y = z\}; \tag{N1.8}$$

$$\mu_{\tilde{m}-\tilde{n}}(z) = \sup_{x,y}\{\min(\tilde{m}(x), \tilde{n}(y)) \mid x - y = z\}; \tag{N1.9}$$

$$\mu_{\tilde{m}\times\tilde{n}}(z) = \sup_{x,y}\{\min(\tilde{m}(x), \tilde{n}(y)) \mid x \times y = z\}; \tag{N1.10}$$

$$\mu_{\tilde{m}+\tilde{n}}(z) = \sup_{x,y}\{\min(\tilde{m}(x), \tilde{n}(y)) \mid x \div y = z\}; \tag{N1.11}$$

Next, we provide another method to derive the fuzzy arithmetic operations based on the concept of *α-cut arithmetic*. Let $\tilde{m} = [m^l, m^m, m^u]$ and $\tilde{n} = [n^l, n^m, n^u]$ be two fuzzy numbers in which the superscripts *l*, *m*, and *u* denote the infimum, mode, and supremum, respectively. The standard fuzzy arithmetic operations can be defined using the concepts of *α-cut* as follows:

$$\tilde{m}(\alpha) + \tilde{n}(\alpha) = [m^l(\alpha) + n^l(\alpha), m^u(\alpha) + n^u(\alpha)]; \tag{N1.12}$$

$$\tilde{m}(\alpha) - \tilde{n}(\alpha) = [m^l(\alpha) - n^u(\alpha), m^u(\alpha) - n^l(\alpha)]; \tag{N1.13}$$

$$\tilde{m}(\alpha) \div \tilde{n}(\alpha) \approx [m^l(\alpha), m^u(\alpha)] \times [1/n^u(\alpha), 1/n^l(\alpha)]; \tag{N1.14}$$

$$\tilde{m}(\alpha) \times \tilde{n}(\alpha) \approx [M, N] \tag{N1.15}$$

where (α) denotes the *α-cut* operation, ≈ is the approximation operation, and

$$M = \min\{m^l(\alpha)n^l(\alpha), m^l(\alpha)n^u(\alpha), m^u(\alpha)n^l(\alpha), m^u(\alpha)n^u(\alpha)\};$$
$$N = \max\{m^l(\alpha)n^l(\alpha), m^l(\alpha)n^u(\alpha), m^u(\alpha)n^l(\alpha), m^u(\alpha)n^u(\alpha)\}.$$

CHAPTER 2: MULTI-OBJECTIVE EVOLUTIONARY ALGORITHMS

Pseudo Code of Genetic Algorithms

```
procedure
GA
begin
    t = 0
    initialize P(o)
    evaluate P(t)
    while not satistfy stopping rule do
    bigin
        t = t + 1
        select P(t) from P(t–1)
        alter P(t)
        evaluate P(t)
    end
end
```

Evolutionary Computations

Evolutionary Algorithms (EAs)

1. Evolutionary programming (EP)
2. Genetic algorithms (GAs): (a) cluster analysis, classification, identification) and (b) programming for optimization
3. DNA (Deoxyribonucleic acid) computing (for pattern forecasting, optimization, and other operations)
4. Genetic programming (GP)
5. Genetic neural programming (GNP)
6. Simulated annealing (SA)
7. Evolutionary strategies (ES)

Computational Intelligence (Soft Computing)

1. Artificial neural networks (ANNs)
2. Fuzzy systems
3. Rough sets

Family Systems of Computational Intelligence

1. Ant systems (ASs) and ant colony systems (ACSs), i.e., ant colony optimization algorithms (ACOAs)
2. Swarm intelligence (SI)
3. Particle swarm optimization (PSO)
4. Immunological systems (ISs)

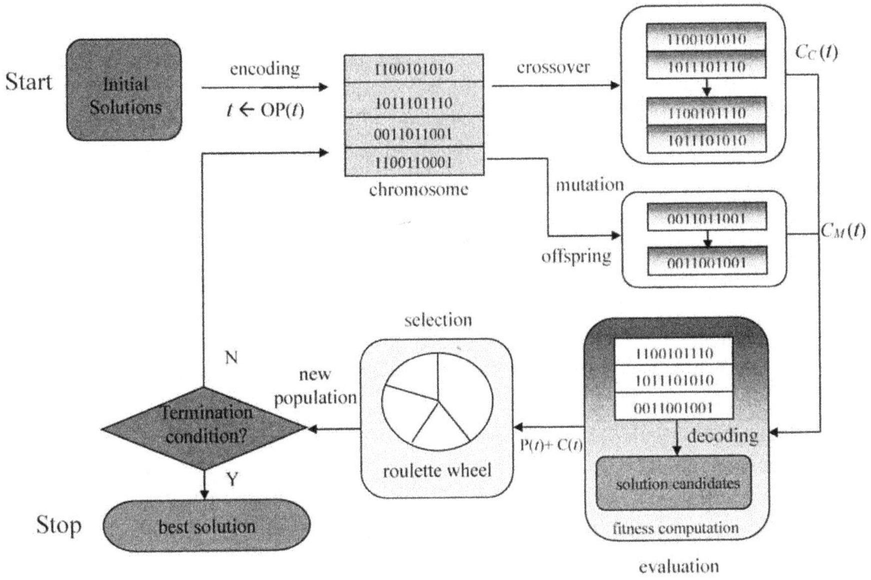

FIGURE N2.1 General structure for genetic algorithm.

5. Cat swarm optimization (CSO)
6. Artificial bee colonies (not including cockroaches, bacterial foraging insects, frogs, flies, stick insects, etc.)
7. Parallel particle swarm optimization (PPSO)
8. Parallel ant colony systems (PACSs)
9. Parallel cat swarm optimization (PCSO)

Basic Concepts of Genetic Algorithms

Figure N2.1 depicts genetic algorithms. Interested readers should consult Gen and Cheng (1997).

Software of MOEA

1. Vector evaluation genetic algorithm (VEGA)
2. Vector-optimized evolution strategy (VOES)
3. Weight-based genetic algorithm (WBGA)
4. Random weighted genetic algorithm (RWGA)
5. Multi-objective genetic algorithm (MOGA)
6. Non-dominated sorting genetic algorithm (NSGA)
7. Niched Pareto genetic algorithm (NPGA)
8. Elitist non-dominated sorting genetic algorithm (NSGA-II)
9. Web sources:
   http://www.iitk.ac.in/kangal/codes.shtml
   http://www.downloadplex.com/Scripts/Matlab/Development-Tools/ev-moga-multi-objective-evlutionary-algorithm_444078.html
   http://www.downloadplex.com/Scripts/Matlab/Development-Tools/godlike-a-robust-single-multi-objective-optimizer_342076.html

CHAPTER 3: GOAL PROGRAMMING

Goal programming (GP) is an analytical approach devised to address decision-making problems in which targets have been assigned to all the attributes and the decision maker is interested in minimizing the non-achievements of the corresponding goals (Romero, 2004). Initially conceived as an application of single objective linear programming by Charnes and Cooper (1955 and 1961), GP gained popularity in the 1960s and 70s from the works of Ijiri (1965), Lee (1972), and Ignizio (1976). GP is ideal for criteria in which target values of achievement are of significance (Steurer, 1986). Goal programming is distinguished from linear programming by:

1. The conceptualization of objectives as goals
2. The assignment of priorities and/or weights to the achievement of the goals
3. The presence of deviational variables
4. Measurement of overachievement and underachievement from target or (threshold) levels
5. Minimization of weighted sums of deviational variables to find solutions that best satisfy the goals

Tamiz and others (1995) show that around 65% of GP applications reported in the literature use lexicographic achievement functions, 21% use weighted achievement functions, and the rest use other types of achievement functions such as min–max structures that minimize maximum deviation.

Multiple Objective Programming

Most multi-objective programming (MOP) problems can be represented mathematically as:

$$\begin{aligned} &\max\left[f_1(\boldsymbol{x}), f_2(\boldsymbol{x}), \ldots, f_k(\boldsymbol{x})\right] \\ &s.t. \quad \boldsymbol{Ax} \le \boldsymbol{b} \\ &\qquad\;\; \boldsymbol{x} \ge \boldsymbol{0} \end{aligned} \tag{N3.1}$$

Weighted GP (WGP) Model

The mathematical programming of a WGP model (Ignizio, 1976) is the following:

$$\begin{aligned} \text{Min} \quad & \sum_i (\alpha_i d_i^- + \beta_i d_i^+) \\ s.t. \quad & f_i(\boldsymbol{x}) + d_i^- - d_i^+ = g_i, \quad i = 1,2,\ldots,k \\ & \boldsymbol{Ax} \le \boldsymbol{b} \\ & d_i^- \cdot d_i^+ = 0, \quad d_i^- \ge 0, \quad d_i^+ \ge 0 \end{aligned} \tag{N3.2}$$

where $\alpha_i = w_i^- / k_i$ if d_i^- is unwanted, otherwise $\alpha_i = 0$ and $\beta_i = w_i^+ / k_i$ if d_i^+ is unwanted, otherwise $\beta_i = 0$.

The parameters w_i^-, w_i^+ and $k_i = f_i^* - f_i^-$ are the weights reflecting preferential and normalizing purposes attached to achievement of the i-th goal; f_i^* can be set as a positive ideal point or as an aspiration level of the i-th goal and f_i^- can act as a negative ideal point or as the worst value of the i-th goal, respectively. We also can rewrite Equation (N3.2) into Equation (N3.3) as follows:

$$\min \sum_{i=1}^{q} \alpha_i (d_i^- + d_i^+) \tag{N3.3}$$

$$\begin{aligned} s.t.\ & f_i(x) + d_i^- - d_i^+ = g_i, \quad i = 1,2,\ldots,q \\ & \boldsymbol{Ax} \le \boldsymbol{b} \\ & d_i^- \cdot d_i^+ = 0 \\ & d_i^- \ge 0, \quad d_i^+ \ge 0 \end{aligned}$$

where $\alpha_i = \frac{w_i}{f_i^* - f_i^-}$ and w_i can be obtained by AHP, ANP, or DANP (DEMATEL-based ANP).

Lexicographic GP (LGP) Model

The achievement function of the LGP model is an ordered vector whose dimension coincides with the Q number of priority levels established in the model. Each component in this vector represents the unwanted deviation variables of the goals placed in the corresponding priority level. The mathematical programming of a LGP model (Ignizio, 1976) is the following:

$$Lex\ Min\ a = \left[\sum_{i \in h_1} \left(\alpha_i d_i^- + \beta_i d_i^+ \right), \ldots, \sum_{i \in h_r} \left(\alpha_i d_i^- + \beta_i d_i^+ \right), \ldots, \sum_{i \in h_Q} \left(\alpha_i d_i^- + \beta_i d_i^+ \right) \right] \tag{N3.4}$$

$$\begin{aligned} s.t.\quad & f_i(\boldsymbol{x}) + d_i^- - d_i^+ = g_i \quad i \in \{1,\ldots,q\} \quad i \in h_r \quad r \in \{1,\ldots,Q\} \\ & \boldsymbol{x} \in \boldsymbol{F}, \quad d_i^- \ge 0, \quad d_i^+ \ge 0 \end{aligned}$$

where h_r represents the index set of goals placed in the r-th priority level and $x \in F$ denotes a feasible solution (decision space). Lexicographic achievement functions imply a non-compensatory structure of preferences. In other words, there are no finite trade-offs among goals placed at different priority levels (Romero, 1991).

Min–Max GP (MGP) Model

The achievement function of the MGP model seeks the minimization of the maximum deviation from any single goal. If we represent this maximum deviation by D, the mathematical programming of the LGP model (Flavell, 1976) is:

$$\min_{\boldsymbol{x}} D \tag{N3.5}$$

$$\begin{aligned} s.t.\quad & \alpha_i d_i^- + \beta_i d_i^+ \le D \\ & f_i(\boldsymbol{x}) + d_i^- - d_i^+ = g_i \quad i \in \{1,\ldots,q\} \\ & \boldsymbol{x} \in \boldsymbol{F}, \quad d_i^- \ge 0, \quad d_i^+ \ge 0 \end{aligned}$$

CHAPTER 4: COMPROMISE SOLUTION AND TOPSIS

Compromise Solution

Multi-objective programming (MOP) problems can be represented mathematically as:

$$\max\left[f_1(\boldsymbol{x}), f_2(\boldsymbol{x}), \ldots, f_k(\boldsymbol{x})\right] \quad \text{(N4.1)}$$
$$s.t. \quad \boldsymbol{Ax} \le \boldsymbol{b}$$
$$\boldsymbol{x} \ge \boldsymbol{0}$$

MOP problems can be solved by Pareto optimal solutions using traditional methods as follows:

Weightings Method

$$\max \sum_{i=1}^{k} w_i f_i(\boldsymbol{x}) \quad \text{(N4.2)}$$
$$s.t. \quad \boldsymbol{Ax} \le \boldsymbol{b}$$
$$\boldsymbol{x} \ge \boldsymbol{0},$$
$$\sum_{i=1}^{k} w_i = 1$$

If $k = 2$ (two objectives), we can assume weights as follows for finding Pareto optimal solutions (Figure N4.1).

w_1	w_2
1.0	0.0
0.9	0.1
⋮	⋮
0.5	0.5
⋮	⋮
0.1	0.9
0.0	1.0

ε- Constraints

If $k = 2$ (two objectives),

$$\max\{f_1(\boldsymbol{x}), f_2(\boldsymbol{x})\} \quad \text{(N4.3)}$$
$$s.t. \quad \boldsymbol{Ax} \le \boldsymbol{b}$$
$$\boldsymbol{x} \ge \boldsymbol{0}$$

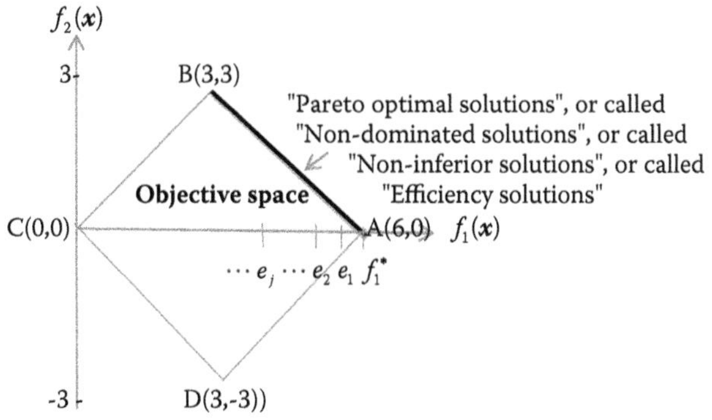

FIGURE N4.1 Objective space and Pareto optimal solutions by ε-constraints.

We can find the ideal point of the first objective f_1^* and set the ε-constraint values $e_1, e_2, ..., e_j, ...$ to the first objectives. Then we can find the Pareto optimal solutions with respective to $f_1(\boldsymbol{x})$ and $f_2(\boldsymbol{x})$.

$$\begin{aligned} &\max f_2(\boldsymbol{x}) \\ &s.t.\ \ f_1(\boldsymbol{x}) \geq e_j,\ \ j = 1,2,... \\ &\quad \boldsymbol{Ax} \leq \boldsymbol{b} \\ &\quad \boldsymbol{x} \geq \boldsymbol{0} \end{aligned} \tag{N4.4}$$

Surrogate Worth Trade-Off (SWT) Method

Compromise Solution

The compromise solution method originally was proposed by Yu and Zeleny in 1972 (Figure N4.2). The concept of the SWT method is depicted In Figure N4.3).

$$\begin{aligned} &\max/\min \{f_1(\boldsymbol{x}), ..., f(\boldsymbol{x})\} \\ &s.t.\ \ \boldsymbol{Ax} \leq \boldsymbol{b} \\ &\qquad \boldsymbol{x} \geq \boldsymbol{0} \end{aligned} \tag{N4.5}$$

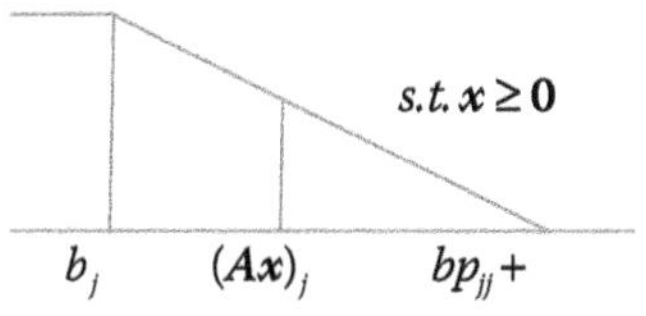

FIGURE N4.2 Basic concept of a compromise solution.

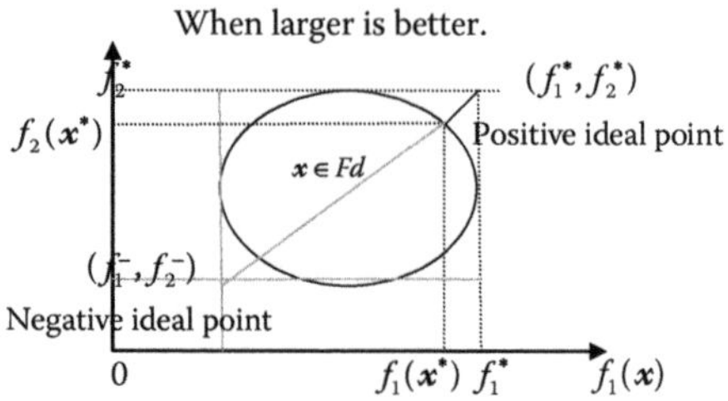

FIGURE N4.3 Basic concept of the constraint set.

The basic concept is demonstrated in the following equations.

$$\min d^p = \left\{ \sum_{i=1}^{k} \left(w_i \left(\frac{|f_i^* - f_i(\boldsymbol{x})|}{|f_i^* - f_i^-|} \right) \right)^p \right\}^{1/p} \tag{N4.6}$$

where

$$d^{p=1} = \sum_{i=1}^{k} \left(w_i \left(\frac{|f_i^* - f_i(\boldsymbol{x})|}{|f_i^* - f_i^-|} \right) \right)$$

$$\vdots$$

$$d^{p=\infty} = \max_i \left\{ \frac{|f_i^* - f_i(\boldsymbol{x})|}{|f_i^* - f_i^-|} \middle| i = 1, 2, ..., k \right\}$$

Problem 1:

$$\min_{\boldsymbol{x}} \lambda \tag{N4.7}$$

$$s.t. \; \frac{|f_i^* - f_i(\boldsymbol{x})|}{|f_i^* - f_i^-|} \leq \lambda, \; i = 1, 2, ..., k \quad \text{or} \quad \min_{\boldsymbol{x}} \max_{i} \lambda$$

$$\boldsymbol{Ax} \leq \boldsymbol{b}$$

$$\boldsymbol{x} \geq \boldsymbol{0}$$

Problem 2:

$$\min_{x} \lambda \tag{N4.8}$$

$$s.t. \frac{|f_i^* - f_i(x)|}{|f_i^* - f_i^-|} \leq \lambda, \quad i = 1, 2, \ldots, k$$

$$\frac{(\boldsymbol{Ax})_j - b_j}{p_j} \leq d, \quad j = 1, 2, \ldots, m$$

$$\boldsymbol{Ax} \leq \boldsymbol{b}$$

$$\boldsymbol{x} \geq \boldsymbol{0}$$

Fuzzy Goal Programming

In fuzzy goal programming problems, we can refer to the concept of TOPSIS for MODM with a compromise solution (Lai et al., 1994) to define the membership function of a fuzzy goal as follows:

$$\begin{aligned} &\max/\min\{f_1(\boldsymbol{x}),\ldots,f_k(\boldsymbol{x})\} \\ &\quad s.t.\quad \boldsymbol{Ax}\le\boldsymbol{b} \\ &\qquad\quad \boldsymbol{x}\ge\boldsymbol{0} \end{aligned} \tag{N4.8}$$

$$\mu_{g_i}(\boldsymbol{x})=\begin{cases} 1, & f_i(\boldsymbol{x})>f_i^*(\boldsymbol{x}) \\ 1-\dfrac{f_i^*(\boldsymbol{x})-f_i(\boldsymbol{x})}{f_i^*(\boldsymbol{x})-f_i^-(\boldsymbol{x})}, & f_i^-(\boldsymbol{x})\le f_i(\boldsymbol{x})\le f_i^*(\boldsymbol{x}) \\ 0, & f_i(\boldsymbol{x})<f_i^-(\boldsymbol{x}) \end{cases}$$

We can transfer the above equation to the following programming:

$$\begin{aligned} &\max_x \lambda \\ &\quad s.t.\ \lambda\le\frac{f_i(\boldsymbol{x})-f_i^-(\boldsymbol{x})}{f_i^*-f_i^-},\ i=1,\ldots,k \quad \text{or} \quad \max_{\boldsymbol{x}}\min_i \lambda \\ &\quad s.t.\quad \boldsymbol{Ax}\le\boldsymbol{b} \\ &\qquad\quad \boldsymbol{x}\ge\boldsymbol{0} \end{aligned} \tag{N4.9}$$

Fuzzy Goal and Fuzzy Constraint Programming

$$\begin{aligned} &\max_x \lambda \\ &\quad s.t.\ \lambda\le 1-\frac{f_i(\boldsymbol{x})-f_i^-(\boldsymbol{x})}{f_i^*-f_i^-},\ i=1,2,\ldots,k \quad \text{or} \quad \max_{\boldsymbol{x}}\min_{i,j} \lambda \\ &\qquad \lambda\le 1-\frac{(\boldsymbol{Ax})_j-b_j}{p_j},\quad j=1,2,\ldots,m \\ &\qquad \boldsymbol{Ax}\le\boldsymbol{b} \\ &\qquad \boldsymbol{x}\ge\boldsymbol{0} \end{aligned} \tag{N4.10}$$

Fuzzy Multiple Objective Linear Programming (FMOLP): General Form:

$$\max\ \tilde{z}_k = \sum_{i=1}^{n} \tilde{c}_{ki} x_i, \quad k = 1, 2, \dots, q_1 \tag{N4.12}$$

$$\min\ \tilde{w}_k = \sum_{i=1}^{n} \tilde{c}_{ki} x_i, \quad k = q_1 + 1, \dots, q$$

$$s.t.\ \sum_{i=1}^{n} \tilde{a}_{ij} x_i \le \tilde{b}_j, \quad j = 1, 2, \dots, m_1$$

$$\sum_{i=1}^{n} \tilde{a}_{ij} x_i \ge \tilde{b}_j, \quad j = m_1 + 1, \dots, m_2$$

$$\sum_{i=1}^{n} \tilde{a}_{ij} x_i = \tilde{b}_j, \quad j = m_2 + 1, \dots, m$$

$$x_i \ge 0, \qquad i = 1, 2, \dots, n$$

The FMOLP problem can be solved by transferring it into a crisp MOLP:

$$\max\ (z_k)_\alpha = \sum_{j=1}^{n} (c_{kj})_\alpha^U x_j, \quad k = 1, 2, \dots, q_1 \tag{N4.13}$$

$$\min\ (w_k)_\alpha = \sum_{j=1}^{n} (c_{kj})_\alpha^L x_j, \ k = q_1 + 1, \ \dots, q$$

$$s.t.\ \sum_{j=1}^{n} (a_{ij})_\alpha^L x_j \le (b_i)_\alpha^U, \ \ i = 1, 2, \dots, m_1, \ m_2 + 1, \dots, m$$

$$\sum_{j=1}^{n} (a_{ij})_\alpha^U x_j \ge (b_i)_\alpha^L, \ \ i = m_1 + 1, \ \dots, m_2$$

$$x_j \ge 0, \quad j = 1, 2, \dots, n$$

where superscripts L and α represent an α-cut in the smaller site; superscripts U and α represent an α-cut in the larger site. This problem can be solved interactively by a fuzzy algorithm. See Zimmermann (1978) and Lee and Li (1993) for details. For applications and extensions, refer to Sakawa (1993), Sakawa et al. (1995), Shibano et al. (1996), Shih et al. (1996), Ida and Gen (1997), and Shih and Lee (1999).

Two-Phase Approach for Solving FMOLP Problem: General Form

$$\max_{x} [\tilde{f}_1(\tilde{c}_1, x), \tilde{f}_2(\tilde{c}_2, x), \ldots, \tilde{f}_{k_1}(\tilde{c}_{k_1}, x)] \tag{N4.14}$$

$$\min_{x} [\tilde{f}_{k_1+1}(\tilde{c}_{k_1+1}, x), \tilde{f}_{k_1+2}(\tilde{c}_{k_1+2}, x), \ldots, \tilde{f}_k(\tilde{c}_k, x)]$$

$$s.t. \quad \tilde{A}x \Uparrow \tilde{b}$$

$$x \geq 0$$

where $\Uparrow$ represents a binary relation defined as $\{\Uparrow\} = \{>\}\vee\{\geq\}\vee\{\leq\}\vee\{<\}\vee\{=\}$ and $\vee$ means *or*.

Step 1:

$$\max_{x} \tilde{f}_1(\tilde{c}^U_{1\alpha}, x), \tilde{f}_2(\tilde{c}^U_{2\alpha}, x), \ldots, \tilde{f}_{k_1}(\tilde{c}^U_{k_1\alpha}, x)\} \tag{N4.15}$$

$$\min_{x} \{\tilde{f}_{k_1+1}(\tilde{c}^L_{(k_1+1)\alpha}, x), \tilde{f}_{k_1+2}(\tilde{c}^L_{(k_1+2)\alpha}, x), \ldots, \tilde{f}_k(\tilde{c}^L_{k\alpha}, x)]$$

$$s.t. \; (A)^L_\alpha x \leq (b)^U_\alpha$$

$$(A)^U_\alpha x \geq (b)^L_\alpha$$

$$x \geq 0, \quad x \in X_\alpha$$

According to Zimmermann (1978), the two important relations between α and β are (1) optimal level of α and β, that is $\alpha = \beta$; and (2) a trade-off relation between α and β.

Step 2:

$$\max_{x} \beta \tag{N4.14}$$

$$s.t. \;\; \beta \leq \mu_{g_{i(\max)}}(x)$$

$$\beta \leq \mu_{g_{i(\min)}}(x)$$

$$x \in X_\alpha$$

where

$$\mu_{g_{i(\max)}}(x) = \frac{f_{i(\max)}(c^U_{i\alpha}, x) - f^-_{i(\max)\alpha}}{f^*_{i(\max)\alpha} - f^-_{i(\max)\alpha}}, \qquad \text{for } i = 1, 2, \ldots, k_1$$

$$\mu_{g_{i(\min)}}(x) = \frac{f^-_{i(\min)\alpha} - f_{i(\min)}(c^U_{i\alpha}, x)}{f^-_{i(\min)\alpha} - f^*_{i(\min)\alpha}}, \qquad \text{for } i = k_1 + 1, k_1 + 2, \ldots, k$$

We can set the positive ideal points $f^*_{i(\max)\alpha}$ (large is better) and $f^*_{i(\min)\alpha}$ (smaller is better) to denote the aspiration level and the negative ideal points $f^-_{i(\max)\alpha}$ (large is better) $f^-_{i(\min)\alpha}$ (smaller is better) as the worst values.

We can then find the optimal solution. Use of the iteration procedure has proven a good approach and when $\alpha \risingdotseq \beta$, iteration is stopped. The second phase is to find λ such that $\lambda = min\{\alpha,\beta\}$. Lee and Li (1993) proposed the following algorithm for solving FMOLP problems:

Step 1. Set tolerable error τ, step width ε, initial α-cut ($\alpha = 1.0$), and iterative frequency $t = 1$.
Step 2. Put $\alpha = \alpha - t\varepsilon$, solve c-LP problem, then obtain β and $\boldsymbol{x}$.
Step 3. If $|\alpha - \beta| \le \tau$, let $\lambda = \min\{\alpha,\beta\}$ and go to step 4; otherwise, return to Step 2. If width ε is too large, let $\varepsilon = \varepsilon/2$ and $t = 1$, and return to step 2.
Step 4. Obtain λ, α, β, and x.

Using the first phase α, β, refer to the algorithm of Lee and Li (1993) and solve the c-LP2 problem based on the following mathematical programming; for more details refer to Ida and Gen (1997).

$$\max \quad \bar{\beta} = \frac{1}{k}\sum_{i=1}^{k} \beta_i \tag{N4.17}$$

$$s.t. \quad \mu_{g_{i(\max)}}(\boldsymbol{x}) = \frac{f_{i(\max)}(\boldsymbol{c}_{i\alpha}^{U}, \boldsymbol{x}) - f_{i(\max)\alpha}^{-}}{f_{i(\max)\alpha}^{*} - f_{i(\max)\alpha}^{-}}, \quad \text{for } i = 1,2,\ldots,k_1$$

$$\mu_{g_{i(\min)}}(\boldsymbol{x}) = \frac{f_{i(\min)\alpha}^{-} - f_{i(\min)}(\boldsymbol{c}_{i\alpha}^{U}, \boldsymbol{x})}{f_{i(\min)\alpha}^{-} - f_{i(\min)\alpha}^{*}}, \quad \text{for } i = k_1+1, k_1+2,\ldots,k$$

$$\boldsymbol{x} \in X_\alpha, \ \beta, \beta_i \in [0,1]$$

TOPSIS

The basic concept of TOPSIS (technique for order preference by similarity to ideal solution) using multiple-attribute decision making (MADM) criteria is $c_1 \ \cdots \ c_j \cdots \ c_n$. The alternatives are:

Alternatives	Criteria				
	c_1	...	c_j	...	c_n
	w_1	...	w_j	...	w_n
a_1	f_{11}	...	f_{1j}	...	f_{1n}
$\vdots$	$\vdots$		$\vdots$		$\vdots$
a_k	f_{k1}	...	f_{kj}	...	f_{kn}
$\vdots$	$\vdots$		$\vdots$		$\vdots$
a_m	f_{m1}	...	f_{mj}	...	f_{mn}
Aspired value	f_1^*	...	f_j^*	...	f_n^*
The worst value	f_1^-	...	f_1^-	...	f_1^-

$$\text{Data matrix } \left[f_{kj}\right]_{m\times n} \xrightarrow{\text{normalization}} \left[r_{kj}\right]_{m\times n} \xrightarrow{\text{weight}} \left[w_j r_{kj}\right]_{m\times n} \longrightarrow \left[v_{kj}\right]_{m\times n}$$

$$r = (| f_{kj} - f_j^- |)/(| f_j^* - f_j^- |)$$

$$\min r_{kj}^{aspire} = (| f_j^* - f_{kj} |)/(| f_j^* - f_j^- |)$$

$$\max r_{kj}^{worst} = (| f_{kj} - f_j^- |)/(| f_j^* - f_j^- |)$$

The distance from point v_{kj} to the positive ideal point v_j^* and negative ideal point v_j^- for $j = 1,2,...,$ is:

$$d_k^+ = \left[\sum_{j=1}^{n}(v_{kj} - v_j^*)^2\right]^{1/2} \quad \text{or} \quad \sqrt{\sum_{j=1}^{n}(v_{kj} - v_j^*)^2} \tag{N4.18}$$

$$d_i^- - k = \left[\sum_{j=1}^{n}(v_{kj} - v_j^-)^2\right]^{1/2} \quad \text{or} \quad \sqrt{\sum_{j=1}^{n}(v_{kj} - v_j^-)^2} \tag{N4.19}$$

The ranking index for achieving the appropriate level (large is better) is:

$$R_k = \frac{d_k^-}{d_k^* + d_k^-} \quad \text{or} \quad R_k = 1 - \frac{d_k^*}{d_k^* + d_k^-} \tag{N4.20}$$

The ranking index for the gap to the positive ideal point (small is better) is:

$$R_k = \frac{d_k^*}{d_k^* + d_k^-} \tag{N4.21}$$

The optimal solution of a MODM problem should be nearest the PIS and farthest from the NIS in the traditional approach. We can use this basic concept of TOPSIS in MADM to get closest to the aspiration level (in a traditional approach, closest to the PIS), that is, minimize the gap and be the most distance from the worst value (in a traditional approach, farthest from the NIS) for each criterion.

$$\min r_{kj}^{aspire} = (| f_j^* - f_{kj} |)/(| f_j^* - f_j^- |) \tag{N4.22}$$

$$\max r_{kj}^{worst} = (| f_{kj} - f_j^- |)/(| f_j^* - f_j^- |)$$

We can let alternative k act as an objective function in objective (criterion) j and $j = 1,2,...,n$; we can also rewrite f_{kj} into $f_j(\boldsymbol{x})$ and the vector of decision variables as $\boldsymbol{x} = (x_1, x_2, ..., x_p)$; the weight w_j of objective (criterion) j can be obtained from AHP, ANP, or DANP based on the relationship of objectives (criteria). We can use the L_p-norm to measure the distance between objective values, the aspiration, and the worst value as follows:

$$d_p^{aspiration}(\boldsymbol{x}) = \sum_{j=1}^{n} w_j r_j^{aspiration}(\boldsymbol{x}) = \sum_{j=1}^{n} w_j(|f_j^* - f_j(\boldsymbol{x})|)/(|f_j^* - f_j^-|) \quad \text{(N4.23)}$$

and

$$d_p^{worst}(\boldsymbol{x}) = \sum_{j=1}^{n} w_j r_j^{worst}(\boldsymbol{x}) = \sum_{j=1}^{n} w_j(|f_j(\boldsymbol{x}) - f_j^-|)/(|f_j^* - f_j^-|) \quad \text{(N4.24)}$$

Then we can transform the concept of TOPSIS for solving the following two-objective programming:

$$\begin{aligned} \min \quad & d_p^{aspiration}(\boldsymbol{x}) \\ \max \quad & d_p^{worst}(\boldsymbol{x}) \\ s.t. \quad & \boldsymbol{Ax} \le \boldsymbol{b}, \\ & \boldsymbol{x} \ge \boldsymbol{0}. \end{aligned} \quad \text{(N4.25)}$$

or the following fractional single objective programming:

$$\begin{aligned} \min \quad & \frac{d_p^{aspiration}(\boldsymbol{x})}{d_p^{worst}(\boldsymbol{x})} \\ s.t. \quad & \boldsymbol{Ax} \le \boldsymbol{b} \\ & \boldsymbol{x} \ge \boldsymbol{0}. \end{aligned} \quad \text{(N4.26)}$$

CHAPTER 5: DE NOVO PROGRAMMING AND CHANGEABLE PARAMETERS

De Novo Programming Method

When dealing with a multiple criteria optimization problem, we usually confront a situation in which it is almost impossible to optimize all criteria in a system. This property requires trade-offs.

Zeleny (1981 and 1986) suggested that trade-offs are properties of an inadequately designed system and thus can be eliminated by designing a better (more optimal) system. Zeleny (1995) proposed the concept of the optimal portfolio of resources that is a design of system resources based on integration, i.e., the levels of individual resources are not determined separately and thus require no trade-offs.

Zeleny also developed de novo programming for designing optimal systems by reshaping feasible sets and suggested an optimum-path ratio to contract a budget to available level.

Shi (1995) discussed budgets from various views and defined six optimum-path ratios for finding alternatives in designing optimal systems. No matter what optimum-path ratio is used, it can provide only a certain path to locate a solution in the decision space of the new system. A multi-criteria problem can be described as follows (Yu, 1985):

$$\begin{aligned} &Max \quad Cx \\ &s.t. \quad \boldsymbol{Ax} \le \boldsymbol{b} \\ &\qquad\quad \boldsymbol{x} \ge \boldsymbol{0} \end{aligned} \tag{N5.1}$$

where $\boldsymbol{C} = C_{q\times n}$ and $\boldsymbol{A} = A_{m\times n}$ are matrices and $\boldsymbol{b} = (b_1,...,b_m)^T \in R^m$.

We use a graph example to define the maximum objective f_1 as profit and f_2 as quality. We reshape the feasible set to include unavailable *good* alternatives due to the trade-offs between profit and quality as shown in Figure N5.1 and in the following equations.

A simple production problem involves two products (suits and dresses) in quantities x_1 and x_2. Each product consumes five different resources (unit market prices are given). De novo programming by maximizing levels of two products can be calculated by mathematical programming:

$$\text{Profit:} \quad \max f_1(x_1,x_2) = 400x_1 + 300x_2 \tag{N5.2}$$

$$\text{Quality:} \ \max f_2(x_1,x_2) = 6x_1 + 8x_2$$

$$\begin{aligned} s.t.\ & 4x_1 \le 20 \\ & 2x_1 + 6x_2 \le 24 \\ & 12x_1 + 4x_2 \le 60 \\ & 3x_2 \le 10.5 \\ & 4x_1 + 4x_2 \ge 26 \\ & x_1, x_2 \ge 0 \end{aligned} \Rightarrow \begin{bmatrix} 4 & 0 \\ 2 & 6 \\ 12 & 4 \\ 0 & 3 \\ 4 & 4 \end{bmatrix} \begin{bmatrix} x_1 \\ x_2 \end{bmatrix} \le \begin{bmatrix} 20 \\ 24 \\ 60 \\ 10.5 \\ 26 \end{bmatrix}, \begin{pmatrix} s.t. & \boldsymbol{Ax} \le \boldsymbol{b} \\ & x_1, x_2 \ge 0 \end{pmatrix}$$

The cost of the given resources portfolio is calculated as (30 × 20) + (40 × 24) + (9.5 × 60) + (20 × 10.5) + (10 × 26) = \$2600. The unit costs of the two products are calculated as:

x_1: (30 × 4) + (40 × 2) + (9.5 × 12) + (20 × 0) + (10 × 4) = \$354

x_2: (30 × 0) + (40 × 6) + (9.5 × 4) + (20 × 3) + (10 × 4) = \$378

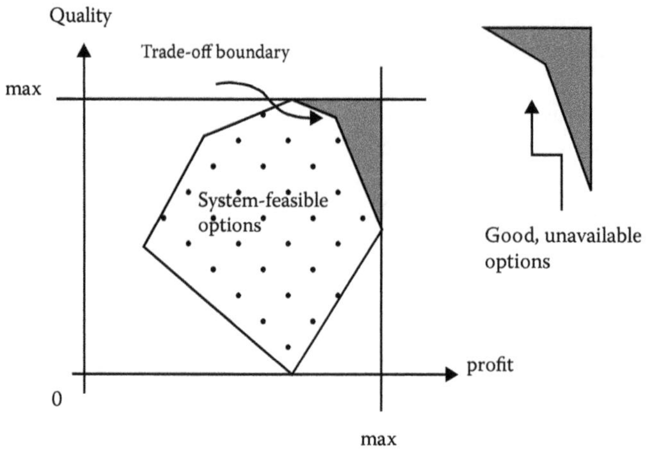

FIGURE N5.1 Trade-off boundary.

Ideal point:
Maximize profit $f_1(x_1, x_2)$ as:

$$\max f_1(x_1, x_2) = 400x_1 + 300x_2 \qquad \text{(N5.3)}$$
$$s.t. \quad \boldsymbol{Ax} \le \boldsymbol{b}$$
$$x_1, x_2 \ge 0$$

Answer: $x_1 = 4.25, x_2 = 2.25;\ f_1^* = 400 \times 4.25 + 300 \times 2.25 = \$2375$
Maximize total quality index $f_2(x_1, x_2)$ as:

$$\max f_2(x_1, x_2) = 6x_1 + 8x_2 \qquad \text{(N5.4)}$$
$$s.t. \quad \boldsymbol{Ax} \le \boldsymbol{b}$$
$$x_1, x_2 \ge 0$$

Answer: $x_1 = 3.75, x_2 = 2.75;\ f_2^* = 6 \times 3.75 + 8 \times 2.75 = \$44.5$

TABLE N5.1
Data Summary

Unit Price ($)	Resources (Raw Material)	Technological Coefficients (Resource Requirements)		No. of Units (Resource Portfolio)
		x_1	x_2	
30	Nylon	4	0	20
40	Velvet	2	6	24
9.5	Silver thread	12	4	60
20	Silk	0	3	10.5
10	Golden thread	4	4	26

Multi-objective programming:

$$\max \{f_1(\boldsymbol{x}),\ldots,f_i(\boldsymbol{x}),\ldots,f_k(\boldsymbol{x})\} \tag{N5.5}$$

$$\begin{array}{lll} s.t.\ \ \boldsymbol{Ax} \le \boldsymbol{b} & s.t.\ \ \boldsymbol{p'Ax} \le \boldsymbol{p'b} & s.t.\ \ \boldsymbol{c'x} \le \boldsymbol{B} \\ \qquad\quad \Rightarrow & \qquad\quad \Rightarrow & \\ \boldsymbol{x} \ge \boldsymbol{0} & \boldsymbol{x} \ge \boldsymbol{0} & \boldsymbol{x} \ge \boldsymbol{0} \end{array}$$

where vector p denotes the unit price of each resource; vector $\mathbf{c}' = \mathbf{p}'\mathbf{A}$ is product unit cost, and B indicates budget.

De novo programming:

$$\begin{aligned} &\min \boldsymbol{cx} \\ &\quad s.t.\ \ f_i(\boldsymbol{x}) \ge f_i^*,\quad i = 1,2,\ldots,k \\ &\qquad\quad \boldsymbol{x} \ge \boldsymbol{0} \end{aligned} \tag{N5.6}$$

Example:

$$\begin{aligned} &\min \boldsymbol{cx} = 354x_1 + 378x_2 \\ &s.t.\ \ f_1(x_1,x_2) = 400x_1 + 300x_2 \ge 2375 \\ &\qquad f_2(x_1,x_2) = 6x_1 + 8x_2 \ge 44.5 \\ &\qquad x_1, x_2 \ge 0 \end{aligned}$$

Maximum profit $f_1(x_1,x_2) = \$2375$

Answer: $x_1 = 4.03, x_2 = 2.54;\ \ f_1^* = 400 \times 4.03 + 300 \times 2.54 = \$2375$

Maximum total quality index $f_2(x_1,x_2) = \$44.5$

Answer: $x_1 = 4.03, x_2 = 2.54;\ \ f_2^* = 6 \times 4.03 + 8 \times 2.54 = \$44.5$

Cost of newly designed system = $2386.74 [**(30 × 16.12)** + **(40 × 23.3)** + **(9.5 × 58.52)** + **(20 × 7.62)** + **(10 × 26.28)** = **$2386.74]**

Changeable parameters:
Extend to the basic concept of the changeable decision space and aspiration level as shown in Figure N5.2.

TABLE N5.2
Data Summary

Unit Price ($)	Resources (Raw Material)	Technological Coefficients (Resource Requirements) x_1	x_2	No. of Units (Resource Portfolio) Relationship of Original and New Designs[a]
30	Nylon	4	0	20 > **16.12**
40	Velvet	2	6	24 > **23.3**
9.5	Silver thread	12	4	60 > **58.52**
20	Silk	0	3	10.5 > **7.62**
10	Golden thread	4	4	26 < **26.28**

[a] Bolding indicates new designs.

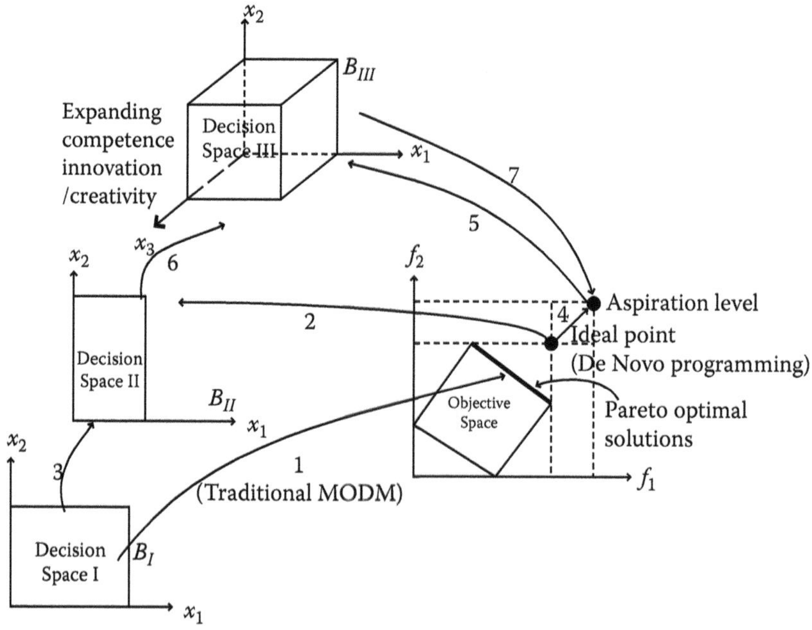

FIGURE N5.2 Change spaces (decision space and objective space) for MODM.

CHAPTER 6: MULTI-STAGE PROGRAMMING

Multi-Stage Decision Making with Multiple Criteria

Yu and Seiford (1981) proposed a general framework for multi-criteria finite stage problems as follows:

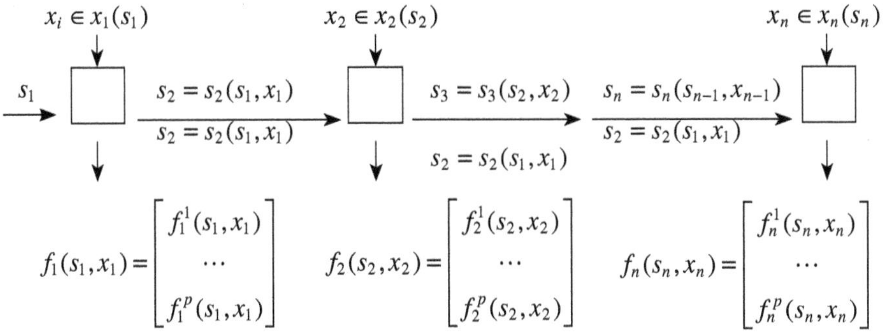

The decision variable is $\boldsymbol{x} = (x_1, \ldots, x_n)$ with each $x_t \in x_t(s_t)$. The state variables $\{s_t\}$ are generated by:

$$s_{t+1} = s_{t+1}(s_t, x_t) \qquad t = 1, \ldots, n-1$$

where $x_t \in x_t(s_t)$ and s_t specifies the set of alternatives when the state is reached: (1) the sequence $\{s_t\}$ generated serially by $\{x_t\}$ is a *path* in the state space or (2) the familiar constraints in mathematical programming are reached:

$$\sum_j s_j \geq c, \quad \prod_j g_j(x_j) \geq c,$$
$$\text{or} \quad \max_j \{g_j(x_j)\} \geq c$$

CHAPTER 7: MULTI-LEVEL MULTI-OBJECTIVE PROGRAMMING

Bi-Level Programming

Single Objective Bi-Level Programming

$$\begin{aligned} &\max_{\boldsymbol{x}} f_1(\boldsymbol{x}, \boldsymbol{y}) = \boldsymbol{c}_1 \boldsymbol{x} + \boldsymbol{d}_1 \boldsymbol{y} \\ &\text{where} \quad y \quad \text{solves} \\ &\max_{\boldsymbol{y}} f_2(\boldsymbol{x}, \boldsymbol{y}) = \boldsymbol{c}_2 \boldsymbol{x} + \boldsymbol{d}_2 \boldsymbol{y} \\ &s.t. \quad \boldsymbol{A}_1 \boldsymbol{x} + \boldsymbol{A}_2 \boldsymbol{y} \leq \boldsymbol{b} \\ &\qquad \boldsymbol{x} \geq \boldsymbol{0}, \boldsymbol{y} \geq \boldsymbol{0} \end{aligned} \tag{N7.1}$$

Let $\hat{x}$ be the decision specified by the leader; then the follower solves a linear programming operation:

$$\max_{y} f_2(\hat{x},y) = d_2y + c_2\hat{x} \tag{N7.2}$$
$$s.t.\quad A_2y \le b - A_1\hat{x}$$
$$y \ge 0$$

We can define the following concepts related to the Stackelberg solution to the above problem:

Constraint region S:

$$S = \{(x,y) \mid A_1x + A_2y \le b, x \ge 0, y \le 0\} \tag{N7.3}$$

Feasible region of follower S(x):

$$S(x) = \{y \mid A_2y < b - A_1x, y \ge 0\} \tag{N7.4}$$

Set of follower rational responses R(x):

$$R(x) = \{y \mid y \in \arg \max_{y \in S(x)} f_2(x,y)\} \tag{N7.5}$$

The set $R(x)$ is often assumed to be a singleton.

Inducible region *IR*:

$$IR = IR = \{(x,y) \mid (x,y) \in S, y \in R(x)\} \tag{N7.6}$$

Stackelberry solution:

$$\{(x,y) \mid (x,y) \in \arg \max_{(x,y) \in IR} f_1(x,y)\} \tag{N7.7}$$

Bi-Level Multi-objective Programming

$$\max_{x} f_{11}(x,y) = c_{11}x + d_{11}y \tag{N7.8}$$
$$\vdots$$
$$\max_{x} f_{1k_1}(x,y) = c_{1k_1}x + d_{1k_1}y$$

where y solves

$$\max_{y} f_{21}(x,y) = c_{21}x + d_{21}y$$
$$\vdots$$
$$\max_{y} f_{2k_2}(x,y) = c_{2k_2}x + d_{1k_2}y$$
$$s.t.\quad A_1x + A_2y \le b$$
$$x \ge 0, y \ge 0$$

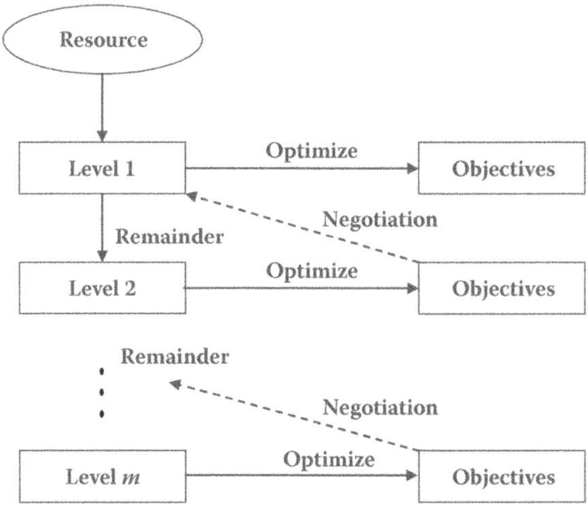

FIGURE N7.1 Basic concept of multiple-level resource allocation problem.

Multi-Level Multi-Objective Programming

We can extend bi-level multi-objective programming into multi-level multi-objective programming. The procedures of multiple-level resource allocation problems are depicted in Figure N7.1. Based on these concepts, multiple-level resource allocation problems can be considered to maximize the following knapsack equation:

$$\begin{aligned}
&\text{Level 1:} \quad \max \quad z_1 = \boldsymbol{c}_{11}\boldsymbol{x}_1 + \boldsymbol{c}_{12}\boldsymbol{x}_2 + \cdots + \boldsymbol{c}_{1p}\boldsymbol{x}_p \\
&\text{Level 2:} \quad \max \quad z_2 = \boldsymbol{c}_{21}\boldsymbol{x}_1 + \boldsymbol{c}_{22}\boldsymbol{x}_2 + \cdots + \boldsymbol{c}_{2p}\boldsymbol{x}_p \\
&\qquad\qquad\quad \vdots \\
&\text{Level } m\text{:} \quad \max \quad z_m = \boldsymbol{c}_{m1}\boldsymbol{x}_1 + \boldsymbol{c}_{m2}\boldsymbol{x}_2 + \cdots + \boldsymbol{c}_{mp}\boldsymbol{x}_p \\
&\qquad\qquad\quad s.t. \quad \boldsymbol{a}_{11}\boldsymbol{x}_1 \le \boldsymbol{b}_1, \\
&\qquad\qquad\qquad\quad \boldsymbol{a}_{21}\boldsymbol{x}_2 \le \boldsymbol{b}_2, \\
&\qquad\qquad\qquad\quad \vdots \\
&\qquad\qquad\qquad\quad \boldsymbol{a}_{m1}\boldsymbol{x}_m \le \boldsymbol{b}_m, \\
&\qquad\qquad\qquad\quad \boldsymbol{x} \ge \boldsymbol{0}
\end{aligned} \tag{N7.9}$$

where $\boldsymbol{c}_{ij}$ and $\boldsymbol{x}_i$ denote given resource parameters at the i-th level and are usually represented as technological coefficients and products, respectively, and $\boldsymbol{b}_i$ denotes the maximum limited resource portfolios at the i-th level.

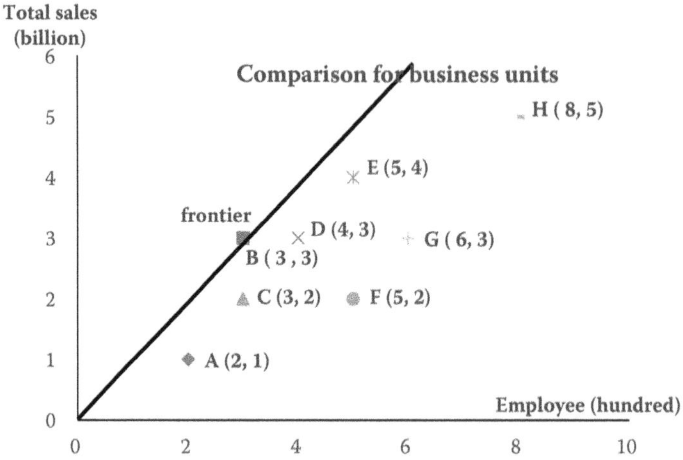

FIGURE N8.1 Comparison of MDUs.

CHAPTER 8: DATA ENVELOPMENT ANALYSIS (DEA)

Methods for Assessment of Efficiency

These methods are based on the work of Charnes, Cooper, and Rhodes (CCR; 1978). Ratio scales for measuring efficiency are depicted in Figures N8.1 through N8.4 and Tables N8.1 through N8.3. (1978). The equation is:

$$\frac{\text{Output (efficiency)}}{\text{Input (resource)}}$$

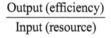

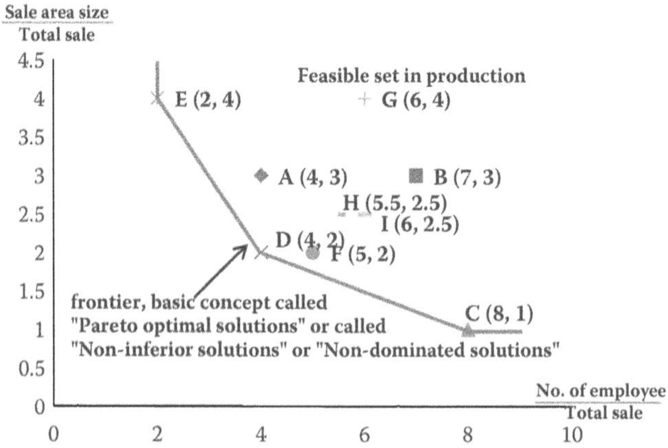

FIGURE N.8.2 Improving efficiency of Store A.

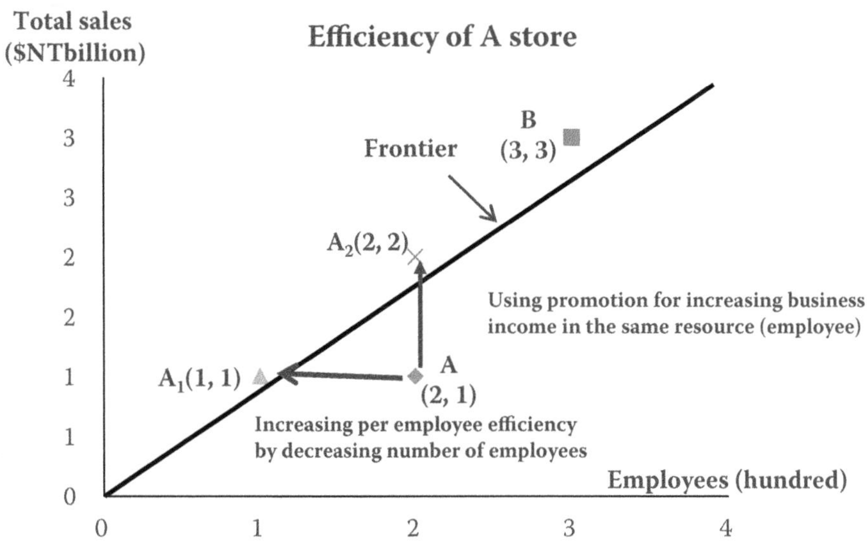

FIGURE N8.3 Case of two inputs and one output.

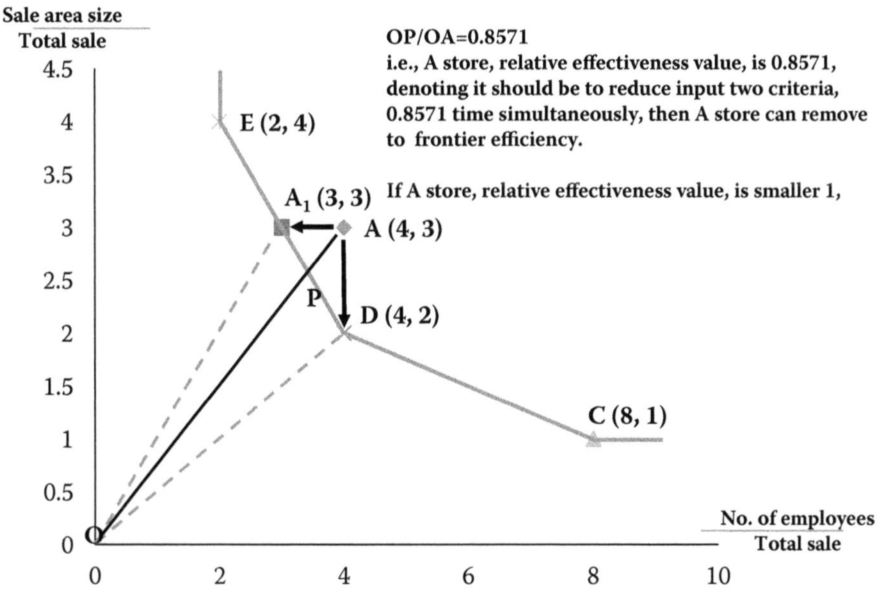

FIGURE N8.4 Improving efficiency of Store A with two inputs.

TABLE N8.1
Example: One Input and One Output

Business Units (Stores)	A	B	C	D	E	F	G	H
Number of employees (hundreds)	2	3	3	4	5	5	6	8
Total sales (NT$ millions)	1	3	2	3	4	2	3	5
Sales per employee (NT$ millions/hundreds)	0.5	1	0.667	0.75	0.8	0.4	0.5	0.625

TABLE N8.2
Improving Efficiency in Non-Efficiency Units (Stores)

Business Units (Stores)	A	B	C	D	E	F	G	H
Efficiency	0.5	1	0.667	0.75	0.8	0.4	0.5	0.625

TABLE N8.3
Example: Two Inputs and One Output

Business Units (Stores)	A	B	C	D	E	F	G	H	I
Number of employees x_1	4	7	8	4	2	5	6	5.5	6
Sales area size x_2	3	3	1	2	4	2	4	2.5	2.5
Total sales y	1	1	1	1	1	1	1	1	1

Extension: Multiple Input and Output Variables

The CCR model is the basic model of DEA and can be demonstrated as a linear programming (LP) problem:

$$
\begin{aligned}
<\mathrm{CCR}_k> \ \min \theta_k \qquad & \qquad (\mathrm{N8.1}) \\
s.t.\ \theta_k \boldsymbol{x}_k - X\lambda \geq \mathbf{0} & \\
\boldsymbol{y}_k - Y\lambda \leq \mathbf{0} & \\
\lambda \geq \mathbf{0} &
\end{aligned}
$$

where θ_k shows all reduced θ_k time in input; min θ_k denotes satisfactory condition by using minimal improvement. $X\lambda$ and $Y\lambda$ show the input and output frontiers, respectively. Vector $\boldsymbol{\lambda}^*$ is an optimal value of vector λ.

Slack-Based Efficiency Measures

Slack variables are input surpluses (over inputs) or output insufficiencies (outputs nevertheless). Input surplus (over input) and output insufficiency (output nevertheless) can be defined as follows:

$$\text{<Input surplus>} \quad \boldsymbol{s}^*_{x_k} = \theta^*_k \boldsymbol{x}_k - \boldsymbol{X}\boldsymbol{\lambda}^* \tag{N8.2}$$

$$\text{<Output insufficiency>} \quad \boldsymbol{s}^*_{y_k} = \boldsymbol{Y}\boldsymbol{\lambda}^* - \boldsymbol{y}_k \tag{N8.3}$$

When a DMU_k can satisfy flowing conditions, we consider it efficient.

$$\theta^*_k = 1$$

For CCR_k all slack variables are optimal solutions when $\boldsymbol{s}^*_{x_k} = \boldsymbol{0}$ and $\boldsymbol{s}^*_{y_k} = \boldsymbol{0}$. If DMU_k can not satisfy these conditions, it is inefficient. Therefore, this DMU_k should face efficient direction to be improved (by an alternative) as follows:

$$\boldsymbol{x}^*_k = \theta^*_k \boldsymbol{x}_k - \boldsymbol{s}^*_{x_k} \tag{N8.4}$$

$$\boldsymbol{y}^*_k = \boldsymbol{y}_k + \boldsymbol{s}^*_{y_k} \tag{N8.5}$$

We can build a dual problem in linear programming CCR_k and use v and u as variables.

DEA Efficiency Measure

Classical Efficiency Measure

If the k-th DMU uses m-dimension input variables x_{ik} $(i = 1,2,...,m)$ to produce s-dimension output variables y_{jk} $(j = 1,2,...,s)$, the efficiency of DMU h_k $(k = 1,...,n)$ can be found from the following model:

$$\max h_k = \frac{\sum_{j=1}^{S} u_j\, y_{jk}}{\sum_{i=1}^{m} v_i x_{ik}} \tag{N8.6}$$

$$s.t. \quad \frac{\sum_{j=1}^{S} u_j\, y_{jr}}{\sum_{i=1}^{m} v_i x_{ir}} \le 1, \quad r = 1,2,...,n$$

$$u \ge \varepsilon > 0; \quad v \ge \varepsilon > 0; \quad i = 1,2,...,m; \quad j = 1,2,...,s; \quad k, r \in \{1,2,...,n\}$$

where x_{ik} is the i-th input of the k-th DMU; y_{jk} indicates the j-th output of the k-th DMU; u_j, and v_i denote the weights of the j-th output and i-th input respectively;

h_k is a relative efficiency value; ε is an Archimedean number equal to 10^{-6}. We next convert the above equation to the LP model:

$$\max\ h_k = \sum_{j=1}^{s} u_j y_{jk} \tag{N8.7}$$

$$s.t.\ \sum_{i=1}^{m} v_i x_{ir} - \sum_{j=1}^{s} u_j y_{jr} \geq 0, \quad r = 1,2,\ldots,n$$

$$\sum_{i=1}^{m} v_i x_{ik} = 1$$

$$u_j \geq \varepsilon > 0,\ k = 1,2,\ldots,s$$

$$v_i \geq \varepsilon > 0,\ i = 1,2,\ldots,m$$

Dual problem:

$$\min\left\{\theta_k - \varepsilon\left[\sum_{i=1}^{m} s_{ik}^{-} + \sum_{j=1}^{s} s_{jk}^{+}\right]\right\} \tag{N8.8}$$

$$s.t.\ \ \theta_k x_{ik} - \sum_{r=1}^{n} \lambda_r x_{ir} - s_{ik}^{-} = 0, \quad i = 1,2,\ldots,m$$

$$y_{jk} - \sum_{r=1}^{n} \lambda_r y_{jr} + s_{jk}^{+} = 0, \quad j = 1,2,\ldots,s$$

$$s_{ik}^{-}, s_{jk}^{+}, \lambda_r \geq 0$$

where s_{ik}^{-} and s_{jk}^{+} are slack variables.

Andersen and Petersen (A&P) Efficiency Measure

The A&P efficiency measure function is shown below (Chiang and Tzeng, 2000).

Revenue Aspect (N8.9)

$$\max\ \Phi_k$$

$$s.t.\ \sum_{r=1}^{n} \lambda_r x_{ir} \leq x_{ik}$$

$$\sum_{r=1}^{n} \lambda_r y_{jr} \geq \theta_k y_{jk}$$

$$\sum_{r=1}^{n} \lambda_r = 1$$

$\lambda_r \geq 0; i = 1,2,\ldots,m; r = 1,2,\ldots,n; j = 1,2,\ldots,s$

Cost Aspect (N8.10)

$$\min\ \theta_k$$

$$s.t.\ \sum_{r=1}^{n} x_{ir} \lambda_r \leq \theta_k x_{ik}$$

$$\sum_{r=1}^{n} y_{jr} \lambda_r \geq y_{jk}$$

$$\sum_{r=1}^{n} \lambda_r = 1$$

$\lambda_r \geq 0; i = 1,2,\ldots,m; r = 1,2,\ldots,n; j = 1,2,\ldots,s$

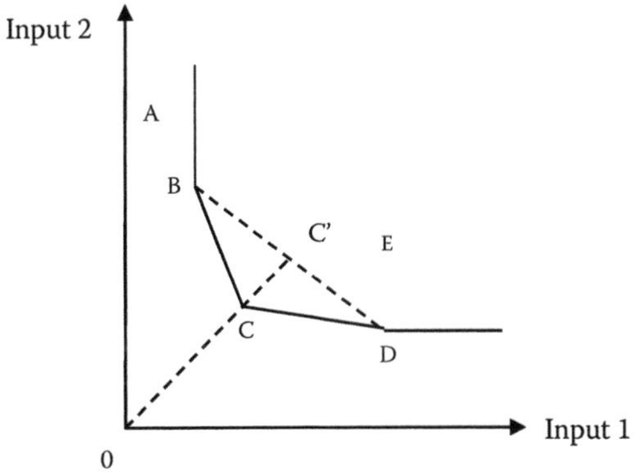

FIGURE N8.5 Unit isoquant spanned by an A&P model.

where Φ_k is the revenue efficiency value ($k = 1,2,\ldots,n$) and θ_k denotes the cost efficiency value ($k = 1,2,\ldots,n$).

The ratio OC′/OC defines the efficiency measure for evaluating unit C. Index values equal to or larger than 1 may be interpreted as the maximum proportional increase in a vector that contains the corresponding input characterization of that observation as an efficiency value (Andersen and Petersen, 1993). The unit isoquant spanned by the A&P model is described as shown in Figure N8.5.

Efficiency Achievement Measure

The overall relative efficiency achievement (Chiang and Tzeng, 2000) calculation is:

$$
\begin{aligned}
\max h_1 &= \frac{\sum_{j=1}^{s} u_j y_{j1}}{\sum_{i=1}^{m} v_i x_{i1}} \\
\max h_2 &= \frac{\sum_{j=1}^{s} u_j y_{j2}}{\sum_{i=1}^{m} v_i x_{i2}} \\
&\vdots \\
\max h_n &= \frac{\sum_{j=1}^{s} u_j y_{jn}}{\sum_{i=1}^{m} v_i x_{in}} \qquad \text{(N8.11)} \\
s.t. \quad & \frac{\sum_{j=1}^{s} u_j y_{jk}}{\sum_{i=1}^{m} v_i x_{ik}} \le 1, \quad k = 1,2,\ldots,n \\
& u_j \ge \varepsilon > 0, \quad j = 1,2,\ldots,s \\
& v_i \ge \varepsilon > 0, \quad i = 1,2,\ldots,m
\end{aligned}
$$

To convert to the pattern of fuzzy multiple objective programming (Chiang and Tzeng, 2000):

$$\max \alpha \tag{N8.12}$$

$$s.t. \sum_{j=1}^{s} u_j y_{jk} - \sum_{i=1}^{m} v_i x_{ik} \leq 0, \quad k = 1,2,\ldots,n \quad \left(\text{or } \frac{\sum_{j=1}^{S} u_j\, y_{jk}}{\sum_{i=1}^{m} v_i x_{ik}} \leq 1,\ k = 1,2,\cdots,n \right)$$

$$\sum_{j=1}^{s} u_j y_{jk} - \alpha \sum_{i=1}^{m} v_i x_{ik} \geq 0, \quad k = 1,2,\ldots,n \quad \left(\text{or } \alpha \leq \frac{\sum_{j=1}^{S} u_j\, y_{jk}}{\sum_{i=1}^{m} v_i x_{ik}},\ k = 1,2,\cdots,n \right)$$

$$0 < \alpha \leq 1; \mathrm{u}_r \geq \varepsilon > 0, \quad r = 1,\ldots,s; \mathrm{v}_i \geq \varepsilon > 0, \quad i = 1,\ldots,m$$

The efficiency achievement measure in each DMU (Figure N8.6) is as follows:

$$\text{First phase:} \quad \alpha_k = \sum_{j=1}^{s} u_j^* y_{jk} \,/\, \sum_{i=1}^{m} v_i^* x_{ik}, \quad k = 1,\ldots,n \tag{N8.13}$$

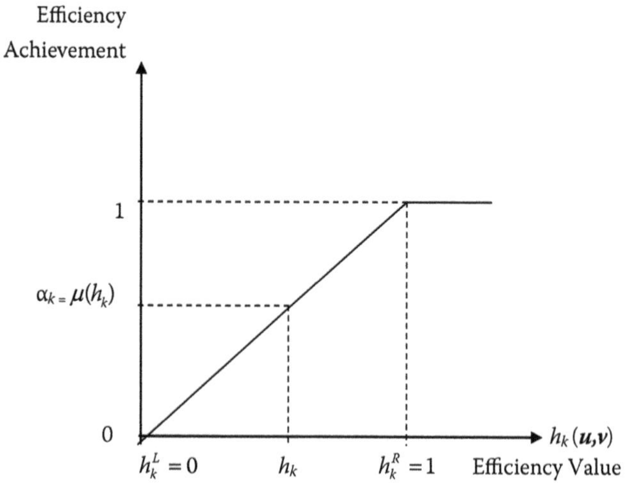

FIGURE N8.6 Identity function of efficiency achievement.

Second phase: $$\max_{u,v} \bar{\alpha} = \frac{1}{n}\alpha_k \tag{N8.14}$$

$$s.t. \quad \sum_{j=1}^{s} u_j y_{jr} - \sum_{i=1}^{m} v_i x_{ir} \le 0, \quad r = 1,2,\ldots,n$$

$$\frac{\sum_{j=1}^{s} u_j y_{jk}}{\sum_{i=1}^{m} \upsilon_i x_{ik}} \ge \alpha_k \ge \alpha, \quad k = 1,2,\ldots,n$$

$$0 < \alpha \le 1; \mathrm{u}_r \ge \varepsilon > 0, \quad r = 1,\ldots,s; \mathrm{v}_i \ge \varepsilon > 0, \quad i = 1,\ldots,m$$

Fuzzy Number for DEA Output Variables

A triangular fuzzy number (TFN) is used for measuring DEA output values. If a triangular fuzzy number is expressed as $\tilde{a} = (a^l, a^m, a^u)$, we can show the following membership function as shown in Figure N8.7.

$$u_{\tilde{a}}(x) = \begin{cases} 0, & x \le a^l \\ \dfrac{x - a^l}{a^m - a^l}, & a^l < x < a^m \\ 1, & x = a^m \\ \dfrac{a^u - x}{a^u - a^m}, & a^m < x < a^u \\ 0, & x \ge a^u \end{cases} \tag{N8.15}$$

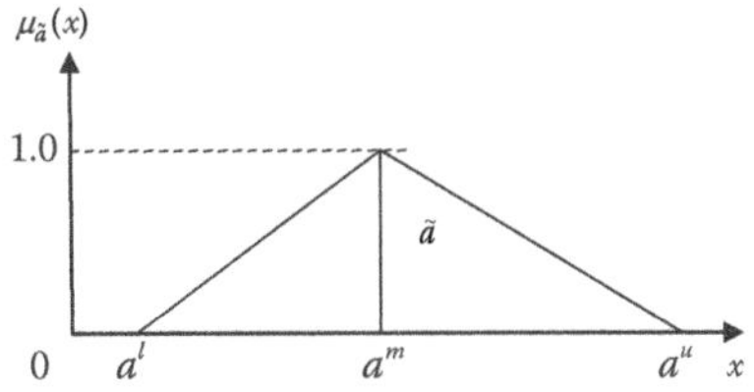

FIGURE N8.7 Membership function of triangular fuzzy number.

Fuzzy operator: Two TFN $\tilde{a} = (a^l, a^m, a^u)$ and $\tilde{b} = (b^l, b^m, b^u)$

Sum of fuzzy number: $\tilde{a} + \tilde{b} = (a^l + b^l, a^m + b^m, a^u + b^u)$

Scalar of fuzzy number: $k\tilde{a} = (ka^l, ka^m, ka^u), \quad k \geq 0$

Fuzzy DEA

If planning output values are given as fuzzy numbers (for example, a fuzzy number can be forecast by fuzzy interval regression), we want to find the output optimization in a DEA model as follows.

$$\max \ \tilde{f}_k = \sum_{j=1}^{s} u_j \tilde{y}_{jk} \tag{N8.16}$$

$$s.t. \ \sum_{i=1}^{m} \upsilon_i x_{ik} = 1$$

$$\sum_{j=1}^{s} u_j \tilde{y}_{jr} - \sum_{i=1}^{m} \upsilon_i x_{ir} \leq 0, \quad r = 1,2,\ldots,n$$

$$u_j \geq \varepsilon > 0, \ \upsilon_i \geq \varepsilon > 0, \ j = 1,2,\ldots,s; \ i = 1,2,\ldots,m$$

The optimal triangular fuzzy objective function value can be shown as: $\tilde{f}_k^* = \left(f_k^{l*}, f_k^{m*}, f_k^{u*}\right)$, we call this value the FDEA efficiency (see Figure N8.8). We first calculate fuzzy objective and fuzzy constraint (Figure N8.9). First, let $r = k$ be fuzzy number DMU_k, then $\max \tilde{f}_k = \Sigma_{j=1}^{s} u_j \tilde{y}_{jk} \leq 1$

$$\max h_{1k} = \frac{\Sigma_{j=1}^{s} u_j y_{jk}^l}{1 - \left(\Sigma_{j=1}^{s} u_j y_{jk}^m - \Sigma_{j=1}^{s} u_j y_{jk}^l\right)} \tag{N8.17}$$

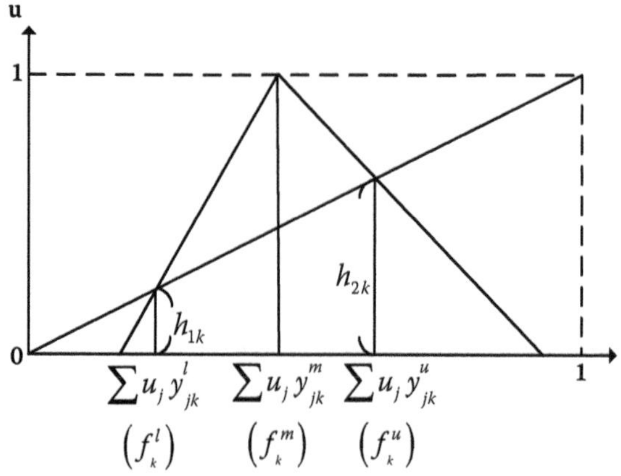

FIGURE N8.8 Fuzzy number DMU_k.

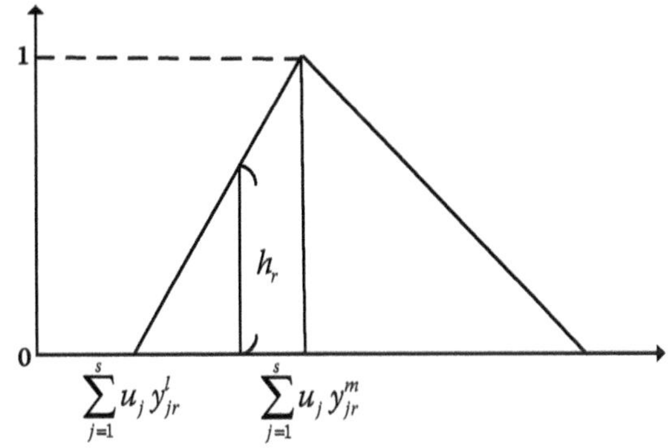

FIGURE N8.9 The concept of the fuzzy set of DMUK.

$$\max h_{2k} = \frac{\sum_{j=1}^{s} u_j y_{jk}^{u}}{1+\left(\sum_{j=1}^{s} u_j y_{jk}^{u} - \sum_{j=1}^{s} u_j y_{k}^{m}\right)} \tag{N8.18}$$

Let $r \neq k$ be fuzzy number DMU_k.

$$\max \ h_r = \frac{\sum_{j=1}^{s} u_j y_{jr} - \sum_{j=1}^{s} u_j y_{jr}^{l}}{\sum_{j=1}^{s} u_j y_{jr}^{m} - \sum_{j=1}^{s} u_j y_{jr}^{l}} \tag{N8.19}$$

Then

$$\left.\begin{aligned}
\max h_{1k} &= \frac{\sum_{j=1}^{s} u_j y_{jk}^{l}}{1-\left(\sum_{j=1}^{s} u_j y_{jk}^{m} - \sum_{j=1}^{s} u_j y_{jk}^{l}\right)} \\
\max h_{2k} &= \frac{\sum_{j=1}^{s} u_j y_{jk}^{u}}{1+\left(\sum_{j=1}^{s} u_j y_{jk}^{u} - \sum_{j=1}^{s} u_j y_{k}^{m}\right)} \\
\max \ h_r &= \frac{\sum_{j=1}^{s} u_j y_{jr} - \sum_{j=1}^{s} u_j y_{jr}^{l}}{\sum_{j=1}^{s} u_j y_{jr}^{m} - \sum_{j=1}^{s} u_j y_{jr}^{l}}
\end{aligned}\right\} \tag{N8.20}$$

$$\begin{aligned}
s.t. \ \sum_{i=1}^{m} v_i x_{ik} &= 1 \\
0 \le h_{1k} &\le 1 \\
0 \le h_{2k} &\le 1 \\
0 \le h_r &\le 1, \ r = 1,2,\ldots,n; r \neq k \\
u_j \ge \varepsilon &\ge 0, \quad j = 1,2,\ldots,s \\
v_i \ge \varepsilon &\ge 0, \quad i = 1,2,\ldots,m
\end{aligned}$$

$$\max \ \lambda_k \tag{N8.21}$$

$$s.t. \ \sum_{i=1}^{m} v_i x_{ik} = 1$$

$$\lambda_k \leq \frac{\sum_{j=1}^{s} u_j y_{jk}^{l}}{1-\left(\sum_{j=1}^{s} u_j y_{jk}^{m} - \sum_{j=1}^{s} u_j y_{jk}^{l}\right)}, \quad k = 1,2,\ldots,n$$

$$\lambda_k \leq \frac{\sum_{j=1}^{s} u_j y_{jk}^{u}}{1+\left(\sum_{j=1}^{s} u_j y_{jk}^{u} - \sum_{j=1}^{s} u_j y_{k}^{m}\right)}, \quad k = 1,2,\ldots,n$$

$$\lambda_k \leq \frac{\sum_{j=1}^{s} u_j y_{jr} - \sum_{j=1}^{s} u_j y_{jr}^{l}}{\sum_{j=1}^{s} u_j y_{jr}^{m} - \sum_{j=1}^{s} u_j y_{jr}^{l}}, \quad r = 1,2,\ldots,n; \ r \neq k$$

$$0 \leq \lambda_k \leq 1$$

$$u_j \geq \varepsilon \geq 0, \ j = 1,2,\ldots,s$$

$$v_i \geq \varepsilon \geq 0, \ i = 1,2,\ldots,m$$

DEA Cross-Period Efficiency Analysis Testing: Malmquist Index

Assume F^t is the production frontier of the t period and F^{t+1} is the production frontier of the $t+1$ period. When the production efficiency changes from F^t to F^{t+1}, the change is called a relative efficiency variation. According to Fare et al. (1992), the Malmquist productivity index is a product of technical change (TC) and efficiency change (EC); see Figure N8.10.

Technical Change (TC)—When $TC > 1$, technical progress is indicated. Conversely, $TC < 1$ indicates technical regress.

$$TC = \left[\frac{D^{t+1}(X^t, Y^t)}{D^t(X^t, Y^t)} \quad \frac{D^{t+1}(X^{t+1}, Y^{t+1})}{D^t(X^{t+1}, Y^{t+1})} \right]^{1/2} \tag{N8.22}$$

Efficiency Change (EC) — This factor compares the efficiency of the production frontier of the t period and that of the $t+1$ period. $EC > 1$ indicates improved efficiency whereas $EC < 1$ indicates reduced efficiency.

$$\begin{aligned} EC &= \left[\frac{A^{t+1}(X^{t+1}, Y^{t+1})}{A^t(X^t, Y^t)} \right] / \left[\frac{D^{t+1}(X^{t+1}, Y^{t+1})}{D^t(X^t, Y^t)} \right] \\ &= \left[\frac{D^{t+1}(X^t, Y^t)}{D^t(X^{t+1}, Y^{t+1})} \right] \times \left[\frac{A^{t+1}(X^{t+1}, Y^{t+1})}{A^t(X^t, Y^t)} \right] \end{aligned} \tag{N8.23}$$

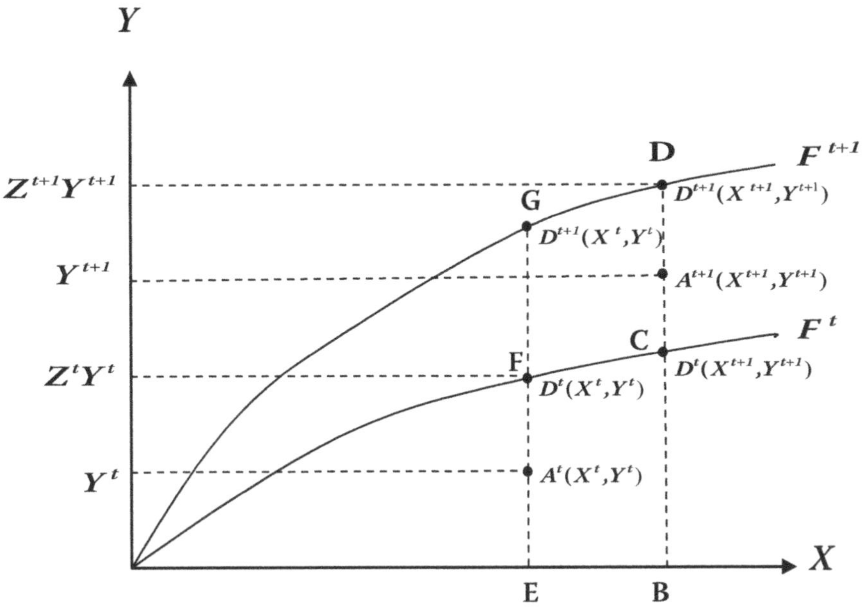

FIGURE N8.10 Measurement of Malmquist productivity index.

Malmquist Productivity Index (M) — The index is product of TC and EC. $M > 1$ indicates improved productivity, whereas $M < 1$ indicates decreased productivity.

$$M_{t,t+1} = TC_{t,t+1} \times EC_{t,t+1}$$

$$= \left[\frac{D^{t+1}(X^t,Y^t)}{D^{t+1}(X^{t+1},Y^{t+1})}\ \frac{D^t(X^t,Y^t)}{D^t(X^{t+1},Y^{t+1})}\right]^{1/2} \times \left(\left[\frac{D^t(X^t,Y^t)}{D^{t+1}(X^{t+1},Y^{t+1})}\right] \times \left[\frac{A^{t+1}(X^{t+1},Y^{t+1})}{A^t(X^t,Y^t)}\right]\right)$$

$$= \left[\frac{D^{t+1}(X^t,Y^t)}{D^{t+1}(X^{t+1},Y^{t+1})}\right] \times \left[\frac{D^t(X^t,Y^t)}{D^t(X^{t+1},Y^{t+1})}\right]^{1/2} \times \left[\frac{A^{t+1}(X^{t+1},Y^{t+1})}{A^t(X^t,Y^t)}\right] \tag{N8.24}$$

Index

D

E

F

G

H

U

V

W

Z

Milton Keynes UK
Ingram Content Group UK Ltd.
UKHW020315111024
449327UK00040B/1158

9 780367 379643